Grover E. Murray Studies in
the American Southwest

Cacti of Texas

A FIELD GUIDE

Cacti of Texas

A FIELD GUIDE

With Emphasis on the Trans-Pecos Species

A. Michael Powell, James F. Weedin, and Shirley A. Powell

Texas Tech University Press

This book is typeset in Sabon LT Std Roman. The paper used in this book meets
the minimum requirements of ANSI/NISO Z39.48–1992 (R1997). ∞

Publication of this book was made possible by the contributors acknowledged
on p. xiv.

Library of Congress Cataloging-in-Publication Data
Powell, A. Michael.
Cacti of Texas, a field guide : with emphasis on the Trans-Pecos species /
A. Michael Powell, James F. Weedin, and Shirley A. Powell.
      p.   cm.   —   (Grover E. Murray studies in the American Southwest)
Summary: "A field guide to the cacti of Texas, with emphasis on the Trans-
Pecos, the region of the Chihuahuan Desert extension into Texas where most of
the species of cacti of Texas may be found"—Provided by publisher.
      Includes bibliographical references and index.
      ISBN-13: 978-0-89672-611-6 (pbk. : alk. paper)
      ISBN-10: 0-89672-611-8 (pbk. : alk. paper)
1. Cactus—Trans-Pecos (Tex. and N.M.)—Identification.   2. Cactus—Texas—
Identification.   I. Weedin, James F.   II. Powell, Shirley A.   III. Title.
      QK495.C11P68   2008
      583'.5609764—dc22
                                                                        2007030415

Printed in Korea
17   18   19   20   21   22   23   24   25   /   10   9   8   7   6   5   4   3   2

Texas Tech University Press
Box 41037
Lubbock, Texas 79409–1037 USA
800.832.4042
www.ttup.ttu.edu

Contents

Preface

The Chihuahuan Desert Region (CDR) holds more species of cacti than any other comparable area in North America. This center of cactus diversity lies mostly in northern Mexico; it extends into the United States only in the Trans-Pecos region of Texas and some adjacent parcels in New Mexico and Arizona. This area west of the Pecos River in Texas measures about 32,000 square miles, or 20.5 million acres, equivalent in size to the state of Maine. The Trans-Pecos shares some species and vegetation types with the Edwards Plateau ("Hill Country") of Central Texas, the Tamaulipan thorn-scrub of the Rio Grande valley, the Great Plains of the Texas Panhandle and eastern New Mexico, and the Apachean floristic region of southeastern Arizona. However, by far the greatest part of the Trans-Pecos vegetation and flora pertains to the Chihuahuan Desert. Consequently, the Trans-Pecos is one of the major centers of cactus distribution in the United States.

In Trans-Pecos Texas, the southern Big Bend area is best known as cactus country. The lowest elevation in the entire CDR lies in the Boquillas Basin near the Rio Grande, just downslope from the southeastern foothills of the Chisos Mountains. Living specimens and subfossil remains of desert plants indicate that the Boquillas Basin served as a major biological refugium during the last Ice Age and thus must have been one of the major dispersal centers for Chihuahuan Desert species.

The definitive work on which the present guide is based, *Cacti of the Trans-Pecos and Adjacent Areas* (Powell and Weedin, 2004), was the first book to emphasize the cacti of the region. Several publications with wider geographic coverage that have included the Trans-Pecos cactus flora are cited in the Bibliography. Most of those sources are out of print, not including recent treatments in *Flora of North America,* vol. 4, pp. 99–257 (2003). The cactus books most widely used to identify Texas cacti have been those by Del Weniger and Lyman Benson, but the extensive nomenclatural differences between those publications were disconcerting to many who wanted to use the scientific names of southwestern cacti. One objective we hoped to achieve in our original book was to reconcile the differences between Weniger and Benson, where possible, and to provide alternate names that reflect contemporary taxonomic research.

The main purpose of the present guide is to provide simpler, less technically detailed treatments of the species for educated general readers. The guide allows preliminary identification by matching unidentified cacti with color photographs of plants in flower and in fruit. Distribution maps, keys, and descriptions provide a second level of information for systematic identification that supersedes the first impression gained from the pictures.

This field guide at least briefly mentions essentially all Texas taxa of Cactaceae and should allow confident identification of cacti throughout Texas, excepting varieties of *Echinocereus reichenbachii* and three eastern species of prickly pears, which are briefly described. We have prepared treatments for all cactus taxa known to occur in the Trans-Pecos except for certain races of *Opuntia leptocaulis* and certain prickly pears. Cactus species of areas outside the Trans-Pecos are not keyed but are discussed along with the Trans-Pecos species or in separate sections at the end of the genera sections. Photographs and distribution maps representing most of these taxa are included, except for some opuntias.

The guide was designed for use by dedicated nonprofessionals, self-taught hobbyists and naturalists, and serious students of cacti. Visitors to the national parks, state parks, and other natural areas in regions adjacent to the Trans-Pecos will find this book useful. The manual will be important to professionals in national and state park resource interpretation, wildlife biology, ecology, range management, and environmental consulting.

Many aspects of format follow that of *Cacti of the Trans-Pecos and Adjacent Areas* and its companion books, *Trees and Shrubs of the Trans-Pecos and Adjacent Areas* (Powell, 1998a) and *Grasses of the Trans-Pecos and Adjacent Areas* (Powell, 2000). Metric system measurements are used in keys, descriptions, and discussions about the species. The English system is used for elevations and distances. Abbreviations for states are two letters in caps, following the U.S. Postal Service and current trends in scientific writing. Directions are in caps (e.g., N, E, S, W, or NW). Only one direction, NE, has the same abbreviation as a state, Nebraska, but the difference will be apparent from the context. Author citations for plant names mostly follow those in the Missouri Botanical Garden, TROPICOS, and Brummitt and Powell (1992).

Most of the morphological data were taken from specimens housed in the Sul Ross State University (SRSC) herbarium and from living specimens at Sul Ross. Some measurements also were taken from pertinent literature. Measurements given in keys, descriptions, and discussions mostly are of the typical range of a particular character and may not include exceptions or extremes outside the normal range.

The generalized distribution maps reflect our current information about the ranges of relevant taxa. To fashion the range maps, we have used a combination of documented localities, remembered localities from our collective field experiences, and our understanding of the habitat requirements of each species. Distributional information recorded in various published sources has been considered on a species-by-species basis but may not have been incorporated, because mistakes are rampant in the literature. These generalized distributions are most accurate for taxa of restricted ranges and are less precise when portraying wider-ranging taxa.

The approximate ranges of Trans-Pecos cactus species in the United States and Mexico, as far as known, were mapped in Powell and Weedin (2004).

The term "cultivated" in figure legends denotes photography of specimens grown in containers, in the experimental *Opuntia* garden, or in the formal Cactus Garden at Sul Ross State University. The photographs were taken by the first author, except for those otherwise specifically acknowledged.

Acknowledgments

Individual recognition and gratitude expressed in *Cacti of the Trans-Pecos and Adjacent Areas* mostly pertains as well to the current abridged version. For support in preparation of the field guide, we wish to thank Amy Valenzuela for prompt, cheerful, and expert office help. Tim Parsons kindly assisted with computer graphics. Martin Terry clarified the currently known range of *Astrophytum asterias*.

The photographs in this guide are the same as those in the original *Cacti of the Trans-Pecos*, where they each appeared with a caption giving the Latin name, common name, population source, and contributor of the photograph, if different from the first author of the book. For providing some of the photos, we are again grateful to Scooter Cheatham, James H. Everitt, David J. Ferguson, Mark Lockwood, Jackie M. Poole, Shirley A. Powell, Marcos A. Rodriguez, James F. Weedin, Brent Wauer, Richard D. Worthington, Dale and Marian Zimmerman, and Allan D. Zimmerman.

Finally, the authors and Texas Tech University Press are deeply grateful to the following contributors, without whose generous and timely support the book from which this work was extracted, *Cacti of the Trans-Pecos and Adjacent Areas,* would not have been possible: The Edwill Fund, Elizabeth Winston Mize, the late Grover Murray and Sally Murray, Colorado Cactus and Succulent Society, Austin Cactus and Succulent Society, Lee J. Miller, and National Capital Cactus and Succulent Society.

Cacti of Texas

A FIELD GUIDE

Introduction

Trans-Pecos Texas is one of the most cactus-rich regions in the United States. Physiographically, it is a mountain and basin region, and it is occupied in part by the Chihuahuan Desert Biome. Two-thirds of the great Chihuahuan Desert Region (CDR) is in Mexico, while the northern one-third extends into the United States, through the Trans-Pecos and into New Mexico, along the Pecos and Rio Grande drainages, and barely into southeastern Arizona.

Cactaceae, the cactus family, includes 1,500–1,600 species in about 100–122 genera, all of which are native to the New World with the exception of epiphytes in the genus *Rhipsalis* Gaertn., found in tropical forests of Kenya, Africa, and Sri Lanka and other islands of the Indian Ocean, including Madagascar. Cacti are widely distributed in the Americas, from Canada south to Patagonia in southern South America. Multiple lines of evidence suggest South America as the continent of origin for the family.

Cacti are well-known components of North American deserts, particularly the Sonoran and Chihuahuan deserts. The Chihuahuan Desert Region in Mexico includes substantially more cacti than either Texas or Arizona, approximately 256 taxa (194 species and 62 varieties; Zimmerman et al., forthcoming). In the country of Mexico there are more than twice as many taxa of cacti (563 species in 48 genera; Hernández and Bárcenas, 1995) than there are in the United States.

In the United States the two major cactus provinces are Arizona and the Trans-Pecos region of Texas, each with similar numbers of cactus species. Benson (1982) tabulated the following numbers of taxa for these areas: Texas, 126 taxa (73 species and 53 additional varieties); Arizona, 130 taxa (70 species and 60 varieties). We estimate that there are currently 136 taxa (95 species and 41 varieties) found in Texas. Approximately 109 taxa (76 species and 33 varieties) of the Texas cacti are distributed in the Trans-Pecos. Thus, about 80% of all Texas cacti are present in the Trans-Pecos. Other cactus-rich states listed by Benson are New Mexico with 88 taxa (53 species and 35 varieties) and California with 57 taxa (32 species and 25 varieties).

Trans-Pecos Texas (Map 1) at lower elevations is mostly desertic. Low basins and arid mountains and mesas extend from El Paso County east to the Pecos River and beyond (Plate 1). The desertic mountains are abundant in the southern Big Bend region. Mid-elevation basins of

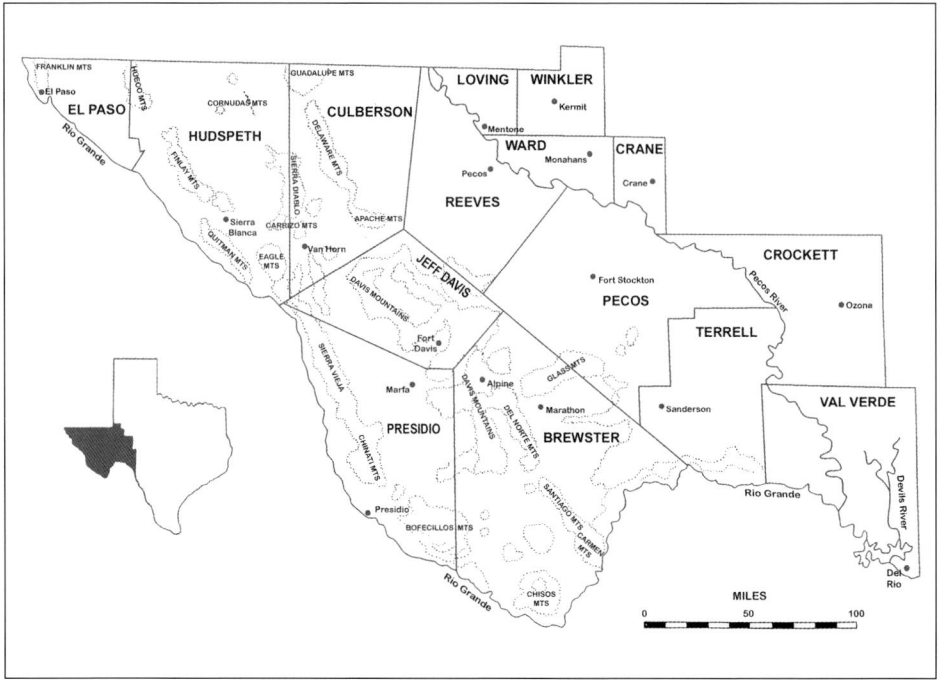

Map 1. Trans-Pecos Texas and adjacent counties, with some major geographic features shown.

mesic mountains support extensive plains grasslands on the Diablo Plateau in Hudspeth County and in the Davis Mountains area in Jeff Davis, Presidio, and Brewster counties (Plates 2 and 3). In the central Trans-Pecos three mesic mountain ranges are dominated at middle and upper elevations by oak-juniper-pinyon woodlands (Plates 4 and 5). Relict montane woodlands or conifer forests are found at the tops of these mountains: 8,749 feet at Guadalupe Peak; 8,382 feet at Mount Livermore in the Davis Mountains (Plate 6); 7,835 feet at Mount Emory in the Chisos Mountains (Plate 7). Another major mountain range in Presidio County is the moderately wooded Chinati Mountains, with its highest point, Chinati Peak, at 7,730 feet.

The relatively arid climate and variable desert and mountain physiognomy in the Trans-Pecos has favored the establishment of a diverse cactus flora. In the Trans-Pecos the average annual precipitation is about 12 inches, with annual precipitation usually increasing from about 8 inches in the west at El Paso to the east where an average 13–20 inches are recorded in the central mountains and nearby areas, depending upon elevation. Wide fluctuations in seasonal and annual precipitation are common, and severe periodic droughts have been recorded in the region since the late nineteenth century. In the twentieth century noted droughts occurred in the 1930s, 1950s, and 1990s.

Miocene volcanic activity and other geologic dynamics have produced extensively varied substrates for the establish-

Plate 1. Chihuahuan desertscrub, looking SE to the igneous Chisos Mts. Habitat of *Opuntia camanchica, O. azurea,* and *Echinocactus horizonthalonius.*

Plate 2. Grassland in N Brewster Co., TX, with the limestone Glass Mts in the background. Habitat of *Opuntia davisii.*

Plate 3. Alluvial basin supporting grassland in N Brewster Co., TX, in contact with igneous mountains. Habitat of *Opuntia tortispina, O. pottsii,* and *Echinomastus intertextus.*

Plate 4. Oak-juniper-pinyon woodland above basin grassland, Davis Mts in central Jeff Davis Co., TX. Habitat of *Opuntia phaeacantha* var. *phaeacantha* and *Echinocereus viridiflorus* var. *cylindricus*.

Plate 5. Oak-juniper-pinyon woodland in the Chisos Mts, southern Brewster Co., TX. Habitat of *Opuntia chisosensis*, *O. engelmannii* var. *engelmannii*, *Mammillaria meiacantha*, and *Echinocereus coccineus*.

Plate 6. Montane woodland in upper Madera Canyon, Mt Livermore, Jeff Davis Co., TX. Habitat of *Opuntia polyacantha* and *Echinocereus viridiflorus* var. *weedinii* (altitude ca. 8,000 ft).

ment of vegetation in the northern CDR. Igneous and sedimentary rock habitats are common, as are those featuring gravel, sand, and clay alluvium (Plates 8–11). Gypsum and salt deposits are exposed in many desert basins. More extensive discussion regarding the physical and vegetative characteristics of the Trans-Pecos can be found in Schmidly (1977), Diamond et al. (1988), and Powell (1998a, 2000).

Actually, only about 20 taxa are strictly endemic to the Trans-Pecos. This means that most of the cactus species that occur in the Trans-Pecos also extend into other parts of Texas and/or into neighboring states and into northern Mexico. Depending upon taxonomic interpretations, only 4 genera, 16 species, and 11 varieties occur in Texas strictly east of the Pecos River. We have no doubt that future study will result in the description or identification of additional taxa for Texas, particularly in the genus *Opuntia.*

Large columnar cacti, such as the saguaro, *Carnegiea gigantea,* of the Sonoran Desert of Arizona do not occur in the northern CDR, probably because it is too cold in the winter. Conversely, Trans-Pecos Texas is known for its diminutive cacti. Some of these plants, such as *Coryphantha minima* and *Echinocereus davisii* (Plate 12) are marble-sized, with showy flowers larger than the stems.

Cactus Morphology

The following brief review of cactus morphology should be helpful in field identification. For much more detail concerning the biology of cacti, including anatomy, physiology, ecology and biogeography, evolution, pollination biology, uses of cacti, horticulture, and conservation, consult *Cacti of the Trans-Pecos and Adjacent Areas* (Powell and Weedin, 2004).

General Morphology. In the Trans-Pecos region, and elsewhere, desert plants with spiny stems or pointed leaves collectively have been confused with "cacti" by some of the uninitiated. Actually, there are several families of desert plants that sometimes are misconstrued as cacti, particularly the yuccas (*Yucca* spp.; Plate 13), agaves or century plants (*Agave* spp.; Plate 14), sotols (*Dasylirion* spp.; Plate 15), and ocotillo (*Fouquieria splendens* Engelm.; Plate 16).

Cacti are distinguished by fleshy (succulent), usually leafless stems, with spines localized in numerous specific areas, called areoles, along the stems. Prickly pears and chollas may produce short-lived slender (cylindroid) leaves on young stems. Some primitive cacti produce broad, persistent leaves (*Pereskia* spp.; Plates 17 and 18), but these species are restricted to tropical and subtropical areas. Cacti also are distinguished by their flowers, which have numerous petal-like structures that intergrade with sepal-like structures to the outside. These parts are referred to as tepals, petaloid and sepaloid tepals, inner and outer tepals, or bracts (Fig. 1; Plate 19). In different cacti the color of the inner tepals may be white, yellow, purple, pink, magenta, red, or a combination of colors. In the flower center the stamens are numerous, and the pistil consists of an inferior ovary and four or more united carpels. The ovary has one seed chamber, usually with many ovules. A pistil exhibits a single style with several stigma lobes at the apex. The ovules develop into seeds in fleshy (berrylike) or dry (capsulelike) fruits.

The largest cactus plants in the Trans-

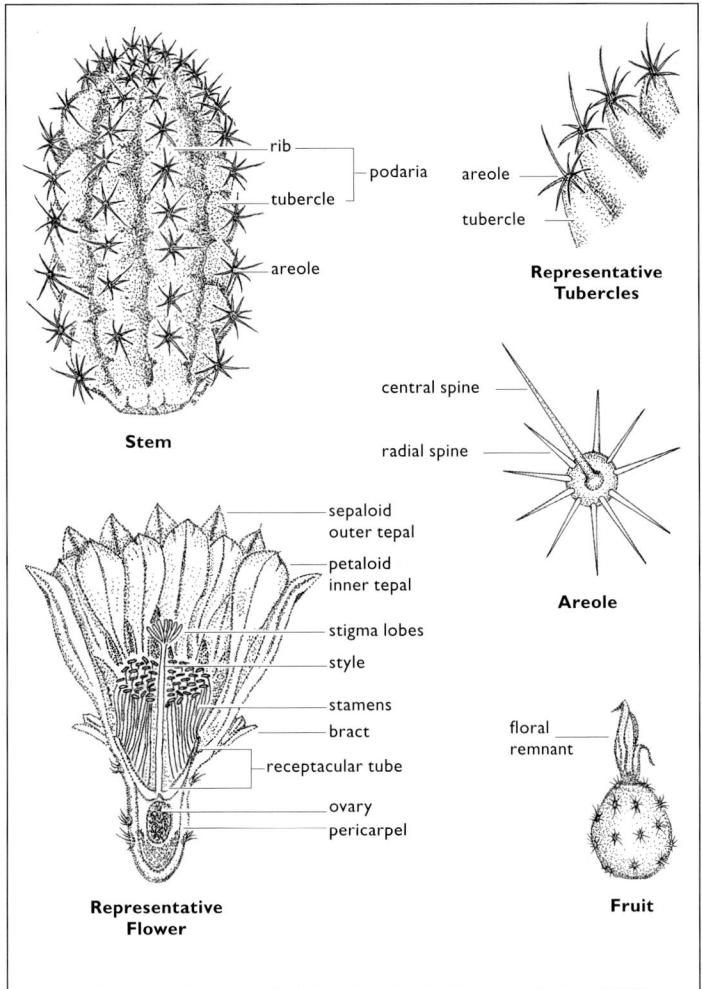

Fig. 1. Vegetative and floral morphology of cacti.

rib
podaria
tubercle
areole

areole
tubercle

Representative Tubercles

Stem

central spine
radial spine

Areole

sepaloid outer tepal
petaloid inner tepal
stigma lobes
style
stamens
bract
receptacular tube
ovary
pericarpel

floral remnant

Representative Flower

Fruit

Pecos are certain prickly pears and chollas of the genus *Opuntia* (subfamily Opuntioideae). Trans-Pecos members of *Opuntia*, such as *O. engelmannii* and *O. imbricata* (Plate 20), may range to 1–3 m (3–10 ft) tall. Different species of *Opuntia* are branched, usually with one or a series of joints (stem segments) forming prostrate (Plate 21), declined, spreading, or upright shrubs (Fig. 2). Prickly pears exhibit flattened joints, and chollas have cylindrical,

club-shaped, or rounded joints. Prickly pears and chollas may or may not have short trunks. In Mexico and some other areas, prickly pears may be treelike, up to 10 m (32 ft) tall, with rather massive trunks.

In the Trans-Pecos the tallest members of the subfamily Cactoideae are the barrel cacti, specifically *Ferocactus wislizeni* (Plate 22) and *F. hamatacanthus*. The larger of the two is decidedly *F. wislizeni*,

Plate 7. Emory Peak, Chisos Mts, southern Brewster Co., TX. Habitat of *Opuntia chisosensis.*

Plate 8. Chihuahuan desertscrub, Hen Egg Mt and Agua Fria in the background, southern Brewster Co., TX.

Plate 9. Chihuahuan desertscrub in igneous-derived substrate, Big Bend Ranch State Park, sedimentary exterior of Solitario Dome in background, southern Presidio Co., TX.

Plate 10. *Agave lechuguilla* and *Euphorbia antisyphilitica* on limestone of Reed Plateau, looking N across sedimentary substrate and alluvial soils of the desert floor to igneous desert mountains, southern Brewster Co., TX.

Plate 11. Sedimentary rock, the Caballos Novaculite formation in Marathon Basin, N Brewster Co., TX. Habitat of the "novaculite endemic" cacti and many others.

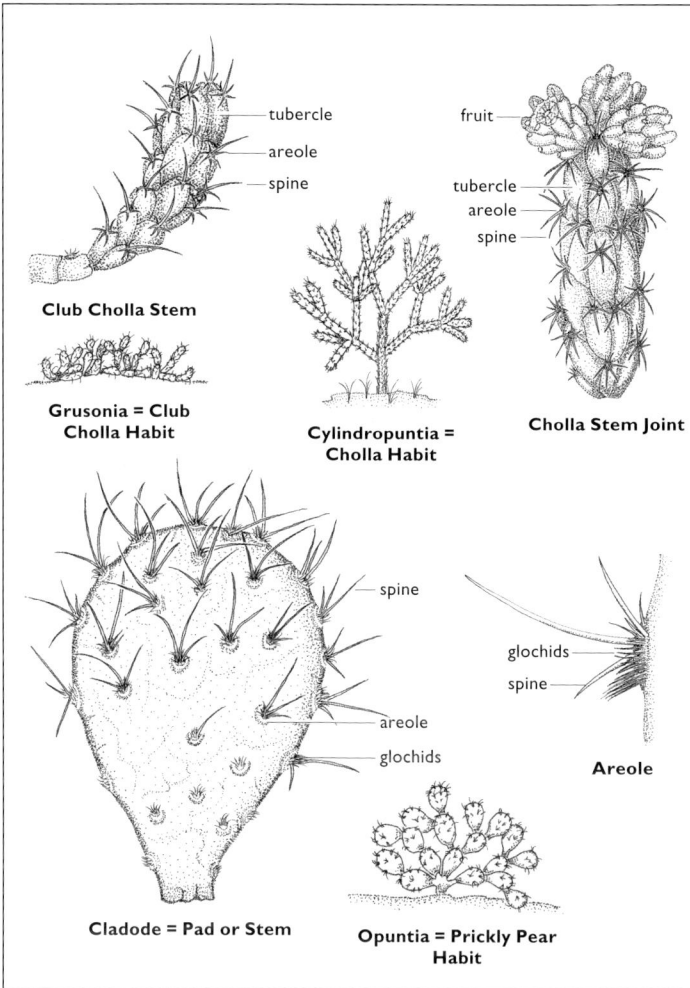

Fig. 2. Morphology of Opuntioideae.

tubercle
areole
spine

Club Cholla Stem

Grusonia = Club Cholla Habit

Cylindropuntia = Cholla Habit

fruit

tubercle
areole
spine

Cholla Stem Joint

spine

glochids
spine

areole
glochids

Areole

Cladode = Pad or Stem

Opuntia = Prickly Pear Habit

with the tallest known specimens, in El Paso County, approaching 2 m (6 ft) high. Some multistemmed and mounded specimens of *Echinocereus* are large in diameter (Plate 23). The biggest plants of *E. coccineus, E. stramineus,* and *E. enneacanthus* may form hundreds of stems and exceed 1 m (3 ft) in diameter.

Trans-Pecos members of the Cactoideae may be branched or unbranched. Branches may form near ground level or at almost any point up the stem. Many species, such as those in *Ferocactus* and *Echinomastus,* are unbranched. Single-stemmed plants of the Cactoideae may be columnar, cylindroid, conical, globular, or hemispheric. Multistemmed plants may form dense mounds or mats.

The plants of many Trans-Pecos species are difficult to detect, either because they are diminutive or because their stems retract almost completely underground

(Plate 24). *Ariocarpus fissuratus* is commonly known as living rock cactus because the flattened stems resemble rocks, except for their pale gray-green color. When dry, the stems of *A. fissuratus* may shrink, pulling the stem apex below ground level and further obscuring the plant in surrounding limestone rubble.

In the life cycle of cactus plants, from seedlings through sexually mature adults, the plants may exhibit different growth habits and spine features. Some of these growth stages or "age classes" are so different that they may appear to represent separate taxa. Cactus literature includes examples of growth stages having been formally described as species.

Roots. In general, cacti are understood to have numerous shallow, horizontal roots that extend in all directions from the plant, out to 2–10 m (6–32 ft) or more. The usually branched shallow roots may be only 1.5–10 cm (0.75–4 in) below the surface, and they can take advantage of light precipitation. Many of the smallish Trans-Pecos cacti have more compact systems, with lateral roots that develop closer to the plant. Some small Trans-Pecos cacti, such as *Coryphantha duncanii* and *Lophophora williamsii,* have fleshy taproots that store water and few lateral roots. Some prickly pears, such as *O. pottsii* and *O. macrorhiza,* produce branched, or "tuberous," fleshy roots. Taproots the size of a basketball or larger are formed by old plants of *Peniocereus greggii.* Cacti strike roots easily when stem surfaces (including whole fruits of chollas) come in contact with the soil. *Opuntia* commonly reproduces asexually by the scattering of stem joints.

Stems. "Shoot" is the anatomical term for the stem and its appendages collectively, as opposed to the root. Cacti produce three different kinds of shoots: sterile long shoots (the stems); fertile long shoots (the flowers); and sterile short shoots (the areoles). The stems of most cacti appear to be leafless, but many species produce almost microscopic rudimentary leaves at the lower edge of areoles developing near the growing tip of the shoot. The rudimentary leaves soon may wither and fall away. Opuntioideae produce ephemeral, cylindrical (to flattened) leaves 0.5–1.5 cm long, usually present only on new stems and fruits. In Cactoideae, areoles, the sterile short shoots, are borne on long shoots in or near the axils of usually rudimentary leaves.

Stem Joints. Species of *Opuntia* produce joints and jointed branches. The flattened stem joints of prickly pears (subgenus *Opuntia*), formally known as cladodes or phylloclads, are commonly called "pads." The prickly pears also are known as the platyopuntias, where the Greek root word *platy,* "flat," is a reference to the pads.

Tubercles and Ribs. Technically, "podaria" (singular, podarium) is the collective term for protuberances of the long-shoot surface. Individually protruding podaria are called "tubercles," and podaria that form longitudinal files on the stem surface are "ribs" (Fig. 1). For example, stems of *Coryphantha* and *Mammillaria* are tuberculate, and stems of *Echinocereus* are ribbed. In some species the tubercles form ribs only near their bases, resulting in low ribs and tuberculate projections in longitudinal files.

Spines. Cactus spines are modified leaves that develop individually or collectively from distinct growing regions of the spines, nearest the lower margin of the

Plate 12. *Selaginella* mat on novaculite substrate, Marathon Basin. Microhabitat of diminutive cactus *Echinocereus davisii*.

Plate 13. *Yucca torreyi*, SW foothills of the igneous Chisos Mts, Brewster Co., TX. In the west *Y. torreyi* is an "indicator species" of the Chihuahuan Desert, but in the east it extends far into the subtropical Tamaulipan scrub, where it is called *Y. treculiana*.

Plate 14. *Agave lechuguilla* in the limestone Dead Horse Mts; Chisos Mts in the background; Brewster Co., TX.

Plate 15. *Dasylirion heteracanthum,* Black Gap Wildlife Management Area, southern Brewster Co., TX.

Plate 16. *Fouquieria splendens,* with its slender, spreading, spiny stems; *Yucca torreyi* in left foreground; Big Bend National Park, Brewster Co., TX.

areole. Very young spines may grow at the tip until they reach approximately 0.1 mm in length, after which they elongate from the base. While they are developing, spines become hardened from the tip down, as the mature cells produce woody cell walls. Completely hardened and mature spine clusters are attached to the areolar tissue and to each other by cork cells that arise near the base of each spine. In many cactus species the specific numbers and arrangements of spines can be considered together in a group, giving rise to the term "spine cluster." Spine clusters are extremely valuable in cactus identification when they are detached from the plant or found as subfossils preserved in pack rat middens or elsewhere.

The shorter and thinner spines of *Opuntia*, the glochids (Fig. 2), are formed in the same manner as larger spines, but they are not persistent because the base does not become hardened. Glochids ultimately are deciduous, and they are easily dislodged from the areole. Sharp glochids readily penetrate human skin, but because they are barbed they are not easily removed.

Spines may be stout, slender, or hairlike; cylindroid, needlelike, angled, flattened and stout, or flattened and thin; smooth, rough, feathery, or corrugated with transverse ridges; straight, twisted, curved, or strongly hooked on the end like a fishhook. Spines may be slightly or much broader at the base. Appressed spines extend parallel to the stem contour nearly in contact with the stem surface, or at most their tips protrude at a tangent with respect to the rounded stem outline, leaving the plants easy to handle without gloves. Such stems appear smooth and safe to the touch.

Erect or ascending spines (shown by many species) angle upward, more or less intermediate between appressed and porrect (perpendicular to the surface). Descending spines angle downward, and deflexed spines are abruptly bent downward from near the base.

Spines may be white, gray, black, yellow, golden, brown, orange, pink or rose, red, or multicolored. Colored spines may be opaque or somewhat translucent. With age spine colors usually fade to dull shades or change colors (for example, from white to gray). The soft, living bases of actively growing (nascent) spines can be brilliantly colored.

The spines of a stem may be a uniform color (concolorous), as in the whitish spines of *Echinocereus pectinatus* var. *wenigeri*, or the central and radial spines may be of different colors. In some taxa, such as the rainbow cacti, including many species of *Echinocereus*, different spine colors make the plant appear to have broad or narrow bands or rings around the stem. Those varying colors reflect the beginning and/or end of each growth season.

The spines of North American cylindropuntias may have sheaths formed from the epidermal layer and separate from the woody spine core during development. The paper-thin sheaths may be yellowish, silver, or whitish and may fit tightly or loosely over the spine. Sheaths may cover the spines completely to the base, partially, or only on the tip. In the Trans-Pecos, sheaths are prominent in *Opuntia tunicata*, *O. imbricata*, and several other taxa.

Radial and Central Spines. The distinction between radial and central spines is evident in most cacti, but sometimes the difference between radial and central spines near the periphery of spine clusters is not clear. Radial spines may be defined as the series that radiates from the periphery of the spine cluster in a plane that is more or less parallel with the stem surface. Radial spines, like spokes of a wheel, often are similar in appearance, although often of different lengths at the top and bottom or sides of the spine cluster. Central spines protrude from the central region of the spine cluster, originating anywhere inside the radial spines, and usually are more robust than radial spines. Even near-peripheral centrals that are similar to radials in size are slightly larger and have other features, such as enlarged basal structure, that allow them to be differentiated when the two spine types are compared under magnification.

When identifying cacti through the evaluation of spine morphology, pay attention first to the central spines, because the most useful comparative spine characters typically are those of the centrals. Also, when distinguishing between central and radial spines, for the purpose of counting spine number or evaluating other characters, first distinguish the centrals, particularly any that may be similar in size and position to the radials. This guide follows the "centrals first" orientation in descriptions and discussion.

In a departure from the traditional cactus literature, we distinguish central and radial spines in our descriptions of the Opuntioideae as well as those of the Cactoideae. We have observed that two or more discrete peripheral spines often are present in species of *Opuntia*, and we refer to these smaller spines as "radials" even though they usually occupy only the lower periphery of the areoles. The radial spines

are not present in certain other specimens or species of *Opuntia*. When discussing *Opuntia*, we refer to the larger spines as centrals. The centrals may be positioned from the center of the areole to near the periphery.

Glochids occur along with the spines in the areoles of most opuntias. One or more bristles, which are slender, relatively long, pointed, and stiff but flexible, may be present as well.

Trichomes. Cactus trichomes are hair-like, multicellular structures, and in the broad sense they potentially could be found on the surfaces or edges of most organs, including spines. Trichomes are most conspicuous in new, actively growing areoles of certain species, where they may be densely matted together like wool, obscuring whole spines and flower buds at stem tips. In areoles the densely arranged trichomes may be straight, curved, or curled, giving rise to the popular term "wool," especially when the trichomes are whitish. Densely packed, straight trichomes may be referred to as "pile," as in a carpet, or as "felt." Areolar trichomes may be white, yellow, brown, gray, nearly black, or other colors, and the color may fade or change with age. In cacti, trichomes also may be conspicuous at shoot apexes, where the closely arranged young areoles collectively form a dense apical mat of hair (Plate 25). In *O. rufida* the stem surfaces are densely covered with short hairs, but these unicellular hairs are totally different from those in the areoles or those on the stem surface of *Astrophytum*.

Glands. Multicellular glands (highly specialized secretory spines) are produced in the areoles (e.g., in some *Opuntia*) or areolar grooves (e.g., in *Coryphantha*,

Plate 17. *Pereskia grandifolia*, a leafy cactus species from **Brazil**.

Plate 18. *Pereskia bleo*, a leafy cactus species from **Panama** and **Colombia**.

Plate 19. Representative cactus flower with yellow inner (petaloid) tepals (*Echinocereus dasyacanthus*).

Plate 20. Individual plants of *Opuntia imbricata* (tree cholla) are the largest native, wild cacti in the Trans-Pecos.

Plate 21. Low, mounded habit of *Opuntia densispina* (Big Bend devil cholla).

Ancistrocactus, Thelocactus, Ferocactus) of some cacti. The glands apparently develop among the trichomes in areolar grooves and/or upper edges of the spine clusters. The glands extend slightly from the areolar grooves, where they are visible to the naked eye as red, orange, yellow, or brownish structures, often surrounded by trichomes.

Stem in Cross Section. The thin, hard outer "skin" of the stem is called the rind. Beneath the rind is a relatively thick cortex, which is made up of several layers of dark green chlorophyll-bearing cells near the rind and a thick tissue of colorless, water-filled cells. In many cactus species, very large, clear, mucilaginous cells can be seen in the cortex. Inside the cortex is a vascular cylinder, which in cross section is seen as a nearly solid or incomplete ring of tissue that appears more compacted than that of the cortex. Inside the vascular cylinder, in the center of the stem, is the pith, a tissue that resembles the cortex and that also is succulent. Woody tissue, called secondary xylem, may be produced from the vascular cambium in an interrupted ring between the pith and the inside of the vascular cylinder. The cambium, a single layer of cells, is only visible through a microscope.

Vascular System. Many cacti produce woody vascular tissues, particularly in the secondary xylem inside the cambium. It may seem unusual that the succulent stems of cacti have hard or soft woody parts inside, but it is the wood that provides much support for the stems. In many types of cacti, after the stem dies and dries, the central woody cylinder is exposed. Among Trans-Pecos cacti prominent woody skeletons are produced in *Opuntia imbricata* and its allies (Plate 26). The wood skeletons differ between species, particularly in the shape and pattern of holes in the wood, and they provide useful characters in cactus taxonomy. Relatively thin, netlike vascular systems develop in prickly pear pads, and more extensive wood forms in the trunks of some prickly pears.

Fasciation. In cacti, as well as plants of other families, fasciated or "crested" stems may develop as a result of injury or genetic anomaly in the growing tip of the shoot. Orientation of growth is altered so that a fanlike, eventually wavy-margined apex forms at the tip of a normal cylindrical stem. Crested cacti are rare and are prized by collectors who appreciate anomalous plant forms. In the Trans-Pecos we have observed or had reports of occasional fasciated individuals in several species of *Echinocereus* and in *Thelocactus bicolor* (Plate 27), *Coryphantha tuberculosa, C. duncanii, Ancistrocactus tobuschii, Mammillaria meiacantha, M. grahamii,* and *Opuntia imbricata.*

Flowers, Fruits, and Seeds. Flowers form in areoles that are in characteristic position on the stem, perhaps very near the apex, one or more centimeters away from the apex, or on the sides of the stem. Flower position in some cases may be influenced by ecological conditions; for example, more flowers may tend to develop on the warmer sides of the stem.

Flowers may be formed from the areolar growing point, or meristem, just above the spine cluster (many cacti); in an areolar groove toward the stem (e.g., *Coryphantha*); in the tubercle-stem axis (*Mammillaria*); or the flower may erupt through the rind of the stem above the areole (*Echinocereus*). In prickly pears, flowers typically form in areoles on the apical margin of pads, but they may develop in upper lateral areoles as well.

A cactus flower is a leafy shoot, with areoles, in which an ovary is embedded and other parts of the pistil are enclosed. The style extends through tissue, known as the "column," above the top of the ovary and into or through the receptacular tube. Stigma lobes or "style branches" are found at the end of the style. At the base of the flower the tissue surrounding the ovary is known as the pericarpel (Plate 28). The nectar chamber is the recess around the base of the style and below the base of the lowest stamens; it receives any sugary nectar from the adjacent tissue. Distal to the pericarpel is the receptacular tube (referred to as the floral tube or hypanthium by some other authors). The receptacular tube, with bracts surrounding it, extends from the floor of the nectar chamber to the base of the outer tepals. The pericarpel plus the receptacular tube are equivalent, respectively, to the inferior and superior floral cup or tube, the terminology favored by Benson (1982). The sepaloid outer tepals are transitional to the usually showy and petaloid inner tepals. In cactus flowers the stamens usually are numerous, all with filaments of about equal length.

The cactus fruit is made up of the ovary and accessory tissues. A mature cactus fruit enlarges in size and changes color. After the flower opens, the pericarpel matures and ripens. Inside a mature fruit the ovary contains seeds. Outside the ovary the pericarpel tissues (cortex and surface tissues), including the superficial areoles and their subtending bracts (Fig. 1), form a rind.

Mature fruits can be either fleshy or dry. A mature fleshy fruit usually, but not necessarily, achieves a characteristic succulence, color, or color change and encloses fully formed seeds. A mature dry fruit changes from a relatively fleshy developing pericarpel to a structure with dry internal tissues and fruit wall, and it contains fully formed seeds. A ripe fruit is one that has achieved its characteristic color and tissue succulence or dryness and is easily removed from the stem, no matter whether the seeds have matured. Seed morphology, particularly the size and shape of the seed and the color and surface structure of the seed coat, is useful in cactus identification.

In Opuntioideae, the seed typically is completely enclosed by the funicular envelope (Fig. 3), the main vein forming the funicular girdle (Stuppy, 2002). The

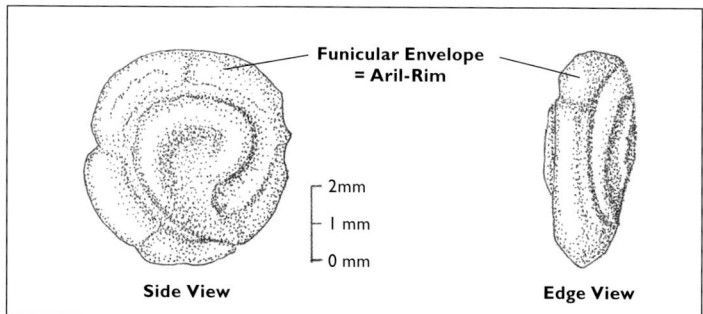

Fig. 3. Seeds of *Opuntia tortispina*, showing the funicular envelope.

Funicular Envelope = Aril-Rim

2mm
1 mm
0 mm

Side View

Edge View

Plate 22. Medium-sized specimen of *Ferocactus wislizeni* (Arizona barrel cactus) from the Franklin Mts, El Paso Co., TX.

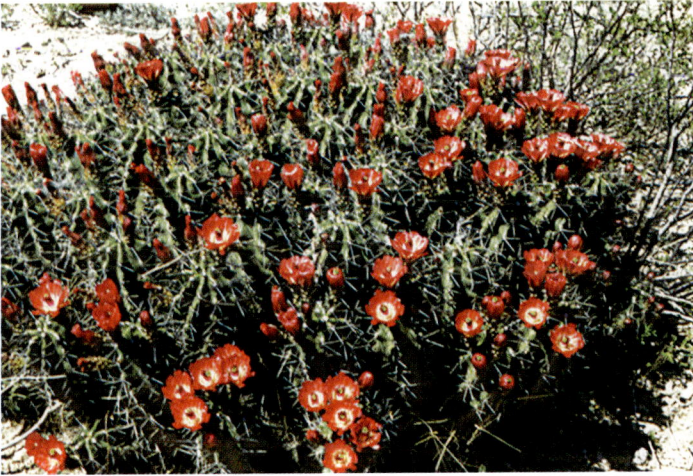

Plate 23. Multistemmed (branched) specimen of *Echinocereus coccineus* (claret-cup cactus) from Pecos Co., TX.

Plate 24. Some cryptic Trans-Pecos cacti. Find *Epithelantha bokei* (Boke's button-cactus), left middle; *Echinomastus mariposensis* (Mariposa cactus), upper right in flower; and *Ariocarpus fissuratus* (living rock cactus), lower right.

Plate 25. Apical mat of hair pulled away from fruit in *Echinocactus horizonthalonius* (eagle-claw cactus).

Plate 26. Woody skeleton of *Opuntia imbricata* (tree cholla).

funicular girdle usually hardens, forming a tan or brownish bony aril-like envelope surrounding the seed (Anderson, 2001). In the taxonomic section for convenience we sometimes use the terms "aril," "arillate," "aril-rim," or "pseudoaril" in reference to the funicular girdle.

Floral Remnant. The shriveled remains at the apex of a cactus fruit are best referred to collectively as the floral remnant (Fig. 1). Following successful pollination and subsequent fertilization in the ovules of the ovary, the flower collapses, dries, and sometimes twists. As the fruit matures, the apical remnant persists as a brownish or blackish structure. The floral remnant is more or less persistent in different species of Trans-Pecos cacti (characteristically deciduous in Opuntioideae). Persistent floral remnants (e.g., in *Ancistrocactus* spp.) may function to provide leverage for the dissemination of fruits and seeds, as might the beaks (necks) in other taxa (e.g., *Peniocereus greggii*). A beak of a fruit, rarely present in Trans-Pecos cacti, is protruding necklike tissue between the main body of the fruit and the floral remnant (see photo of *P. greggii* fruit), which is attached to the apex of the beak.

Coloring. The betalains are special water-soluble pigments found only in Cactaceae and some other plant and mushroom families. Betalains are similar in color to the pigmented compounds common in most plant species. Betalains form two color groups, betacyanins (red, magenta, pink, etc.) and betaxanthins (yellows and orange-red).

In cacti, betacyanins are responsible for the reddish colors of the stems, flowers, and fruits. Most yellow colors in cacti probably result from betaxanthins. The purple stems of the long-spined (purple) prickly pear (*Opuntia macrocentra*) are the result of betacyanins (Plate 29), and its yellow flowers are colored by betaxanthins.

Ecology

Although some cactus species can be found almost everywhere in the Trans-Pecos, more taxa and more individuals tend to be concentrated in certain habitats. Cacti tend to occur in thin, rocky soils or in soil-filled rock crevices on south and west exposures. These conditions tend to favor the establishment of cacti and restrict the growth of other, competing vegetation.

We speculate that desert cacti in the Trans-Pecos, if not those in the grasslands and mountains as well, become established mostly in the protection of nurse plants or similarly effective microhabitat modifiers, such as boulders, ledges, or other sources of shade. Nurse plants seem to protect and nourish cactus seedlings, which probably are more vulnerable to temperature extremes, desiccation, and nutrient deficiencies than are adults. Juveniles and young adults characteristically are associated with other plants (Plate 30), which we presume function as nurse plants.

Conservation

There is no unanimity about how to conserve cacti. Legislative attempts include (1) the Convention of International Trade in Endangered Species of Wild Fauna and Flora (CITES), an international treaty that applies only to international trade in species listed for protection under the treaty; (2) the Lacey Act, a statute that can impose federal, civil, and criminal penalties for violations of state "wildlife" conservation laws, including plant laws, and involving, among other entities, the U.S. Fish and Wildlife Service and state governments in legal protection for listed cactus species; (3) the Endangered Species Act (ESA), relating to the protection of federally listed ("Threatened" or "Endangered") species of cacti; (4) state laws by which many states, including Texas, protect state-listed species (Arizona, for example, restricts the removal of all cacti and other specialized wild plants from natural populations); and (5) regulations in various parks and preserves.

Success of formal regulations depends largely upon the cooperation of those who are involved in commercial trade, selling and buying the plants. Many people interested in cactus conservation believe that cacti should be propagated and made available in such large numbers and at such reasonable prices that the incentive for commercial harvesting will diminish. In Arizona and elsewhere cacti are being propagated in large numbers.

Overharvesting

Wholesale cactus harvesting has taken place mostly in the deserts of the southwestern United States and in northern Mexico, where cacti are most abundant and diverse. In Trans-Pecos Texas, especially the Big Bend area, uncountable tens of thousands of cactus plants have been harvested for wholesale trade (Plate 31). Many of the cacti gathered for sale in the Big Bend or for shipment out of the Big Bend were harvested in adjacent Chihuahua and Coahuila, Mexico. Individuals and teams of laborers have been digging large numbers of cacti in the desert and mountain habitats since before 1930, and probably for much longer.

Plate 27. Fasciated (crested or cristate) individual of *Thelocactus bicolor* var. *flavidispinus* (Marathon Basin thelocactus).

Plate 28. Representative cactus flowers showing pericarpel, receptacular tube, sepaloid tepals, and petaloid tepals (*Echinocereus* X *roetteri* var. *neomexicanus,* Lloyd's hedgehog cactus).

Plate 29. Betacyanin pigmented stems of *Opuntia azurea,* a purple prickly pear.

Plate 30. *Echinocereus chisoensis* (Chisos hedgehog cactus) in "nurse plant" association with *Larrea tridentata* (creosote bush) and *Opuntia aggeria* (clumped dog cholla).

Plate 31. Harvested cacti awaiting sale in southern Brewster Co., TX.

Classification of Cacti

Recent estimates place the number of legitimate species in the family at 1,500–1,600 in 100–122 genera (Nobel, 1988; Hershkovitz and Zimmer, 1997). Traditionally, three subfamilies of Cactaceae—Pereskioideae, Opuntioideae, and Cactoideae—were recognized, but more recently additional subfamilies, Maihuenioideae and Blossfeldioideae, have been proposed (Anderson, 2001; Nyffeler, 2002; Crozier, 2004). Members of the Opuntioideae and Cactoideae are represented in the Trans-Pecos. About 80% of the cactus species, and most of the Trans-Pecos cacti, are classified in subfamily Cactoideae.

This guide includes keys, descriptions, abbreviated taxonomic discussion, distribution maps, and color photographs for the Trans-Pecos taxa. It also includes color photographs, distribution maps, and short descriptions/discussions of Texas cacti from north and east of the Pecos River. Texas cactus taxa most briefly treated in the present manuscript include *Echinocereus reichenbachii* varieties *baileyi/albispinus, perbellus,* and *albertii*; the cylindroid "setaceus" race of *Hamatocactus bicolor;* the endemic race *roemeri* of claret-cup cacti in the granite area of Central Texas; and *Echinocereus milleri* in Coke County. For each of the Trans-Pecos cactus taxa, more detail can be found in Powell and Weedin (2004). The Trans-Pecos cacti are listed by Latinized scientific names under their respective genera. The species names are binomials, consisting of the genus name, always capitalized (e.g., *Opuntia*) and specific epithet, always in lowercase (e.g., *aggeria*). Taxa below the rank of species, which zoologists would call "subspecies" but which plant nomenclature has traditionally called "varieties," bear trinomials. The designation of scientific names follows the International Code of Botanical Nomenclature. The Code dictates the recognition of a single legitimate binomial name for each plant species or a single trinomial for each infraspecific variety. Any other scientific names that have been applied to a species, through nomenclatural superfluity or taxonomic misinterpretation, are regarded as synonyms. Pertinent synonyms are listed under each species.

Often several or numerous common names, sometimes more accurately referred to as popular names, vernacular names, or English names, have been applied to the same Texas cactus species. The common name that we deem most appropriate, based upon historical, taxonomic, or other reasons, appears in capital letters at the right of each Latin name. Other common names are listed for some species.

Key to the Genera

A taxonomic key is an artificial device constructed to help identify an unknown organism, such as an unknown cactus plant. The keys in this book provide a choice between two different "leads," arranged in pairs (here indicated by numbers: 1,1; 2,2; 3,3; etc.). Each pair of leads is known as a couplet. Some of the keys are limited to a single couplet when identification of only two taxa is required, and other keys are composed of a more extensive series of couplets, depending upon the number of taxa that have been grouped for identification. Where possible, our cactus keys contrast macromorphological characters, but the leads often include microscopic and sometimes highly technical features that are required for the distinction of closely related or morphologically similar taxa.

To use the keys, evaluate the characters of the unknown cactus plant that are required to make a choice between leads in the first couplet (number 1). If the second lead is chosen, a parenthetical number at the end of that lead directs the user to the next couplet. A number in parentheses at the beginning of a first lead reminds the user about which couplet was previously consulted.

Because leads of a couplet differ from each other in very discriminating ways, they must include scientific terms with specific meanings. Throughout the text scientific terminology is simplified, as much as is feasible, but as is true in any discipline, the use of specific terms not only preserves accuracy but also saves space. For definitions of terms, see the Glossary.

1. Plants erect, sprawling shrubs or low mats (cholla and prickly pear); stems jointed, the segments abruptly separated, cylindroid, globular, or laterally compressed (pads); glochids present, usually in tufts, in areoles of stems, flowers, and fruits; leaves present on new growth, usually terete, fleshy, and caducous; seeds often discoid, always enclosed by a bony tan or pale pseudoaril; cotyledons foliaceous, ca. 1 cm long, usually deciduous; subfamily Opuntioideae

Opuntia, p. 33

1. Plants cespitose, few-branched, or of solitary stems; stems cylindroid, globular, dorsiventrally flattened, or angular, with ribs or tubercles, not jointed (except seemingly in *Peniocereus*); glochids absent; leaves absent (or stems perhaps with microscopic scales or protrusions); seeds generally spherical or elongated in shape, not discoid, pseudoaril absent; cotyledons deltate to hemispheric and short, much less than 1 cm long to vestigial, persistent; subfamily Cactoideae (2).

2(1). Stems slender, 10–50 times longer than thick, erect or clambering in shrubs, prominently 3–6 angled; spines 1–5 mm long, appressed and arachnid-like on the stem angles; epidermis velvetlike with microscopic unicellular trichomes; flowers nocturnal, white, 11–17 cm long; root tuberlike, often deep underground; tribe Pachycereeae

Peniocereus, p. 124

2. Stems thicker, erect, globular, or dorsiventrally flattened, rarely elongate (inherently prostrate in some *Echinocereus*); stems ribbed or tuberculate, these features sometimes inconspicuous or small and hidden by spines; spines various or absent; epidermis glabrous or merely papillate; flowers diurnal (but in some remaining open at night), white or colored, shorter; roots various (3).

3(2). Receptacle surface bearing five or more spiny areoles, these comparable to those of the stem but smaller; stems 5–19 ribbed; stem tissue strongly mucilaginous; flower buds erupting through epidermis above areoles; seeds black, testa cells strongly convex, in some species forming irregular ridges; tribe Echinocereeae

Echinocereus, p. 128

3. Receptacle surface spineless, bearing spine-tipped bracts with woolly axils in *Echinocactus*; stems ribbed or tuberculate; stem tissue mucilaginous or not; flower buds forming in areoles with or without spines, in areolar grooves, or in tubercle-stem axils (the origin often obscured under wool at stem apexes); seeds black, brown, reddish, or yellowish, testa cells convex, flat, concave, or papillate but not forming ridges; tribe Cacteae (4).

4(3). Plants spineless, or appearing so (short or rudimentary spines often sporadic and hidden by trichomes, or in seedlings) (5).

4. Plants with spines in the areoles (often short, hairlike, or flattened, but usually many in a characteristic pattern and readily visible) (6).

5(4). Plants soft to the touch; stems weakly tuberculate or weakly ribbed, the areoles with tufts of trichomes to 1 cm long

Lophophora, p. 223

5. Plants hard to the touch (cuticle many times thicker than epidermis); stems strongly tuberculate, the tubercles pyramidal, conical, cylindroid-acute, or dorsiventrally flattened, fissured, forming a rosette

Ariocarpus, p. 227

6(4). Spines, at least one or some of them, hooked (7).

6. Spines not (any of them) hooked (12).

7(6). Stems of adults usually greatly exceeding 15 cm in diameter, the larger barrel cacti in the Trans-Pecos (*F. wislizeni*, Franklin Mts, and *F. hamatacanthus* east of the Franklin Mts; juveniles of *F. hamatacanthus* most likely to be confused with *Glandulicactus*)

Ferocactus, p. 205

7. Stems of adults usually less than 15 cm in diameter (except 10–30 cm wide in *Echinocactus*) (8).

8(7). Tubercles, if evident, without areolar grooves (9).

8. Tubercles with areolar grooves (10).

9(8). Stems tuberculate; flowers rose to magenta (*M. grahamii* and *M. wrightii*)

Mammillaria, p. 267

9. Stems ribbed; flowers yellow with red centers

Hamatocactus, p. 214

10(8). Lower three radial spines hooked, in addition to the hooked central spine; flowers dark reddish

Glandulicactus, p. 233

10. No radial spines hooked; flowers yellow, white, or greenish with brown or rose-pink (11).

11(10). Tubercles usually 1.5–3 cm long; central spine terete, straight, curved, or hooked; pith and cortex nonmucilaginous; flowers yellow (*C. scheeri* var. *uncinata*)

Coryphantha, p. 287

11. Tubercles usually less than 1 cm long; central spines dorsiventrally flattened, the lower one hooked; pith and cortex strongly mucilaginous; flowers white or greenish with brown or rose-pink (yellow in *A. tobuschii*)

Ancistrocactus, p. 237

12(6). Spines robust and strongly annulate or cross-ribbed (spines of some *Ferocactus* also annulate); floral bracts spinose-tipped, their axils conspicuously woolly

Echinocactus, p. 199

12. Spines relatively slender or less robust, not annulate or cross-ribbed; floral bracts not spinose-tipped or with woolly axils (13).

13(12). Central spines, at least some of them, strongly flattened and bladelike (14).
13. Central spines, terete, angled, or flattened usually on one side, not bladelike (15).

14(13). Central spines 1–4 per areole, 1–3 upper ones (in near radial position) flat and straight or curved, 1.3–7.5 cm long, 0.2–1.5 mm wide; radial spines 10–20, largest ones 1–1.5 cm long; flowers rose-pink to magenta

Thelocactus, p. 217

14. Central spines 1–3 per areole, usually one porrect, flat, and often curled (resembling a dried grass leaf blade), 1.2–2.7 cm long, 0.4–2 mm wide; radial spines 6–9, largest ones 2–5 mm long; flowers white

Toumeya, p. 246

15(13). Plants small, adult stems usually 1.5–3.5 cm in diameter near ground level, to ca. 3 cm long, rarely longer; tubercles less than 2 mm long, ungrooved; spines all equally thin; dense covering of whitish spines obscuring the stem, the spines 2–3 mm long on the sides of the stem but appressed and collectively forming a smooth surface, spines slightly longer (to 7 mm) at the stem apex; flowers pink to white, rarely yellowish, produced immediately adjacent to spine clusters in a woolly apical region

Epithelantha, p. 261

15. Plants usually larger, with adult stems exceeding 3.5 cm in diameter and 3 cm long, except for *Mammillaria lasiacantha*, which resembles *Epithelantha* in habit, and a few smaller species of *Echinomastus* and *Coryphantha* with stems covered by whitish spines but these different in habit; tubercles longer than 2 mm, if present (3–6 mm long in *M. lasiacantha*), grooved or ungrooved; spines usually tapering to a point; spines whitish or not, covering the stem or not, usually longer than 5–6 mm on sides of stem, forming a bristly surface except in *M. lasiacantha*; flowers of various colors, borne apically or laterally (16).

16(15). Tubercles without areolar grooves; flowers and fruits lateral (i.e., away from the stem apex, and often conspicuously on the sides of the stems, as in *M. pottsii*); flowers emerging from tubercle axils, often subtended by trichomes, spines, or bracts, which also are present in sterile condition

Mammillaria, p. 267

16. Tubercles with areolar grooves; flowers and fruits apical; flowers emerging from the areoles or areolar grooves, although sometimes at the base of the groove near the tubercle axil, spines or bracts not present in the tubercle axils (17).

17(16). A thin, yellowish, mucilaginous layer beneath older stem surface tissue (bark), usually evident in longitudinal sections of living plants; pith and cortex not mucilaginous; flowers pink to magenta; fruits dry, thin-walled, green to tan or whitish when mature; seeds like those of *Ariocarpus*

Neolloydia, p. 230

17. A thin, yellowish, mucilaginous layer not present beneath old bark in living plants; pith and cortex mucilaginous or not; flowers white, yellow, pink, or magenta; fruits scarcely succulent to succulent, green to red when mature; seeds very different (18).

18(17). Stems ribbed, the ribs well or poorly defined, the podaria decurrent and confluent to some degree; areolar glands absent; fruits green, scarcely succulent at first, quickly drying after ripening

Echinomastus, p. 249

18. Stems strongly tuberculate; areolar glands present except in subgenus *Escobaria*; fruits green or red, usually succulent (often juicy), in most species remaining succulent after ripening

Coryphantha, p. 287

Other genera of cacti in Texas outside the Trans-Pecos are listed following p. 350.

Descriptive Cactus Flora

OPUNTIA

Club Chollas, Chollas, Prickly Pears

Opuntia Mill. sensu lato is a genus of ca. 200 taxonomic species distributed over much of the Western Hemisphere from southern Canada to southern South America. Some species of *Opuntia* occurs in every state of the contiguous United States except Maine, Vermont, and New Hampshire (Benson, 1982). Many species are native to Mexico, Central and South America, and the West Indies; one species-group of prickly pears occupies the Galápagos Islands. *Opuntia* species have been introduced worldwide in warm climates, including the Hawaiian Islands. Glochids are produced in all *Opuntia*, but they are most pernicious in prickly pears. Stamens of many species are sensitive to touch, quickly closing around the style. The genus name is from Opus, an ancient town in Greece, where the name was inspired either from a cactuslike plant or from a prickly pear of early introduction there (Meyer and McLaughlin, 1981).

Opuntia is notorious for its taxonomic complexity. Natural hybridization between species and vegetative reproduction are responsible for at least some of the complexity, and there are relatively few exomorphic characters to reflect the underlying phylogenetic relationships. Because prickly pears often are difficult to identify from herbarium specimens, most type specimens are difficult to compare with present-day collections or populations. To reliably interpret the type collections, often it is necessary to visit type localities to study the habit and other characters that are evident only from living plants and/or populational samples.

The taxonomy of *Opuntia* remains poorly understood today, although it has been greatly clarified in recent years through the efforts of D. Pinkava and his students and D. Ferguson. Our treatment of *Opuntia* is tentative, pending much additional populational, chromosomal, genetic, and other biological investigation of the taxa. We suspect that we have overlooked as many as 10 to 20 potentially recognizable taxa of cholla and prickly pear in the Trans-Pecos.

Classically, two great subgenera are recognized in *Opuntia* (Britton and Rose, 1919–23), along with a number of series and sections (Benson, 1982) that further organize the genus. One traditional subgenus is *Cylindropuntia*, the species with cylindroid stem joints, and the other is *Opuntia*, with flat stem joints. The cylindropuntias are known collectively as the chollas; the platyopuntias (flat opuntias) are known as prickly pears in English or *nopales* in Spanish.

The Cactaceae Working Party of the International Organization for Succulent Plant Study (IOS) has decided to subdivide the subfamily Opuntioideae into 14 genera, mostly restricted to South America (Anderson, 1999b). In the United States and northern Mexico, the prickly pears make up part of *Opuntia*, the chollas compose *Cylindropuntia*, and the club/dog chollas (with miscellaneous others) are placed into a newly expanded concept of *Grusonia*. The elevation of traditional subgenera, such as *Cylindropuntia*, to generic level was proposed earlier by Robinson (1973). Generic status for the prickly pears (*Opuntia*), *Cylindropuntia*, and *Grusonia* was accepted by D. J. Pinkava in the *Flora of North America* (Pinkava, 2003a, b, c, d), but the latter two groups continue to be regarded as closest relatives of one another. The Trans-Pecos club/dog chollas might best be placed in the genus *Corynopuntia* F. M. Knuth (P. Griffith, 2002).

In this guide three subgenera of *Opuntia* are recognized for the Trans-Pecos species: subgenus *Grusonia* (F. Rchb. & K. Schum.) Bravo (club/dog chollas), subgenus *Cylindropuntia* Engelm. (chollas), and subgenus *Opuntia* (prickly pears).

Key to the Trans-Pecos Species

1. Stem segments (joints) cylindroid; unweathered young fresh spines enclosed by evident paperlike sheaths, or the sheaths only at the tips of some unweathered spines in the mat- or mound-forming dog chollas; seeds not encircled by a projecting rim (2).
1. Stem segments (cladodes or pads) strongly flattened (except in O. *polyacantha* var. *arenaria* and in seedlings of all species); spines without sheaths; seeds encircled by a projecting (or visibly) thickened specialized rim, the funicular envelope, 0.1–1 mm or more high (subgenus *Opuntia*) (11).

2(1). Spine sheaths rudimentary; plants mat- or mound-forming; (subgenus *Grusonia*) (3).
2. Spine sheaths full-size and persistent for at least one year; plants erect or low, compact, intricately branched shrubs, not mat- or mound-forming, except for O. *tunicata* and stunted O. *davisii* (with especially conspicuous papery sheaths) (subgenus *Cylindropuntia*) (7).

3(2). Stem segments 7–17 cm long, 2.5–5 cm in diameter; tubercles 2.5–3.5 cm long; largest spines 1.5–3 mm in diameter; NW Presidio Co., near the Rio Grande

O. *emoryi*, p. 48

3. Stem segments 3–7 cm long, 1.5–3.5 cm in diameter; tubercles 0.8–2 cm long; largest spines 0.4–2 mm in diameter; widespread in Trans-Pecos (4).

4(3). Roots tuberous (including the large adventitious roots, at least those more than a year old); central spines usually whitish-gray, stramineous to pinkish, less often red-brown, bladelike to nearly terete (5).

4. Roots all diffuse (including even the primary roots of the oldest plants); central spines red-brown or tan (to pinkish or white), bladelike (6).

5(4). Radial spines 2–4; central spines usually whitish-gray, 0.8–1.3 mm in diameter; stem segments all firmly attached, clavate or narrowly obovoid

O. aggeria, p. 40

5. Radial spines 6–8; central spines usually stramineous to pinkish or red-brown, 0.4–0.8 mm in diameter; stem segments (at least distal ones) weakly attached, ovoid or obovoid to cylindroid

O. schottii var. *grahamii*, p. 44

6(4). Central spines red-brown; glochids of second-year and older areoles rarely more than 14 per areole, in tufts 5 mm or shorter; plants usually trailing or creeping, forming open mats or loose chains; stems readily disarticulating and rooting

O. schottii var. *schottii*, p. 42

6. Central spines tan (or pinkish to white); glochids of second-year areoles typically 15–40, in tufts 6–8 mm long; plants usually in dense mats or mounds with an obvious central root system; stems relatively woody, not disarticulating

O. densispina, p. 46

7(2). Plants densely branched, either low and compact or erect, the stems obscured by dense spines; spine sheaths loose; flowers yellowish-green or brownish, pinkish in some *O. davisii* (8).

7. Plants erect, openly branched, the stems not obscured by spines; spine sheaths usually tight; flowers magenta, purplish, or yellow-green (9).

8(7). Plants low, compact; spine sheaths silvery-white; restricted distribution mostly near S slopes of Glass Mts

O. tunicata, p. 50

8. Plants erect, usually at least one or more main axes bearing horizontal branches held aloft above the ground; spine sheaths tan, yellowish, or golden; wider distribution, mostly in grassland of mountain basins

O. davisii, p. 52

9(7). Plants short-trunked treelets or relatively large shrubs; ultimate stem segments 1.5–3 cm in diameter; flowers bright rose-pink to magenta; fruits yellow, 2–3.5 cm in diameter

O. imbricata, p. 54

9. Plants usually smaller shrubs; ultimate stem segments 0.3–1.5 cm in diameter; flowers cream-colored or yellowish-green to dull pink or magenta; fruits red or orange, 0.9–2 cm in diameter (10).

10(9). Ultimate stem segments 0.6–1.5 cm in diameter; flowers magenta to purple-red, violet, brownish, bronze, or greenish-red; fruits 1–2 cm in diameter, tuberculate

O. kleiniae, p. 61

10. Ultimate stem segments 0.3–1 cm in diameter; flowers cream-colored or yellowish-green, never pinkish; fruits 0.9–1.5 cm in diameter, smooth

O. leptocaulis, p. 58

11(1). Plants spineless (but glochids abundant); epidermis (including that of the fruit) microscopically hairy (velutinous)

O. rufida, p. 65

11. Plants bearing at least a few persistent spines, not merely glochids; all surfaces glabrous (12).

12(11). Plants usually less than 30 cm high; spines usually whitish to gray, less often darker colored proximally, distally, or throughout (in O. mackensenii var. minor), rarely yellowish (13).

12. Plants usually more than 30 cm high; spines brown, reddish-brown, dark purple, blackish, yellow, golden, bi- or multicolored, or whitish (17).

13(12). Pads not wrinkling; plants slightly larger, taller (see text); main spines usually 2–5 per areole; tetraploid

O. mackensenii (in part, when less than 30 cm tall), p. 88

13. Pads wrinkling transversely under stress; plants lower, usually less than 20 cm tall (14).

14(13). Branches usually upright from a caudex, not forming long chains; healthy pads glaucous blue-green; usually 1–3 whitish spines in distal areoles, absent proximally; flowers (in our region) red; roots tuberous-thickened

O. pottsii, p. 86

14. Branches longer (chains of pads usually prostrate or sprawling); healthy pads usually brighter green; 1–17 whitish spines in distal areoles or in all but the lower areoles; flowers (in our region) yellow or yellow with red centers; roots diffuse or weakly/sporadically tuberous (15).

15(14). Spines 6–17 per areole; fruits dry, tan, very spiny; largest seeds of any Trans-Pecos *Opuntia*; all diploid in the Trans-Pecos
O. polyacantha, p. 109

15. Spines usually 1–5 per areole; fruits fleshy, reddish, usually spineless or with few spines; seeds smaller; polyploids; relatively similar to *O. phaeacantha*, except for smaller and wrinkling habit (16).

16(15). Main spines usually 1–3 in distal areoles; NE periphery of Trans-Pecos, in Ward Co. and adjacent counties; tetraploid; most similar to *O. pottsii*
O. macrorhiza, p. 84

16. Main spines usually 2–5 in all but the lower areoles; throughout much of the Trans-Pecos (central Trans-Pecos west to El Paso Co.); hexaploid
O. tortispina, p. 82

17(12). Pads "officially" purplish (can be glaucous when fresh); flowers red-centered (18).

17. Pads typically greenish (purplish only during stress, if at all); flowers all yellow or red-centered (21).

18(17). Pads more purple, orbicular to broadly ovate or broader than long (19).

18. Pads less purple, basically blue-green or blue-gray, obovate to orbicular (20).

19(18). Distal pads usually orbicular or broader than long; distribution widespread in the Big Bend region, N at least to Jeff Davis Co.; diploid
O. azurea var. *diplopurpurea*, p. 74

19. Distal pads usually broadly ovate, sometimes orbicular; Trans-Pecos distribution El Paso Co. E to at least Culberson Co.; diploid/tetraploid
O. macrocentra, p. 71

20(18). Pads mostly obovate, especially on upper branches; largest spines 5–12 cm long; diploid
O. azurea var. *parva*, p. 76

20. Pads mostly orbicular to obovate; spines usually 3–5 cm long; tetraploid
O. mackensenii (in part, when both tall and purplish), p. 91

21(17). Spines to 10.5 cm long, whitish, yellow, golden to reddish, often curved and twisted
O. azurea var. *discolor*, p. 80

21. Spines to 7.5 cm long, usually shorter, yellow, reddish, reddish-brown, brown, to blackish or white, sometimes bi- or multicolored (22).

22(21). Spines to 4 cm long, reddish to blackish proximally, yellow distally; pericarpel relatively small and spheroidal; inner tepals uniformly yellow; flowers widely opening; diploids (23).

22. Spines usually fewer than 7–13 per areole, with one or more main central spines; pericarpel narrower or longer; inner tepals either uniformly yellow or red basally; flowers less widely opening; polyploids except for O. *chisosensis* and O. *azurea* var. *aureispina* (24).

23(22). Spines 7–13 per areole, with one central longer, typically straight main spine; Terrell Co. northward and westward

O. *strigil*, p. 67

23. Spines usually 1–2 per areole, 1–2 longer, typically with one central downcurved or straight; central Terrell Co. eastward to SW Edwards Plateau

O. *atrispina*, p. 69

24(22). Plants low, spreading, trailing, or weak-stemmed, 30–60 cm high, with 1–3 whitish spines

O. *phaeacantha* var. *phaeacantha*, p. 97

24. Plants usually erect or spreading with stout stems, usually 30–60 cm or more high, with 1–11 spines of various colors, if mostly whitish then usually showing some brownish at least basally (25).

25(24). Spines typically yellow (red-spined color-phases or forms occur in some populations) (26).

25. Spines various colors, white to reddish, reddish-brown, to nearly black, not yellow; polyploids (28).

26(25). Plants usually over 1 m high; lower Pecos (Sheffield, Pandale) and Boquillas area downstream to the Gulf of Mexico; hexaploids

O. *engelmannii* var. *lindheimeri*, p. 107

26. Plants usually less than 1 m high; distribution mostly otherwise; diploids (27).

27(26). Spines 1–5 per areole, yellow in younger pads but darker with age; fresh flowers pale yellow, lacking red centers; fruits succulent, juicy, spineless; Chisos Mts, in the oak zone

O. *chisosensis*, p. 93

27. Spines 4–11 per areole, yellow to orange, brown, or nearly black; fresh flowers relatively bright yellow, with sharply defined red centers; fruits fast-drying, spiny; near Rio Grande, Mariscal Mt to Boquillas, in rocky desert

O. *azurea* var. *aureispina*, p. 78

28(25). Spines typically whitish (may have dark bases) or color-banded, but not all-brown; pads averaging relatively large (spines appearing relatively short; hexaploids) (29).

28. Spines (at least the largest ones) usually reddish-brown to nearly black, other colors; pads averaging smaller (spines appearing relatively long); tetraploids and hexaploids (30).

29(28). Spines usually 3–4 per areole arranged in a "bird's-foot" pattern; inner tepals all-yellow (may redden prior to wilting); hairy seedlings

O. engelmannii var. *engelmannii*, p. 104

29. Spines usually 2(–4) per areole, not in a "bird's-foot" pattern; inner tepals yellow with red bases; ordinary (bristly) seedlings

O. dulcis, p. 101

30(28). Fruits typically spiniferous distally, drying rapidly; erect shrubs; near Rio Grande, vicinity of Hot Springs, S Brewster Co.; tetraploid

O. spinosibacca, p. 95

30. Fruits spineless or rarely few-spined near the apex, remaining succulent long after ripening; low shrubs; widespread in the Trans-Pecos; hexaploid

O. camanchica, p. 99

Opuntia aggeria
Clumped Dog Cholla

PLATES 32, 33

Opuntia aggeria is a recently named member of a small group of closely related taxa, the *O. schottii* complex. In the Trans-Pecos these low, matted, mound-forming, or sprawling plants are known as dog chollas: "club cholla" is an international name for these and the larger "devil chollas" in the taxon *Corynopuntia*, currently included in the subgenus or segregate genus *Grusonia*. The stem joints of *O. schottii* and its closest relatives easily break apart and disperse as dangerously spiny nuisances that become embedded in shoes, tires, or hapless animals. The stem joints of *O. aggeria*, however, do not break off easily, and it is not at all a pest.

The type locality is Tornillo Flats at 2,800 feet, Big Bend National Park, Brewster County, Texas. The specific epithet refers to the aggregated growth habit of *O. aggeria*, after the Latin *aggestus*, "mound."

Distribution. Loosely consolidated desert alluvium, gravel to silt, often gypseous, igneous, or limestone. Presidio Co., extreme SE portion; Brewster Co., southern half, most common within 10–20 mi of the Rio Grande. 1,800–3,500 ft. Mexico: S into Coahuila, presumably in NE Chihuahua. Map 2.

Vegetative Characters. Distinguished by a combination of characters: roots thickened and tuberous; joints (stems) clavate, strongly persistent; new joint growth from lateral or lower areoles; central spines 3–4, usually grayish-white, less often reddish-brown, flattened or terete, divergent, cross-striated, 3–9 cm long, the widest 0.8–1.3 mm wide near the base; spine bases enlarged-bulbous; radial spines 2–4, grayish-white, slender, deflexed, 0.6–2.5 cm long; glochids numerous, 0.4–1 cm long. Occasional plants spineless.

Flowers. Flowering late Mar–Apr. Flowers of all club chollas open for a single day and are virtually identical except for filament color. Flowers of *O. aggeria* yellow, 5–7 cm long, 4–5 cm wide when fully open, with tepals in 3–4 whorls. Five or more stigma lobes cream-yellow to pale green, ca. 5 mm long. Relatively thick style very pale green to cream-colored, to 2.5 cm long. Anthers yellow to cream; filaments ca. 1 cm long. Innermost filaments yellow-orange to reddish in all Trans-Pecos club chollas, so far as known, except in *O. emoryi* and most flowers of *O. schottii* var. *schottii*, where all filaments are greenish. In at least some plants of *O. aggeria*, even outer filaments are reddish.

Fruits. Fruits light yellow, narrowly obconic or clavate, to 5.5 cm long, becoming gray and dry with age, areoles white-woolly with tufts of glochids and bristles 3–7 mm long. Floral remnant persistent as in all club chollas. Subdiscoid seeds brown to cream-colored, smooth, 5–6 mm in diameter, weakly pointed on one margin.

Full Name and Synonyms. *Opuntia aggeria* Ralston & Hilsenb. [= *Grusonia aggeria* (Ralston & Hilsenb.) E. F. Anderson]. *Corynopuntia aggeria* (Ralston & Hilsenb.) M. P. Griffith.

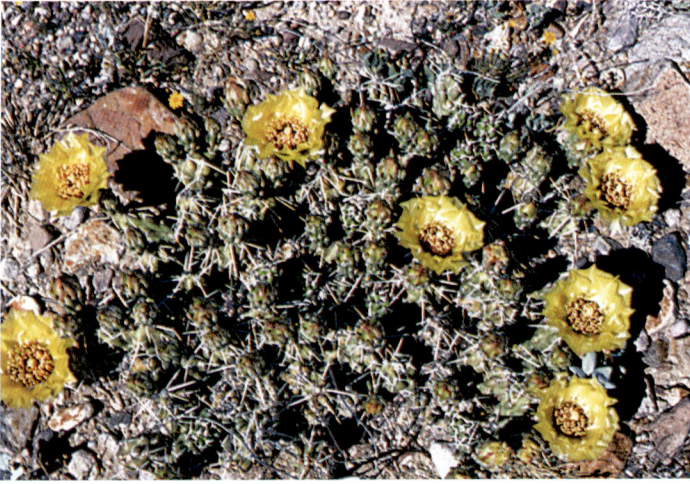

Plate 32. *Opuntia aggeria* (clumped dog cholla) near Study Butte, Brewster Co., TX.

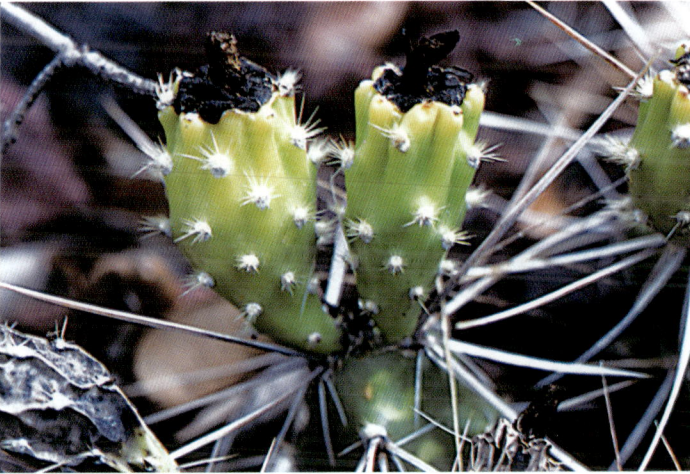

Plate 33. *Opuntia aggeria* (clumped dog cholla); maturing fruits.

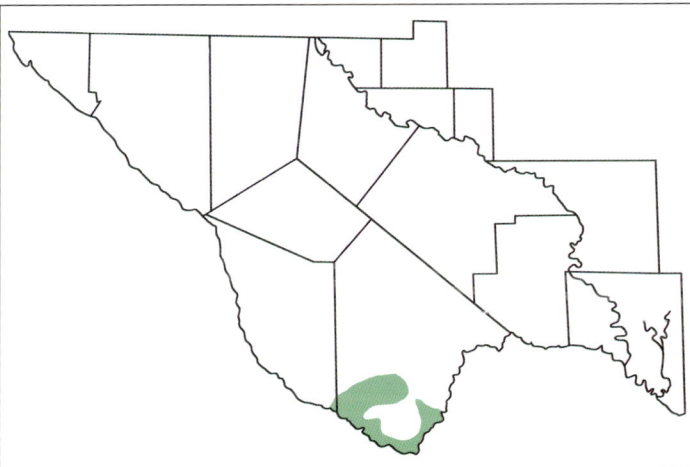

Map 2. *Opuntia aggeria* (clumped dog cholla).

Opuntia schottii
Common Dog Cholla

The type locality is "on the arid hills near the Rio Grande, between the San Pedro and Pecos Rivers (Wright, Schott)" presumably somewhere in Val Verde County. The specific epithet is after Arthur Schott, who collected the type of this species and many others during the United States and Mexican Boundary Survey of 1851–53.

Full Name and Synonym. *Opuntia schottii* Engelm. [= *Grusonia schottii* (Engelm.) H. Rob.].

Key to the Varieties of *Opuntia schottii*

1. Easternmost in distribution: Roots all adventitious (never taprooted); always-disarticulating chains of joints; major spines dark- or reddish-brown, flat and bladelike with translucent edges

    O. schottii var. *schottii*, p. 42

1. Westernmost in distribution: Usually taprooted, roots becoming tuberous after the first year; easily disarticulating joints three in a file; spines gray or brown, terete or moderately flattened, not bladelike

    O. schottii var. *grahamii*, p. 44

Opuntia schottii var. schottii
Schott's Dog Cholla

PLATES 34, 35

Dense populations of var. *schottii* in some parts of the southeastern Trans-Pecos, particularly in Val Verde County, hinder cross-country foot traffic. It appears that increasing desertification enhances the spread of these plants, as readily separated and dispersed spiny joints quickly strike root in bare soil.

Distribution. Alluvial substrates, desertscrub to Tamaulipan scrub. Southeastern Brewster Co., Terrell Co., Pecos Co., Crockett Co., and Val Verde Co. 1,000–4,000 ft. Southeast to Cameron Co., Rio Grande valley; eastern "outposts" in Schleicher and Brown counties (Benson, 1982). Mexico: adjacent in Coahuila; presumably on the Mexican side of the Rio Grande in Nuevo León and Tamaulipas. Map 3.

Vegetative Characters. Sprawling, ankle-high, distinguished primarily by diffuse roots and secondarily by clavate joints (new ones originating from lateral areoles). Central spines 6–8, reddish-brown, flattened, divergent, cross-striated, 3.8–7 cm long, the widest 1.5–2 mm in diameter near the expanded but not bulbous base. Radial spines four, grayish-white, slender, deflexed, 1–2 cm long. Glochids 10–14, ca. 5 mm long.

Flowers. Flowering Jun–Jul. Flowers vary considerably in size. Like those of *O. aggeria,* they may be slightly larger than those of var. *grahamii,* at least in the southern Big Bend region.

Fruits. Fruits and seeds of var. *schottii* similar to those of *O. aggeria,* except fruits

Plate 34. *Opuntia schottii* var. *schottii* (Schott's dog cholla), Langtry, Val Verde Co., TX.

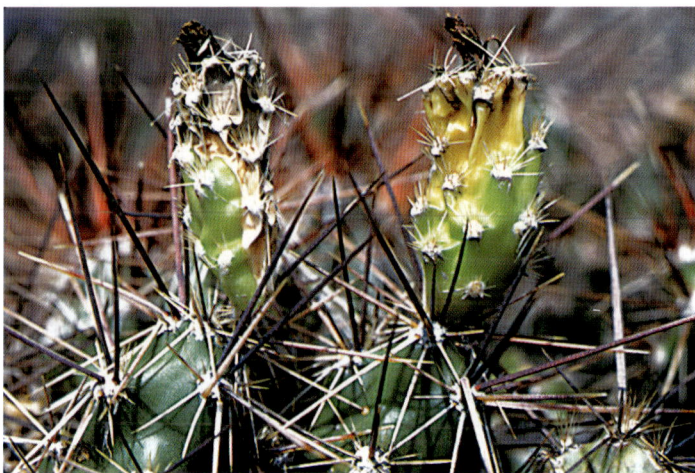

Plate 35. *Opuntia schottii* var. *schottii* (Schott's dog cholla); maturing fruits.

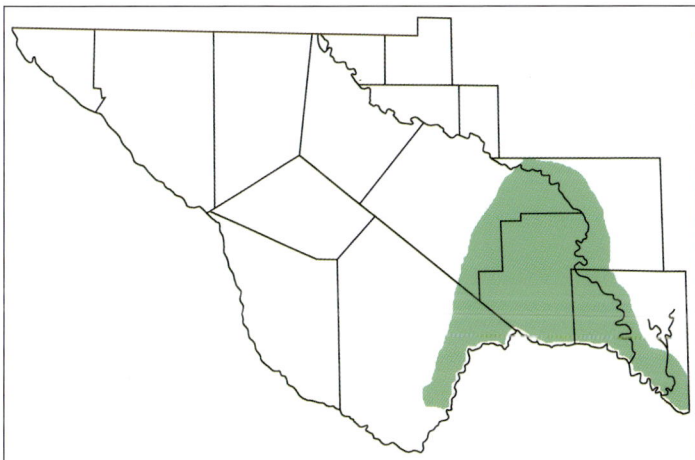

Map 3. *Opuntia schottii* var. *schottii* (Schott's dog cholla).

measured by the authors in var. *schottii* to 4.5 cm long.

Other Distinctions. In southern Brewster County a form of var. *schottii* (evidently not the "pure" form found farther east) is sympatric with *O. aggeria;* var. *schottii* has more, flatter central spines that are wider at the base and typically reddish-brown in color, and fewer glochids. Variety *schottii* also co-occurs with var. *grahamii* in southern Brewster County.

Full Name and Synonym. *Opuntia schottii* var. *schottii. Corynopuntia schottii* (Engelm.) F. M. Knuth.

Opuntia schottii var. grahamii
Graham's Dog Cholla
PLATES 36, 37

Opuntia schottii var. *grahamii* is one of the three dog chollas described from collections made during the United States and Mexican Boundary Survey of 1851–53 (Engelmann, 1856). The type locality is from "near El Paso," specifically "bottoms of the Rio Grande, and downriver

about 100 miles in sandy soil." The specific epithet honors Colonel James D. Graham, head of the scientific corps of the United States and Mexican Boundary Commission that provided many plant collections for study by Engelmann.

Distribution. Loosely consolidated alluvium, igneous or limestone, in desertscrub. El Paso Co. SE to (depending upon classification of intermediates) Brewster Co. 1,800–4,500 ft. South-central NM, E of the Rio Grande. Mexico: from eastern Chihuahua S possibly to Durango, Zacatecas, and San Luis Potosí, depending upon taxonomic circumscription of related endemic taxa. Map 4.

Vegetative Characters. Distinguished by thickened and tuberous roots (like those of *O. aggeria*) and also joints obovoid, ovoid, or cylindroid (new growth originating from apical or upper lateral areoles). Central spines 4–7, stramineous to brown, often tinged pinkish or reddish, to dull reddish, gray with age, terete or only slightly flattened, divergent, inconspicuously cross-striated, 2–4 cm long, the widest 0.4–0.8

Plate 36. *Opuntia schottii var. grahamii* (Graham's dog cholla) from Solitario Uplift, Presidio Co., TX; cultivated.

Plate 37. *Opuntia schottii* var. *grahamii* (Graham's dog cholla); maturing fruits.

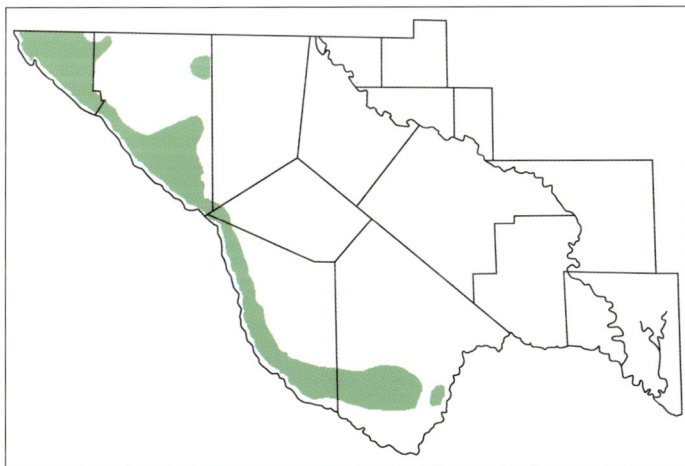

Map 4. *Opuntia schottii* var. *grahamii* (Graham's dog cholla).

mm in diameter near the bulbous base. Radial spines 6–8, grayish-white, slender, deflexed, 1–1.2 cm long. Glochids numerous, 5–7 mm long.

Flowers. Flowering May–Jun. In some plants, flowers slightly smaller than those of *O. aggeria*, but this size difference is not yet substantiated by populational measurements.

Fruits. Fruits similar to those of *O.*

aggeria and *O. schottii* var. *schottii*, except ones measured by the authors were ovoid or oblong and only 2–3.5 cm long.

Full Name and Synonyms. *Opuntia schottii* var. *grahamii* (Engelm.) L. D. Benson [= *Grusonia grahamii* (Engelm.) H. Rob.]. *Corynopuntia grahamii* (Engelm.) F. M. Knuth.

Opuntia densispina
Big Bend Devil Cholla
PLATES 38, 39

Opuntia densispina, the most recently described of the five *Grusonia* taxa in the Trans-Pecos, has a limited distribution. The type locality is on the River Road near Solis Ranch, Big Bend National Park, Brewster County, Texas. The specific epithet refers to the dense appearance of the spine clusters, after the Latin *densus*, "dense," and *spina*, "spine."

Distribution. Bare clay deposits and clay overlaid by gravel or sand, endemic, along and near the Rio Grande, between Mariscal Mt and the Dead Horse Mts, Big Bend National Park, Brewster Co. 1,975–2,200 ft. Not yet reported from adjacent Mexico. Map 5.

Vegetative Characters. Fibrous roots immediately distinguish it from sympatric *O. aggeria* and allopatric *O. schottii* var. *grahamii*, both of which appear visibly smaller. Joints of *O. densispina* strongly persistent, obovate to clavate, with central spines ca. nine, these usually white to pinkish-white or tan, flattened, divergent, cross-striate, 4–8 cm long, the widest 1–2 mm in diameter near the expanded base. Radial spines 2–4, whitish, to 2.5 cm long or longer. Glochids relatively numerous, up to 11 mm long. New stems emerge from lateral areoles.

Plate 38. *Opuntia densispina* (Big Bend devil cholla), Solis, Big Bend National Park, Brewster Co., TX; cultivated.

Plate 39. *Opuntia densispina* (Big Bend devil cholla); maturing fruits.

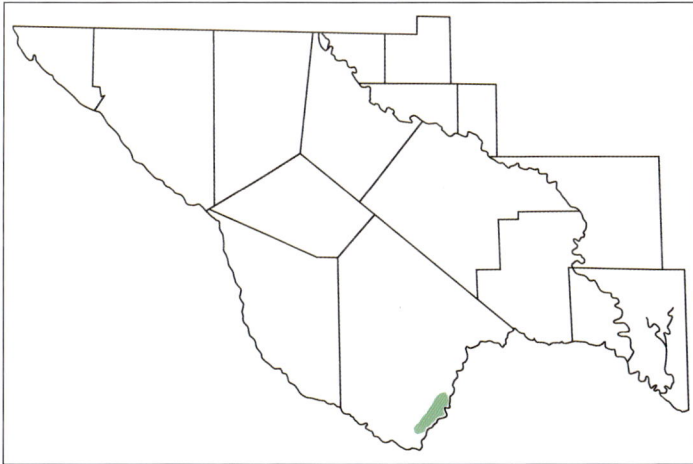

Map 5. *Opuntia densispina* (Big Bend devil cholla).

Flowers. Flowering May–Jun. Flowers like those of other Trans-Pecos club chollas. Filaments, particularly inner ones, reddish-orange or reddish, as in most plants of O. *aggeria*.

Fruits. Normal ripe fruits clavate, uniformly pale lemon-yellow inside and out but quickly drying and browning, with white wool, and with accrescent tufts of bristlelike pale glochids 3–9 mm long in bulging areoles. Subdiscoid seeds cream-colored to tan, smooth, 5–6.5 mm in diameter, 2.4–2.9 mm thick, weakly pointed at one margin.

Other Distinctions. In its natural habitat and typical glistening-mound habit, *O. densispina* is not likely to be confused with any other cactus species except the regionally sympatric *O. aggeria*. *Opuntia densispina* more closely resembles allopatric *O. schottii* var. *schottii,* but the central spines of *O. densispina* on average are longer, more numerous, more often twisted or curved, and of pale pink to tan color, different from the ashy-white aspect of *O. aggeria* and the bright chestnut-brown of typical (eastern) *O. schottii* var. *schottii*.

Full Name and Synonym. Opuntia densispina Ralston & Hilsenb. [= *Grusonia densispina* (Ralston & Hilsenb.) comb. nov., forthcoming].

Opuntia emoryi
Common Devil Cholla

PLATES 40, 41

Opuntia emoryi is easily distinguished from the other Trans-Pecos dog chollas by the larger size of all vegetative parts, except that *O. densispina* has longer spines. Benson (1982) and Weniger (1984) treated this taxon as *O. "stanlyi* Engelm.," an unaccepted provisional epithet (Pinkava and Parfitt, 1988). The type locality is between "the sandhills" (that is, the Samalayuca dunes) and Lake Santa Maria, south and west of El Paso, in Chihuahua, Mexico. The specific epithet honors Colonel W. H. Emory, U.S. Army Corps of Topographical Engineers, a major explorer of the southwestern United States, in charge of the United States and Mexican Boundary Survey and collector of many species of cacti during the earliest expeditions through the Southwest.

Distribution. Sand or gravel flats, washes, and low hills, Presidio Co., below the Sierra Vieja rim, near the Rio Grande between Candelaria and Porvenir. 2,300–3,300 ft. Main population in SE AZ and SW NM. Mexico: type locality in northern Chihuahua, Mexico. Map 6.

Vegetative Characters. Fibrous rooted; can be distinguished from *O. schottii,* sensu lato, by woody, persistent stems, large clavate joints 7–17 cm long, 2.5–5 cm in diameter. Central spines 6–7, tan to yellowish or reddish-brown, flattened, divergent, cross-striate, 4.5–7 cm long, the widest 1.5–3 mm in diameter (broader than those of any other Trans-Pecos *Grusonia* taxa) near the expanded or bulbous base. Radial spines 5–6, reddish-brown or pale yellow, slender, deflexed or divergent, 1.4–2.5 cm long. Glochids 3–30 in number and to ca. 5 mm long. New stems emerge from lateral areoles.

Flowers. Flowering May–Jun. Yellow flowers closely resemble those of the smaller Trans-Pecos dog chollas, except pericarpel may be slightly more elongate, 4–6 cm long, 1–1.3 cm in diameter, and flower slightly larger, 5.5–8 cm long, 4–7 cm wide. Stigma lobes 6–7, cream-colored.

Fruits. Fruits light yellow, clavate or cylindroid-turbinate, 5–6 cm long and 1.5–2.5 cm in diameter. Fruit surfaces display numerous yellowish glochids 5–8 mm long and rarely a few short spines. Fruits may be somewhat larger but are otherwise similar to those of *O. densispina* and the common dog chollas. Seeds slightly smaller, 4.5 mm in diameter, 1.5–2 mm thick, than those of related species.

Plate 40. *Opuntia emoryi* (common devil cholla) near Candelaria, Presidio Co., TX.

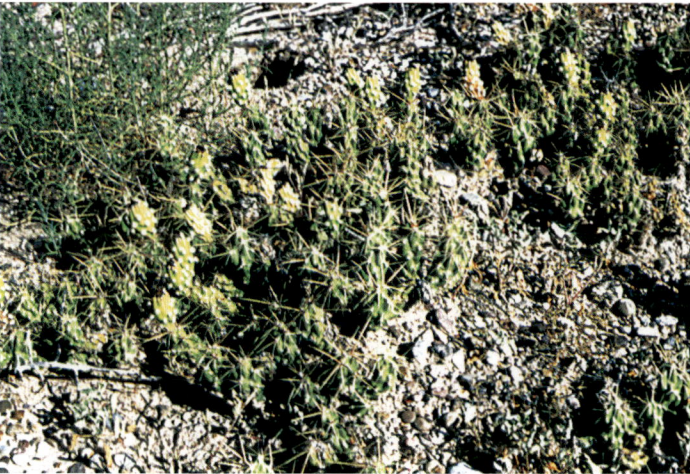

Plate 41. *Opuntia emoryi* (common devil cholla) near Candelaria, Presidio Co., TX; mature and maturing fruits.

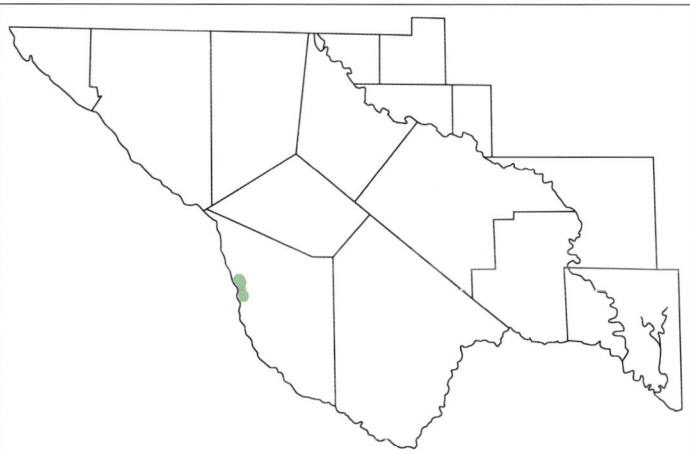

Map 6. *Opuntia emoryi* (common devil cholla).

Full Name and Synonyms. Opuntia emoryi Engelm. [= *Grusonia emoryi* (Engelm.) Pinkava]. *Opuntia stanlyi* Engelm. var. *stanlyi*; *Corynopuntia stanlyi* (Engelm.) F. M. Knuth; *C. emoryi* (Engelm.) M. P. Griffith; *Grusonia stanlyi* (Engelm.) H. Rob.

Opuntia tunicata
Icicle Cholla

PLATES 42, 43

Populations of *O. tunicata* in Mexico are much more extensive than the one most conspicuous population in the United States, visible from the highway in probably less than one square mile of the southeastern Glass Mountains, Pecos County. The largest plants form low-spreading clumps to ca. 1 m in diameter and ca. 35 cm high. On bright days, or when backlit through the glistening mass of spine sheaths, the plants glow as if coated with ice, as reflected in the common name. The type locality is "In Mexico." The specific epithet is after the Latin *tunica*, "garment," a reference to the sheathed (clothed) spines.

Distribution. Localized populations, upper margin of basin grassland, rocky slopes, associated with grasses, shrubs, and *Pinus remota* (Little) D. K. Bailey & Hawksw., eastern Glass Mts, S slope, Pecos Co.; smaller colonies in Brewster Co., outlier of the S Glass Mts, and limestone mesa W of Sanderson near the Brewster-Pecos county line. 4,500–5,000 ft. Mexico: S margins of the Chihuahuan Desert Region at 6,000–6,900 ft. Also Cuba and South America (Ecuador, Peru, and Chile). Map 7.

Vegetative Characters. Plants typically compact masses of 50 or more branching main stems covered with sheathed spines (per areole usually 3–5 sheathed central spines, 2.5–5.5 cm long, to 1 mm wide) almost obscuring stems. Papery-white or silvery-white spine sheaths broader (2–3 mm) than the spines themselves, collectively exhibiting a sheen or glistening mass, as if every spine were individually wrapped in a translucent toothpick sleeve. In the vicinity of the Glass Mountains no other cactus species is likely to be confused with *O. tunicata*, except perhaps *O. davisii*, which is clothed with equally baggy but plain tan, yellowish, or golden spine sheaths. Plants of *O. davisii* occur in the alluvial basin grasslands adjacent to and south of the Glass Mountains and elsewhere in the Trans-Pecos. Typically, some plants in each population or colony of *O. davisii* are erect, with numerous low branches.

Flowers. Flowering May–Jul. Greenish-yellow flowers 3.5–5 cm in diameter. Both outer and inner tepals are greenish-yellow. Stigma lobes 3–6, greenish to greenish-yellow, thick, and ca. 3 mm long. Style greenish, sometimes with a rose tinge. Anthers yellow, and greenish filaments ca. 6 mm long.

Fruits. Fruits somewhat fleshy and yellow at maturity. Tuberculate, obconic fruits ca. 3 cm long, 1.2–1.3 cm across, concave at the apex. Felty areoles support brownish glochids, but usually no spines. Seeds tan, somewhat obovate, 3–3.5 mm long, ca. 2 mm wide, ca. 1 mm thick, with thin aril. Sometimes fruits develop without seeds.

Full Name and Synonyms. Opuntia tunicata (Lehm.) Link & Otto [= *Cylindropuntia tunicata* (Lehm.) F. M. Knuth]. *Opuntia tunicata* (Lehm.) Link & Otto var. *tunicata*.

Plate 42. *Opuntia tunicata* (icicle cholla), Glass Mts, Pecos Co., TX (photo by Marcos A. Rodriguez).

Plate 43. *Opuntia tunicata* (icicle cholla), Glass Mts, Brewster Co., TX; maturing fruits.

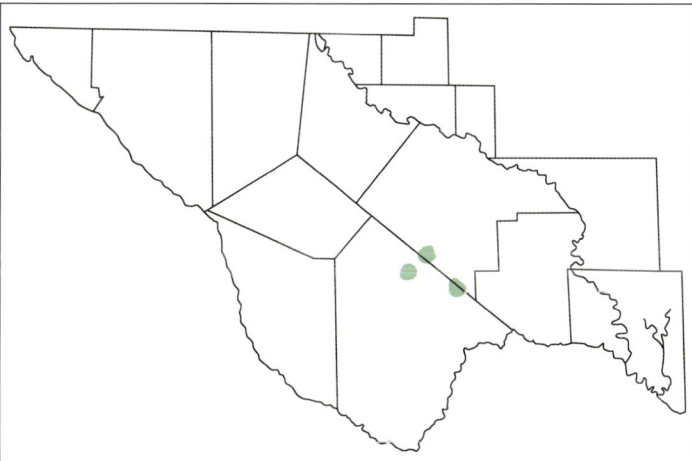

Map 7. *Opuntia tunicata* (icicle cholla).

Opuntia davisii
Davis' Cholla
PLATES 44, 45

In the Trans-Pecos O. *davisii* most likely is to be observed from paved roads through the plains grasslands in the mountain basins near Marfa, Fort Davis, and Marathon. The erect shrublets or mounds, often filled with tall grass, are so densely covered with sheathed spines as to hide the branching pattern of slender stems. When backlit, the plants are conspicuous from great distances owing to their glistening golden spines. The type locality is the "Upper Canadian, about Tucumcari Hills, near the Llano Estacado," probably in northeastern New Mexico. The specific epithet honors Jefferson Davis, the U.S. Secretary of War at the time the Pacific railway surveys were conducted and also the namesake of Jeff Davis County.

Distribution. Widespread but scattered and uncommon, or several plants in local colonies, alluvial mountain basins in plains grasslands, other sites in alluvial soils, especially sand or loam. Hudspeth, Culberson, Presidio, Jeff Davis, Brewster, and Terrell counties. 3,500–5,000 ft. E to eastern Edwards Plateau, N through the Panhandle. Extreme W OK; NM. Mexico: expected in NE Chihuahua. Map 8.

Vegetative Characters. Described as a densely branched erect shrub 40–85 cm tall, but some plants or stems of some plants may collapse into dense, untidy mounds 30–40 cm high. At least some plants produce tuberous roots and rhizomes. Slender, short trunk woody and many-branched, potentially with several branches at each node. Larger terminal joints 6–12 cm long, 1–1.5 cm in diameter, with closely spaced areoles. Stem joints

readily detached, as in O. *tunicata,* but older joints firmly attached. Spine surfaces themselves not readily visible, being covered by loose, flattened sheaths (1–2.5 mm across) pale yellowish-tan or pale golden in color. Central spines in two groups, one with 4–5 wider (0.8–1.5 mm), longer (3–5.5 cm) flattened spines and the other with five additional, more slender, shorter spines. Within its natural range, O. *davisii* is not likely to be confused with any other species.

Flowers. Flowering Jun–Jul. Flowers with a firm, waxy appearance, similar in this respect to those of *Echinocereus coccineus.* Flowers about 5 cm long and ca. 4 cm in diameter when fully open. Tepals green to greenish-yellow or pale green. Stigma lobes 4–7, cream-colored. Style reddish. Anthers yellow, filaments pale purplish distally and greenish below. Flowers of plants in the central Trans-Pecos may have rose-pink stigma lobes and brown tepals.

Fruits. Fruits slightly fleshy and yellow at maturity. Narrowly turbinate or obconic fruits 2.5–3.5 cm long, ca. 1.5 cm wide, concave at the apex. Fruit surface tuberculate with spherical or oval areoles. Whitish areoles have light brown glochids around periphery along with 1–2 short bristles, 0.7–1.5 mm long and deflexed like radial spines. Seeds 3.5 mm in diameter with a thin, beaked aril-margin. Only a few (1–3) seeds per fruit, or in some fruits no viable seeds.

Full Name and Synonyms. Opuntia davisii Engelm. & Bigelow [=*Cylindropuntia davisii* (Engelm. & Bigelow) F. M. Knuth]. *Opuntia tunicata* (Lehm.) Link & Otto var. *davisii* (Engelm. & Bigelow) L. D. Benson.

Plate 44. *Opuntia davisii* (Davis' cholla), N of Roswell, Chaves Co., NM.

Plate 45. *Opuntia davisii* (Davis' cholla) near Marfa, Presidio Co., TX; mature fruit.

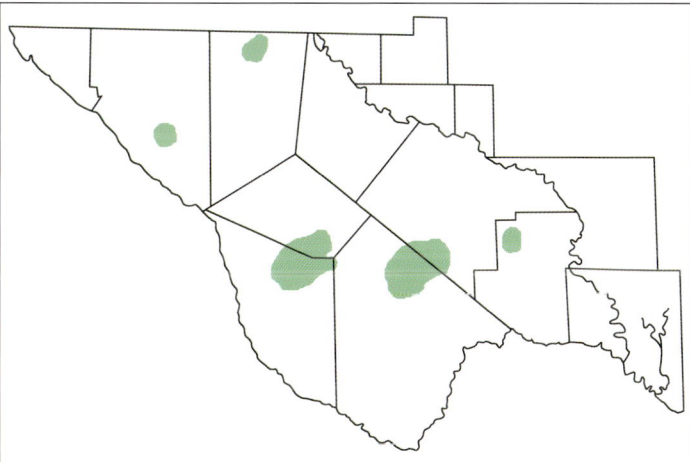

Map 8. *Opuntia davisii* (Davis' cholla).

Opuntia imbricata
Tree Cholla

After Anthony (1956) described *O. imbricata* var. *argentea*, most workers have recognized two varieties of *O. imbricata* to occur in the United States. Additional varieties occur in Mexico (Bravo-Hollis, 1978), one of which is grown in horticulture in the United States (Weniger, 1984). *Opuntia imbricata* is one of the most easily identifiable cacti in the Trans-Pecos, with its relatively large, many-branched shrublike or small treelike habit. The specific epithet alludes to the prominent overlapping appearance of the stem tubercles.

Full Name and Synonym. *Opuntia imbricata* (Haw.) DC. [= *Cylindropuntia imbricata* (Haw.) F. M. Knuth].

Key to the Varieties of *Opuntia imbricata*

1. Plants 1–3 m tall; spine sheaths yellowish, tan, to silver; distribution widespread
   O. imbricata var. *arborescens*, p. 54
1. Plants usually less than 1–1.2 m tall; spine sheaths silvery; distribution restricted to vicinity of Mariscal Mt
   O. imbricata var. argentea, p. 56

Opuntia imbricata var. arborescens
Tree Cholla

PLATES 46, 47

The plants of *O. imbricata* var. *arborescens* are the tallest of any cactus species in the Trans-Pecos. Some rangeland sites in the Davis Mountains and elsewhere have become infested with thickets of var. *arborescens*, probably in most cases where human activities have facilitated the dissemination and growth of dislodged stem segments. The type locality, for Engelmann's *O. arborescens* in effect, is at Santa Fe, where the first specimens with definite locality data were secured, clearly the same variety common from Texas to Colorado. The varietal epithet, after the Latin *arbor*, "tree," and -*escens*, "becoming," is descriptive of the habit.

Distribution. Mountains and desert, igneous and sedimentary substrates, in every county of the Trans-Pecos, most common from Brewster, Jeff Davis, and Presidio counties, W to El Paso Co. 2,400–7,300 ft. South-central TX N through the Panhandle and W through the Trans-Pecos. Panhandle of OK, extreme SW KS, S CO, most of NM, rare and poorly documented in E and SE AZ. Mexico: common in the CDR. Map 9.

Vegetative Characters. Arborescent, cylindroid-stemmed var. *arborescens* not likely to be confused with any other Trans-Pecos cactus species. Plants smaller, densely silver-spined and restricted in distribution. Stems with larger terminal joints 2–3 cm or more in diameter are much thicker than those of *O. kleiniae* and profoundly larger than those of *O. leptocaulis*.

Flowers. Flowering May–Jun. Flowers 5–6 cm long, 5–7.5 cm in diameter. Tepal color magenta, purplish, reddish-purple, rose-pink, and even lavender. Pale yellow or

Plate 46. *Opuntia imbricata* var. *arborescens* (tree cholla), N Brewster Co., TX.

Plate 47. *Opuntia imbricata* var. *arborescens* (tree cholla) near Blue Mt, Jeff Davis Co., TX; mature fruit.

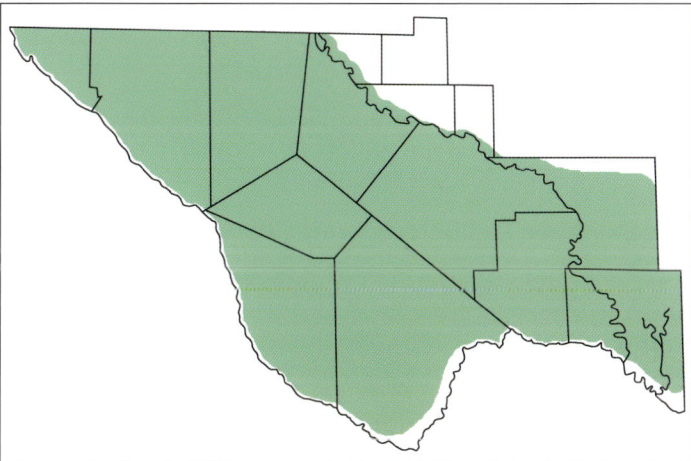

Map 9. *Opuntia imbricata* var. *arborescens* (tree cholla).

colorless flower-forms have been observed near Fort Davis, where the outer tepals are yellowish but the inner tepals are basically without pigments. Filaments 0.8–1 cm long, greenish to greenish-red proximally, reddish distally. Anthers yellow to cream-yellow and ca. 2 mm long. Style 2.5–2.9 cm long, 2–3 mm thick, cream to reddish. Stigma cream-colored, with 5–8 lobes.

Fruits. Fruits yellow, persistent on the plants through the fall and winter, somewhat fleshy at maturity, spineless, turbinate, short-cylindric or hemispheric, 2.5–4.5 cm long, 2.2–3.5 cm in diameter, strongly tuberculate, with a deep umbilicus, and readily proliferate, especially after being detached while still green. Tan seeds discoid, 3–3.5 mm in diameter, 1.5 mm thick, with a narrow, slightly beaked aril-rim.

Other Distinctions. In the Trans-Pecos different spine forms of var. *arborescens* are recognizable. These include a darker-spined form in the central Trans-Pecos mountains and basins and in many surrounding desert habitats and a silvery- or yellowish-spined form (the spine sheaths contribute to the colors) in populations of var. *arborescens* along and near the Rio Grande, from near Presidio in Presidio County west to El Paso County. In var. *arborescens* there are 6–17 central spines and 3–10 radial spines in the areoles of mature stems. Subulate leaves are borne in the upper areoles of young stems.

In southwestern New Mexico and adjacent Arizona, *O. imbricata* is geographically replaced by the more densely spined *O. spinosior* (Engelm.) Toumey, which it closely resembles. Apparently *O. spinosior* does not extend into Texas.

Full Name and Synonyms. Opuntia imbricata var. *arborescens* (Engelm.) A. D. Zimmerman, comb. nov. (forthcoming). [= *Cylindropuntia imbricata* var. *arborescens* (Engelm.) comb. nov., forthcoming]. *O. imbricata* (Haw.) DC. var. *arborescens* Weniger, nom. nud.; *O. imbricata* var. *vexans* (Griffiths) Weniger.

Opuntia imbricata var. argentea
Big Bend Cholla
PLATES 48, 49

In the Trans-Pecos, var. *argentea* is almost restricted to its type locality, Mariscal Mountain, Big Bend National Park, Brewster County. The plants have thrived in experimental cultivation at 4,600 feet. The varietal epithet alludes to the silvery aspect of the spine sheaths and the plant in general, after the Latin *argenteus*, "of silver."

Distribution. Mariscal Mt, mostly on N- and W-facing limestone slopes, fewer on alluvial Rio Grande plain in *Prosopis* thickets, W of Solis Ranch, and NE to Rooney's Place, Big Bend National Park, extreme S Brewster Co. 2,000–2,400 ft. Mexico: adjacent Coahuila and Chihuahua. Map 10.

Vegetative Characters. In its natural habitat, the most distinctive aspect of var. *argentea* is a dense clothing of silvery spine sheaths, contributing to a generally gray aspect to the entire plant. Stems are relatively densely and uniformly covered with spines, because the areoles are comparatively close together. In habit, plants resemble the "bristly-stemmed" form of var. *arborescens*, particularly bristly-stemmed plants with more silvery spines. Variety *arborescens* does not occur sympatrically with var. *argentea*, except for

Plate 48. *Opuntia imbricata* var. *argentea* (Big Bend cholla), Mariscal Mt, Big Bend National Park; cultivated.

Plate 49. *Opuntia imbricata* var. *argentea* (Big Bend cholla), Mariscal Mt, Big Bend National Park; mature and immature fruits; cultivated.

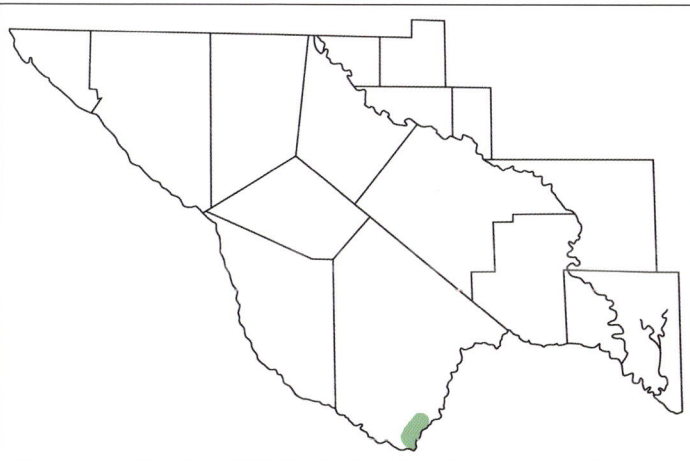

Map 10. *Opuntia imbricata* var. *argentea* (Big Bend cholla).

Chollas and Prickly Pears | 57

an occasional waif. Plants of var. *argentea* are smallest in the species, tending to be chubby, usually less than 1.2 m tall, to ca. 1 m wide. Mature stems have 6–14 central spines per areole; uppermost areoles bear subulate leaves on young stems.

Flowers. Flowering early Apr. Dark magenta, 4.5–5 cm long and 5 cm across, with tepals in four whorls or fewer. Outer tepals pinkish with olive midregions, oblong, apiculate. Filaments deep reddish-purple, 7–9 mm long. Anthers light to cream-yellow, 1.5 mm long. Style reddish-purple, to 1.8 cm long, bulbous above the base, supporting 7–9 cream-colored stigma lobes to 5 mm long.

Fruits. Persistent (for several months) ripe fruits yellow, turbinate, 2.5–4 cm long and 1.8–2.4 cm wide, with deep umbilicus, strongly tuberculate. Areoles circular, ca. 3 mm in diameter, densely covered with gray-white felt, with a compact row of glochids along the upper margin, these ca. 1 mm long and pale yellow. Seeds tan, discoid, ca. 3 mm across, 1–1.5 mm thick, with a narrow, slightly beaked aril.

Full Name and Synonym. Opuntia imbricata var. argentea M. S. Anthony [= Cylindropuntia imbricata var. argentea (M. S. Anthony) Backeb.].

Opuntia leptocaulis
Christmas Cholla

PLATES 50–53

The habit of *O. leptocaulis*, growing among other low shrubs, has been discovered inadvertently by many hikers in the Trans-Pecos. The type locality is "In Mexico." The common name is taken from the tendency of the copious red fruits to persist on plants through the winter. The specific epithet appropriately refers to the slender stems of this species, after the Greek *leptos*, "slender," and *caulis*, "stem."

Opuntia leptocaulis hybridizes with *O. kleiniae* (or backcrosses with it depending upon semantics, if *O. kleiniae* itself is considered a hybrid) in the Davis Mountains and in Mexico. *Opuntia leptocaulis* also hybridizes with *O. spinosior* and other cholla cacti in Arizona.

Distribution. Widely distributed in desert and semidesert habitats, every county of the Trans-Pecos. 1,900–5,000 ft. Throughout most of TX, especially the western two-thirds. Southern OK, NM, and AZ. Mexico: S to Puebla. Map 11.

Vegetative Characters. Plants compact low shrubs with many primary stems, or erect, subarborescent shrubs with one or two primary stems. Pencil-thin secondary branches arise ca. 8 cm above the ground. Typically, a single spine in each areole along the stems. Ultimate joints usually spineless. Plants are not likely to be confused with any other cactus species of the Trans-Pecos except perhaps dehydrated plants of *O. kleiniae* or hybrids between *O. leptocaulis* and *O. kleiniae*.

Flowers. Flowering May–Sep. Greenish-yellow flowers borne on distal halves of longer joints. Flowers 1.5–2 cm long, 1–2.2 cm in diameter. Tepals open widely, and flowers are subrotate. Anthers ca. 1 mm long, pale yellow to cream-yellow. Filaments ca. 6 mm long, pale green to yellow-green. Style ca. 9 mm long. Stigma lobes 3–6, short, thick, cream to cream-white.

Fruits. Fleshy fruits bright red, less often pale to medium red, rarely yellow, obovoid, pear-shaped, clavate, or obconic, 1.4–2.4 cm long, 0.8–1.3 cm in diameter, usually with a shallow umbilicus. Plants with yellow fruits are rare in South Texas,

Plate 50. *Opuntia leptocaulis* (Christmas cholla); short-spine form; near Dryden, Terrell Co., TX.

Plate 51. *Opuntia leptocaulis* (Christmas cholla), S Big Bend National Park, Brewster Co., TX; mature fruits.

Plate 52. *Opuntia leptocaulis* (Christmas cholla); long-spine form; S Brewster Co., TX.

Plate 53. *Opuntia leptocaulis* (Christmas cholla), River Road, Big Bend National Park, Brewster Co., TX; mature fruits.

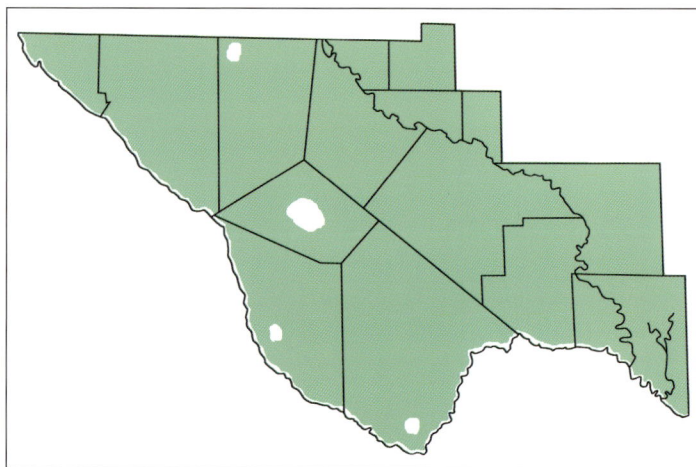

Map 11. *Opuntia leptocaulis* (Christmas cholla).

but evidently not in the Trans-Pecos. Fruit surface smooth between scattered circular to oval areoles, often with a few or a tuft of brown glochids at upper margin, occasionally also with 0–4 glochidlike bristles. Only the fruit wall is red; the cavity is filled with seeds. Seeds tan, discoid, angular, 3–4 mm in diameter, ca. 1.5 mm thick, with slender aril. Fruits may proliferate, especially while still green.

Other Distinctions. Two spine forms of *O. leptocaulis* were recognized in Powell and Weedin (2004). One was referred to as the "long-spined" form, with slightly thicker stems and spines to 4.8 cm long and to 0.8–1 mm wide. The spines are clothed with conspicuous golden-yellow or silver and golden sheaths. The other, "short-spined" form, with thinner stems, has spines typically 0.6–3.2 cm long and 0.3–0.4 mm wide. The spines are inconspicuously sheathed (the sheaths pale yellow or silver) or not sheathed. For more information on the systematics and different daily flowering times of *O. leptocau-lis*, see Champie (2003) and Powell and Weedin (2004).

Full Name and Synonyms. *Opuntia leptocaulis* DC. [= *Cylindropuntia lepto-caulis* (DC.) F. M. Knuth]. *O. vaginata* Engelm.; *O. frutescens* Engelm. var. *brevi-spina* Engelm.

Other Common Names. Pencil cactus; pencil cholla; tasajillo.

Opuntia kleiniae
Candle Cholla

PLATES 54, 55

A taxon similar to *O. kleiniae*, some-times called *O. kleiniae* var. *tetracantha* (Toumey) W. T. Marshall, is based upon Arizona plants, presumably from hybrid-ization of *O. leptocaulis* with a western species instead of *O. imbricata*. The type locality of *O. kleiniae* is "In Mexico." According to Anthony (1956), the specific epithet was inspired by the resemblance of *O. kleiniae* to *Cacalia kleinia* L., a suc-culent in the family Asteraceae.

Distribution. Rocky slopes or alluvial flats, typically with *Prosopis* and other shrubs, Davis Mts, and rocky slopes, alluvial flats, or sandy loam, Rio Grande floodplain. Davis Mts in Jeff Davis and NW Brewster counties, SE Brewster Co. along and near the Rio Grande to El Paso Co.; also N Culberson Co. 1,800–5,000 ft. Central TX, Eastland and Lampasas coun-ties. Scattered introductions in central and western OK. Central and SE NM, in the region of overlap between *O. leptocaulis* and *O. imbricata*. Mexico: S to Hidalgo. Map 12.

Vegetative Characters. Plants larger (1–2.5 m tall) than those of *O. leptocaulis* and smaller than those of *O. imbricata*. In fact, *O. kleiniae* exhibits complete morphological intermediacy between *O. leptocaulis* and *O. imbricata*. The ultimate branches of *O. kleiniae* are more or less the diameter of a fountain pen, as opposed to pencil thick in *O. leptocaulis* and broom handle thick (or thicker) in *O. imbricata*. In *O. kleiniae* there are typically 1–4

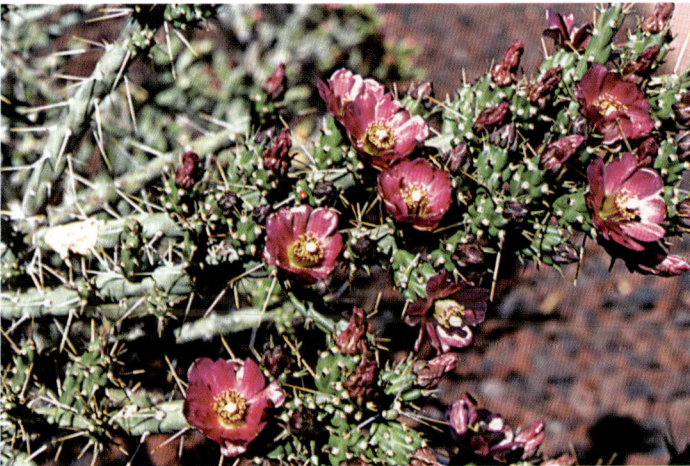

Plate 54. *Opuntia kleiniae* (candle cholla), Musquiz Canyon, Jeff Davis Co., TX.

Plate 55. *Opuntia kleiniae* (candle cholla), Musquiz Canyon, Jeff Davis Co., TX; mature fruits.

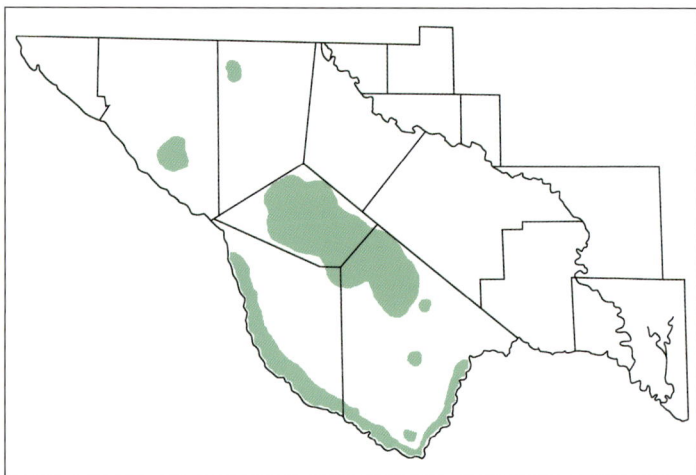

Map 12. *Opuntia kleiniae* (candle cholla).

loosely sheathed central spines per areole, compared to one central per areole in *O. leptocaulis* and 6–17 centrals in *O. imbricata*. Sheaths golden or silver and golden. In Davis Mountains populations, commonly four centrals per areole. Populations along the Rio Grande characteristically have one central spine per areole and exhibit other characters suggesting that they should be given taxonomic recognition (Powell and Weedin, 2004).

Flowers. Flowering Apr–Aug. Pinkish, purplish, to magenta or purple-tinged flowers of *O. kleiniae* contrast sharply with the smaller, yellow-green flowers of *O. leptocaulis* and are smaller and usually paler than the larger magenta flowers of *O. imbricata*. In *O. kleiniae*, flowers 3–3.5 cm long and 2.5–4 cm in diameter. There are floral differences in backcross hybrids (Plates 56 and 57) and between the two forms (Plates 54 and 58) of *O. kleiniae* in the Trans-Pecos. Davis Mountains populations have larger flowers with the inner tepals pinkish, purplish, to purple-red or magenta. Populations along and near the Rio Grande typically have smaller flowers that are not brightly colored, but instead

Plate 56. *Opuntia kleiniae* (candle cholla); putative backcross; near Mitre Peak, Jeff Davis Co., TX.

Plate 57. *Opuntia kleiniae* (candle cholla); putative backcross; near Mitre Peak, Jeff Davis Co., TX; same population as in Plate 56.

the inner tepals often are whitish-pink or greenish-cream and maroon-tinged, more intensely so distally, or exhibit some other combination of pale greenish and cyanic colors. In flowers of *O. kleiniae* there are about five inner tepals and about five outer tepals with another series of outer bracts.

Filaments are greenish to pinkish-red, anthers yellow and ca. 1.5 mm long. Style pale green to pinkish or pale maroon, 1–2 cm long, ca. 1.5 mm in diameter. Stigma lobes 5–7, whitish or cream to pinkish, ca. 3 mm long and thick, persistently folded together in a headlike cluster.

Plate 58. *Opuntia kleiniae* (candle cholla); the "Rio Grande *kleiniae*"; near Castolon, Big Bend National Park, Brewster Co., TX.

Plate 59. *Opuntia kleiniae* (candle cholla) near Castolon, Big Bend National Park, Brewster Co., TX; mature fruit of the "Rio Grande *kleiniae*."

Fruits. Fully ripe fruits red to red-orange, fleshy, obovoid, ovate to subclavate, tuberculate or smooth, 2–3.5 cm long, 1–2.2 cm in diameter. Fruit areoles have glochids and sometimes short, whitish bristles, but are spineless. Fruits persistent through winter and proliferous when in contact with a favorable substrate. Seeds tan, irregularly discoid, 3.5–5 mm in diameter, 1–2 mm thick.

We have observed different fruit morphologies in the two forms of *O. kleiniae*.

Plants in the Davis Mountains (Plate 55) typically have tuberculate fruits, and the Rio Grande plants (Plate 59) have smooth or inconspicuously tuberculate fruits. So far we have carefully observed the fruits of the Rio Grande form only, near Santa Elena Canyon and Heath Canyon.

Full Name and Synonym. *Opuntia kleiniae* DC. [= *Cylindropuntia kleiniae* (DC.) F. M. Knuth].

Other Common Names. Klein pencil cholla; candle cactus.

Opuntia rufida
Blind Prickly Pear

PLATES 60, 61

In the Trans-Pecos *O. rufida* is most common along south-facing cliffs in the hottest desert habitats near the Rio Grande in southern Brewster County, downstream at least to the Reagan Canyon drainage. The type locality is "About Presidio del Norte, on the Rio Grande." The specific epithet is after the Latin *rufulus,* "reddish," in reference to the tufts of reddish-brown glochids that dominate the spineless pads. The English name is associated with the blinding of cattle by dislodged glochids (Benson, 1982), corresponding to the Mexican name, *nopal cegador* [also applied to *O. microdasys* (Lehm.) Pfeiff.].

Distribution. Desert habitats, limestone and igneous, alluvium or rocks, flats, hills, canyons, mountain slopes. Hudspeth Co., S Quitman Mts; Culberson Co., Van Horn Mts; Presidio Co., below the rim, near the Rio Grande. Brewster Co., mostly S of Chisos Mts but N to near Agua Fria Mt, Chalk Bluff, and Black Gap. 1,800–4,000 ft. Mexico: Chihuahua, Coahuila, NE Durango. Map 13.

Vegetative Characters. This is the only truly spineless native *Opuntia* species in the Trans-Pecos. Others are basically spiny except for spineless freak individual plants, and there are spineless clonal colonies of *O. leptocaulis. Opuntia rufida* forms erect shrubs 0.5–1.5 m tall, with or without short trunks. Gray-green pads are circular, obovate, or elliptic, 10–20 cm long, dotted with circular to oval spineless areoles. Areoles crowded with whitish or gray hairs and bulging tuft of reddish-brown glochids. Areoles 1.5–2.5 cm apart on distal portion of pad. Pad surfaces densely short-pubescent (velutinous—that is, velvety with straight, short hairs), unlike any other native Texas *Opuntia*, a character detectable either by touch (carefully between areoles) or visually (using a powerful hand lens).

Flowers. Flowering Apr–May. Yellow flowers 5–7 cm long and 4.5–7 cm in diameter. Inner tepals bright yellow to golden-yellow, or sometimes orange-yellow,

Plate 60. *Opuntia rufida* (blind prickly pear), S Big Bend National Park, Brewster Co., TX.

Plate 61. *Opuntia rufida* (blind prickly pear), S Big Bend National Park, Brewster Co., TX; maturing fruits (cultivated).

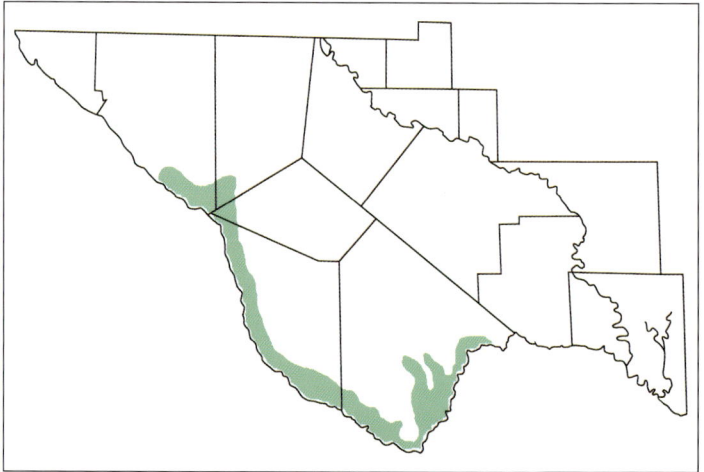

Map. 13. *Opuntia rufida* (blind prickly pear).

without any red pigment at the base. Orangish flowers are probably always older ones. Anthers yellow to cream, filaments whitish. Style 2–2.4 cm long, colorless, with 6–10 dark green stigma lobes.

Fruits. Fruits fleshy, subspherical to obovate, greenish-red to red, sometimes fading yellowish before turning red (1–3 months later), 2–2.5 cm long, 2–2.5 cm in diameter. Fruit surface has circular areoles

2–3 mm in diameter, with white hairs and at least a few red-brown glochids. Fruit pulp greenish. Seeds discoid, irregular in outline, 2–3 mm in diameter, with narrow arillate rim.

Full Name and Synonym. *Opuntia rufida* Engelm. *O. rufida* Engelm. var. *tortiflora* M. S. Anthony.

Opuntia strigil
Marble-Fruit Prickly Pear
PLATES 62, 63

Opuntia strigil has a relatively limited distribution. It extends southeast in the Trans-Pecos to ca. 25 miles north of Langtry (Val Verde County) and northeast across the Pecos River. Reportedly *O. strigil* hybridizes with the closely related *O. atrispina* in the southeast portion of its range. With its characteristic spination *O. strigil* is one of the most easily identified prickly pears in the eastern Trans-Pecos. The type locality is "western Texas, west of the Pecos, in crevices of flat limestone rocks." The word *strigil* (short *i*, soft *g*) is based upon the Latin *strigilis*, "scraper"

or "skin brush," used at baths in the days before soap, probably in reference to the appearance of the pads with their copious deflexed spines; thus, their vesticure is *strigose* (long *i*, hard *g*).

Distribution. Shallow soils, limestone hills, mesas, and canyons, W Stockton Plateau. Western Reeves, Pecos, Terrell, and Val Verde counties. 2,400–4,100 ft. Also Upton Co., Crockett Co., and Nolan Co. (Benson, 1982). Mexico: probably adjacent Coahuila. Map 14.

Vegetative Characters. Plants upright and compact, 0.5–1 m tall, less commonly sprawling. Pads usually obovate. Closely set areoles, with spines in areoles over entire pad, except sometimes at the base. Spines 7–13, strongly deflexed, especially in the upper areoles, almost flattened against the pad except for the main central (2–2.4 cm long, 0.6–0.9 mm wide), which is porrect or deflexed. Single main central spine reddish proximally and yellow distally.

Flowers. Flowering Apr–Jun. Yellow flowers 5–5.5 cm long and 5–6 cm in

Plate 62. *Opuntia strigil* (marble-fruit prickly pear), near Fort Stockton, Pecos Co., TX.

Plate 63. *Opuntia strigil* (marble-fruit prickly pear), S of Fort Stockton, Pecos Co., TX; immature and mature fruit.

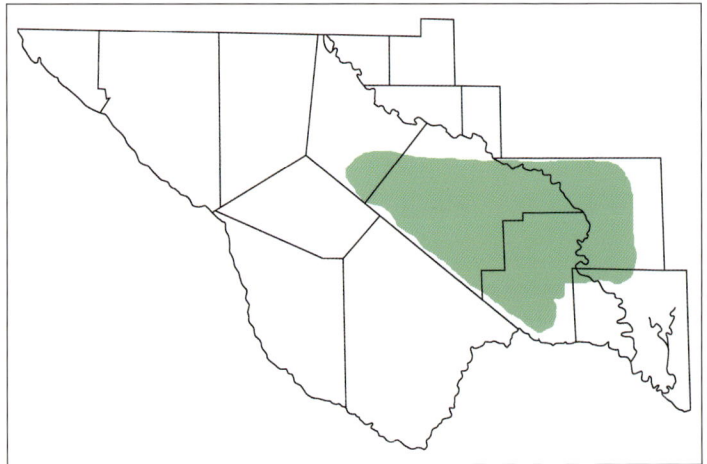

Map 14. *Opuntia strigil* (marble-fruit prickly pear).

diameter. Inner tepals lemon yellow to pale yellow, fading to salmon. Tepal midregions may be pale reddish proximally, but overall appearance of the flower usually is one without a target center. Filaments ca. 1.5 cm long, pale green proximally, cream-colored distally. Anthers ca. 1.5 mm long, yellow to cream-colored. Pale style 1.5–2.4 cm long, slender-urceolate. Stigma lobes 6–8 in number, 2–4 mm long, pale green to cream-colored.

Fruits. The fruits are among the smallest of those of any *Opuntia* in the southwestern United States. Fruits fleshy, not very juicy, subspherical to turbinate, 2–2.7 cm long, 1.3–2.2 cm in diameter. Outer layer red at maturity, but fruit wall and pulp are light green. The limited juice is clear and not sweet. Umbilicus usually shallow, but relatively deep in fruits of some plants. Areoles armed with glo-chids and a few bristles to 4–8 mm long, especially at the fruit apex. Tannish-white seeds irregularly reniform, 2.5–4 mm in diameter, ca. 1.5 mm thick, narrow-margined, not beaked at the hilum, with a slender aril.

Full Name. Opuntia rufida Engelm.

Opuntia atrispina
Black-and-Yellow-Spined Prickly Pear
PLATES 64, 65

Opuntia atrispina is a taxon of limited distribution in Texas, apparently always on limestone, in the extreme southeastern Trans-Pecos southeast to Uvalde County, according to Weniger (1984) in a strip only about 20 miles wide. The type locality is "Near Devil's River, Texas," in Val Verde County, about 35 miles to the east of the lower Pecos River. The specific epithet is after the Latin *atra* or *atrum*, "black," and *spina*, "spine," in reference to the basal color of the central spines.

Distribution. Limestone hills, mesas, and canyons, east of the desert. Val Verde Co., just W of the Pecos River, SE across the county across the Devils River to the Anacacho "Mountains." Reported in Terrell Co. W as far as Dryden. 1,000–2,100 ft. Mexico: eastern Coahuila, reported S to near Monclova. Map 15.

Vegetative Characters. Compact or spreading shrubs 0.5–1 m tall. Pads obovate to subcircular and green to yellow-green. The most distinguishing vegetative feature involves the central spines: 1–2 per areole, black or dark red-brown proximally, yellow distally, but sometimes just the tip is yellow, and often the bases are more brown than black. Centrals usually

Plate 64. *Opuntia atrispina* (black-and-yellow-spined prickly pear), Pecos River high bridge, Val Verde Co., TX; cultivated.

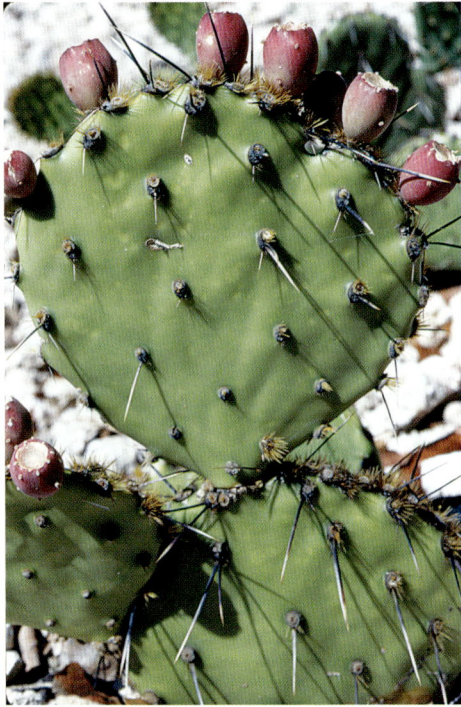

Plate 65. *Opuntia atrispina* (black-and-yellow-spined prickly pear), Pecos River high bridge, Val Verde Co., TX; nearly mature fruit; cultivated.

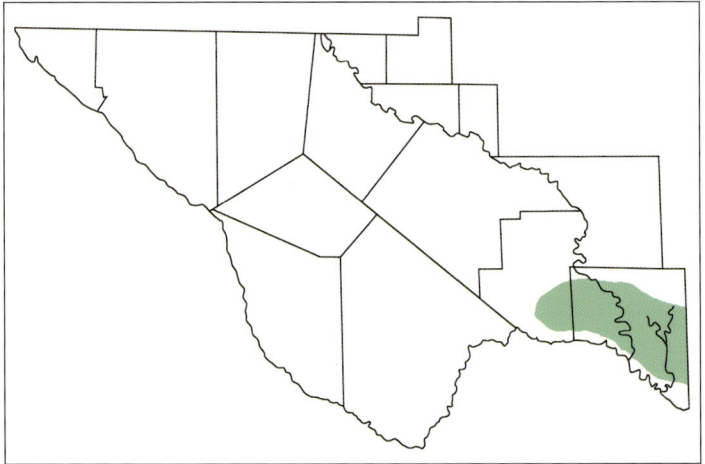

Map 15. *Opuntia atrispina* (black-and-yellow-spined prickly pear).

2.5–3.7 cm long, 1–1.3 mm wide, needle-shaped or weakly flattened near the base. In aspect, the areoles appear to be bulging with glochids, particularly those crowded on the apical margin of the pad.

Flowers. Flowering probably April.

Bona fide O. *atrispina* in cultivation produces yellow flowers that open again for 2–3 days (unusual in *Opuntia*, where typically flowers last one day, maybe two), with tepals turning peachy after the first day. Flowers pure yellow, chrome yellow, or with greenish centers before fading to apricot; relatively small, to 2.5–6.5 cm in diameter. Filaments and anthers cream-colored, colorless, or greenish. Style whitish, fading rose, and 7–8 stigma lobes cream-white to very pale green.

Fruits. Fruits fleshy, bright red to reddish-yellow, subspherical to obovate, variable sizes but mostly small, 1.2–2.5 cm long, 1–1.5 cm in diameter. Pericarpel "rind" red with some clear juice; pulp relatively dry, greenish. In putative hybrids the fruit wall green or partially red, with or without clear juice. Umbilicus shallow to deep. Seeds 3–4 mm in diameter, flattened but relatively thick, with a narrow margin.

Full Name. Opuntia atrispina Griffiths.

Other Common Name. Dark-spined opuntia.

Opuntia macrocentra
Long-Spined Purplish Prickly Pear
PLATES 66, 67

Opuntia macrocentra is one of relatively few prickly pears noted for purple stems, the product of a betalain (betacyanin) pigment most evident when the plants are stressed by cold or drought. See Powell and Weedin (2004) for a review of the taxonomic history of the *O. macrocentra* complex, especially as it pertains to the Trans-Pecos and adjacent areas. The review suggests that there are at least eight taxonomic entities mentioned in the literature concerning this complex, including Mexican and Sonoran Desert fringe taxa. One taxon, previously known as *O. macrocentra* var. *minor,* was paired by us with *O. mackensenii* var. *mackensenii.* In the present treatment we recognize *Opuntia macrocentra,* sensu stricto, and five related taxa now placed in *O. azurea.*

Some populations of *O. macrocentra,* sensu lato, conspicuously fit the popular names of "long-spined prickly pear" and "purple prickly pear." The syntype locality for *O. macrocentra* first appeared in print as "Sand-hills on the Rio Grande near El Paso." Specifically (quote from *Cactaceae of the Boundary*), the syntypes were from "Sandy ridges in the bottom of the Rio Grande near El Paso, also on the Limpia (Wright)," later restricted (by lectotypification) to the El Paso locality. Engelmann's

Plate 66. *Opuntia macrocentra* (long-spined purplish prickly pear), Hueco Mts, El Paso Co., TX (photo by Richard D. Worthington).

Plate 67. *Opuntia macrocentra* (long-spined purplish prickly pear), 35 mi N of Sierra Blanca, Hudspeth Co., TX; mature fruits.

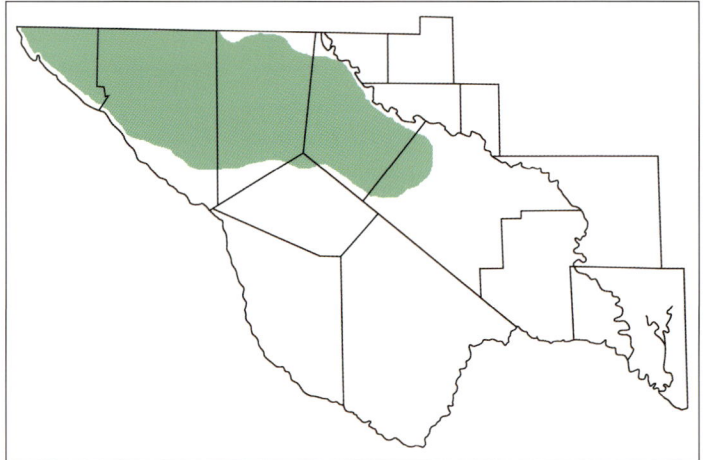

Map 16. *Opuntia macrocentra* (long-spined purplish prickly pear).

diagnosis might be based on a mixture of both El Paso and Limpia Creek populations. The specific epithet alludes to the long central spines, after the Greek *makros*, "long," and *kentron*, "spine."

Distribution. Various substrates and habitats, desert to mountain grasslands and intermediate slopes. El Paso Co. E through Hudspeth, Culberson, and Reeves counties, NE Brewster Co. E of Alpine and adjacent Pecos Co. 3,000–5,000 ft. Southern NM and SE AZ. Mexico: NE Chihuahua S to near Ciudad Chihuahua, and barely into NE Sonora from AZ. Map 16.

Vegetative Characters. Spreading to nearly upright shrubs usually 30–60 cm tall, but to 1 m high, only rarely with a short trunk. "Smooth" blue-gray, blue-green, or purplish pads obovate to orbicular, 10–20 cm long, 10–20 cm in diameter, or slightly wider than long. Areoles 1.5–3 cm apart; spines produced in areoles on upper one-fourth to one-half of the pad, or only on the upper margin. Largest spines notably directed upward. In most upper areoles, 1–3 central spines, with the upper projecting ones usually 5–10 cm long, and black, dark brown, or reddish-brown, at least basally, sometimes tipped more or less extensively with yellow or white. In some plants, the spines may be white for part or most of their length. Spineless or near spineless forms are at least occasional. Reddish-brown to yellowish glochids abundant, especially in upper areoles.

Flowers. Flowering Apr–May. Yellow flowers with sharply defined bright red centers. Flowers 6–8 cm long, 5.5–8 cm wide, in general not opening as widely as most other opuntias. Red bases of yellow tepals extend upward to near midtepal, often tapering distally so that the red center is "star-shaped," and in some plants red pigment extends along the midvein almost to the tepal apex like a "midstripe" (rare in *Opuntia*). Filaments ca. 1.5 cm long, pale green proximally, cream-colored distally. Anthers 1.8–2 mm long, yellow. Cream-colored style 1.7–2 cm long. Stigma lobes ca. six, ca. 5 mm long, cream-colored or pale green.

Fruits. Fruit at ripening reddish-purple or purple and succulent, obovoid, ovoid, or ellipsoid, 3–4.3 cm long, 1.5–3 cm in diameter, and not much, if at all, constricted below the apical rim. Umbilicus deeply concave. Fruits usually with 12–16 areoles. Fruit rind purple; juice and pulp pale purple to clear. Seeds flattened, tan, 3.5–4.5 mm in diameter, 1.5–1.9 mm

thick, irregular in outline, with a broad notch on one side and prominent raphe to ca. 1 mm wide.

Full Name and Synonyms. Opuntia macrocentra Engelm. *O. violacea* Engelm. in Emory (invalid provisional name); *O. violacea* Engelm. ex B. D. Jacks.; *O. violacea* var. *macrocentra* (Engelm.) L. D. Benson; *O. violacea* var. *castetteri* L. D. Benson; *O. phaeacantha* Engelm. var. *nigricans* Engelm.

Other Common Names. Long-spined prickly pear; purple prickly pear.

Opuntia azurea
Purple Prickly Pear, Coyotillo

Opuntia azurea was described as a Mexican species; the name has not been used for any United States opuntia except by Anthony (1956). Our taxonomic concept of *O. azurea* is derived partly from information provided by D. Ferguson (pers. comm.). Herein we interpret *O. azurea* as a species with multiple forms and geographic races; four of them occur in the Big Bend region of the Trans-Pecos, and the typical variety is a Mexican taxon that closely approaches the international border south of Big Bend National Park. The type locality for *O. azurea* is in northeastern Zacatecas, Mexico. The specific epithet is after the Latin *azureus,* "blue," in reference to one of the stem colors of this geographically and seasonally variable species.

Full Name. Opuntia azurea Rose.

Key to the Varieties of *Opuntia azurea*

1. Purple only at the areoles, if at all; spines usually yellow, golden, or reddish (2).
1. Uniformly purple in drought or winter, or greenish-purple between the areoles after moisture in warm seasons; spines usually blackish with white tips, reddish-black, or dark reddish, often with pale tips (3).

2(1). Spines slender (0.7–1.3 mm wide at the base), numerous (4–12); pads small; fruit spiny
    O. azurea var. *aureispina*, p. 78
2. Spines robust (1–1.5 mm wide at the base), few (1–4); pads large; fruit spineless
    O. azurea var. *discolor*, p. 80

3(1). Plants taller (usually 1–2 m); Mexican
    O. azurea var. *azurea*, p. 76
3. Plants shorter (usually 0.3–0.9 m); Trans-Pecos (4).

4(3). Distal pads larger (14–20 cm long, 14–20 cm wide), orbicular
    O. azurea var. *diplopurpurea*, p. 74
4. Distal pads smaller (6.5–19 cm long, 6–14 cm wide), obovate
    O. azurea var. *parva*, p. 76

Opuntia azurea var. diplopurpurea
Diploid Purple Prickly Pear

PLATES 68, 69

The pads of this taxon turn completely reddish-purple or purple under stress. The type locality is Sul Ross Hill, Brewster County, Texas. The varietal epithet is descriptive of a diploid taxon with purple stems, after the Latin *purpureus*, "purple."

Distribution. Common purple prickly pear of Big Bend region of the Trans-Pecos, from mountain grasslands down to the desert. Hudspeth, Jeff Davis, Presidio, Brewster, Pecos, and Crane counties. 2,500–5,500 ft. Mexico: adjacent to the Rio Grande, N Coahuila, and NE Chihuahua. Map 17.

Vegetative Characters. Relatively low, open, sprawling shrubs 30–60 cm high. Pads reddish-purple (for most of year), orbicular to broadly obovate, 14–20 cm long, 14–20 cm in diameter, or broader than long. Usually 2–6 dark spines per areole, more in the distal and marginal areoles. Spiniferous areoles occur on the distal half to one-fourth of the pad, often restricted to the distal margin, sometimes absent. Longest spines 5–11 cm, nearly black, reddish-black, or reddish-brown, and white or at least lighter-colored distally. Occasional "white-spined" forms occur, in which only the bases are dark.

Flowers. Flowering Apr-Jun. Yellow flowers with red centers, virtually identical to those of *O. macrocentra* and other

Plate 68. *Opuntia azurea* var. *diplopurpurea* (diploid purple prickly pear), between Presidio and Ruidosa, Presidio Co., TX (possibly introgressed from var. *parva*).

Plate 69. *Opuntia azurea* var. *diplopurpurea* (diploid purple prickly pear), 10 mi SE of Ruidosa, Presidio Co., TX; mature fruits.

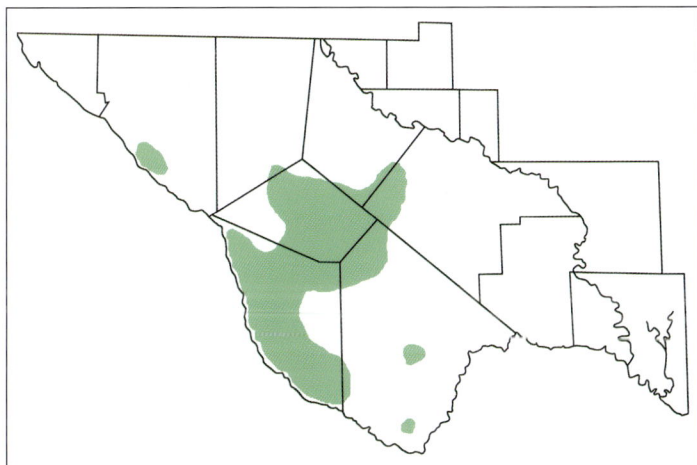

Map 17. *Opuntia azurea* var. *diplopurpurea* (diploid purple prickly pear).

varieties of *O. azurea*. Flowers 6–7.5 cm long and wide. Filaments 1.1–1.5 cm long, supporting yellow anthers 1.5–2 mm long. Style 1–2 cm long with cream-colored to pale green stigma lobes ca. 5 mm long.

Fruits. Upon ripening, fruit turns red to reddish-purple; either remaining fleshy and juicy or drying and shrinking, fading to tan or pale red. Fruit ovate to obovate, 1.5–3 cm long, 1.5–2.8 cm in diameter, somewhat constricted below the apical rim. Umbilicus rather deep. Fruits (in at least some populations) with green pulp and flesh, with very little clear, moderately sweet juice. Seeds flattened, tan, 3.8–4.5 cm in diameter, ca. 1.5 mm thick, and irregular in outline.

Other Distinctions. Variety *diplopurpurea* is not always easy to distinguish from var. *parva*. In the field var. *diplopurpurea* usually can be recognized by its low, sprawling habit that appears "open." The plants are branched at the base, but they produce relatively fewer, larger pads, resulting in a more open plant. Plants of var. *parva* are dense, whether sprawling or upright. The relatively more numerous, smaller, obovate pads collectively contribute to a dense habit.

A few miles south of the Rio Grande, var. *diplopurpurea* is replaced by *O. azurea* var. *azurea* (D. Ferguson, pers. comm.), a Mexican taxon that occurs mostly in desert flats in Coahuila, eastern Chihuahua, Durango, and Zacatecas.

Variety *diplopurpurea* is most common at mid-elevations in the Davis Mountains, in the western foothills of the Chisos Mountains, and in southwestern Presidio County; probably these are the predominant purple prickly pears along the Rio Grande valley northwest to Hudspeth

County. Benson (1982) included var. *diplopurpurea* in *O. violacea* var. *macrocentra*, which he treated as extending from the Big Bend region northwest through southern New Mexico and into southeastern Arizona.

The few-spined or spineless form of var. *diplopurpurea* is more prevalent in the Davis Mountains and on the western slopes of the Chisos Mountains, but plants with essentially spineless pads are also found sporadically in the desert. Likewise, spiny plants like those in the desert are found sporadically in the mountains. Stems of the "spineless" form may be completely spineless or have 1–2 spines in a few areoles at the distal margin of the pads.

Full Name. *Opuntia azurea* var. *diplopurpurea* A. M. Powell & J. F. Weedin.

Opuntia azurea var. parva
Big Bend Purplish Prickly Pear
PLATES 70, 71

Variety *parva* is one of the most common prickly pears in desert habitats around the Chisos Mountains in Big Bend National Park. It extends north in the park to near Persimmon Gap, and north of the park in Brewster County to near Nine Point Mesa. Plants of var. *parva* are among the earliest-blooming opuntias in the Trans-Pecos. The type locality is two miles north of Study Butte, Brewster County, Texas. The varietal epithet is after the Latin *parvus*, "small," denoting the usually small distal pads of the taxon, relative to all of the other purple prickly pears.

Distribution. Restricted to desert habitats surrounding the Chisos Mountains, Brewster Co. 1,900–3,700 ft. Mexico: inferred to be in adjacent Coahuila, but replaced by var. *azurea*. Map 18.

Plate 70. *Opuntia azurea* var. *parva* (Big Bend purplish prickly pear), River Road, Big Bend National Park, Brewster Co., TX.

Plate 71. *Opuntia azurea* var. *parva* (Big Bend purplish prickly pear), Big Bend National Park, Brewster Co., TX; mature fruits.

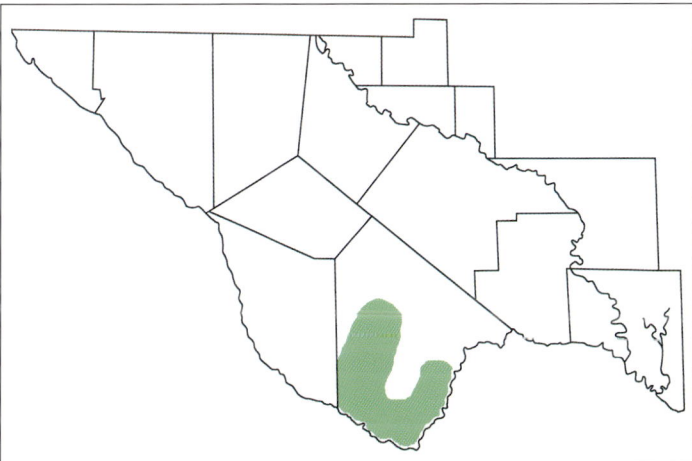

Map 18. *Opuntia azurea* var. *parva* (Big Bend purplish prickly pear).

Vegetative Characters. Usually many-branched, sprawling or upright shrubs 30–90 cm high, sometimes with short trunk. Whole upright plants or individual branches of sprawling plants have a compact appearance. Pads blue-green or blue-gray, glaucous, with an underlying purplish tinge most obvious around the areoles. In apparently stressed plants, reddish-purple betacyanin pigments increase to dominate the whole pads. Pads, at least on terminal branches, typically obovate, cuneate basally, 8–19 cm long, 8–14 cm in diameter. Lower pads may be suborbicular and somewhat larger. Pads typically support long black to dark red-brown spines to 10–12 cm or more (ca. 17 cm) long, with white tips, in areoles on upper one-half of the pad. Usually 1–3 spines per areole, potentially with a few more spines in marginal areoles.

Flowers. Flowering Mar–May. Yellow flowers with red centers, essentially like those in *O. macrocentra* and in other members of the *O. azurea* complex. Flowers 6–7.5 cm long, to 8 cm wide. After anthesis, flowers fade from yellow to orange-red by late afternoon, by which time they are almost closed. It is common to see both young yellow flowers and older orangish flowers open or partially open on the same plant. Stamens with filaments 1.3–1.8 cm long and yellow anthers 1.5–2 mm long. Style ca. 1 cm long, with cream-colored to pale green stigma lobes 3–5 mm long.

Fruits. Mature fruits pale red to purple, tending to dry and shrink with age. Fruit ovate to obovate, 1.5–3 cm long, 1.5–2.7 cm in diameter, constricted below the apical rim. Umbilicus deep. Seeds discoid, irregular in outline, large, 5–5.5 mm in diameter, 1–1.4 mm thick. Seeds of var. *parva* and var. *aureispina* similar in size

and larger than those of var. *diplopurpurea* and var. *discolor.*

Other Distinctions. Plants of var. *parva* throughout the northern part of their range tend to be profusely branched with blue-green to purplish obovate pads. Those in the northeast (i.e., from Panther Junction northward) are relatively low and sprawling. In the northwestern part of its range, mostly south of Study Butte, var. *parva* tends to be upright and compact. These populations, in at least the northeastern and northwestern part of the range, appear to be introgressed from var. *diplopurpurea.* Introgression of these taxa might extend into Presidio County at the lowest altitudes. The ecogeographic range of var. *parva*, near the Rio Grande, stands between var. *azurea* in nearby Mexico and var. *diplopurpurea* in most of the Big Bend region.

Full Name. Opuntia azurea var. *parva* A. M. Powell & J. F. Weedin.

Opuntia azurea var. aureispina
Golden-Spined Prickly Pear
PLATES 72, 73

The type locality for var. *aureispina* is near the Rio Grande, Big Bend National Park, Brewster County, Texas. The specific epithet is descriptive of the yellow spines, after the Latin *aureus*, "golden," and *spina*, "spine."

Distribution. Fractured limestone, desert hills and canyons along and near the Rio Grande. Brewster Co. near Mariscal Mt and in Boquillas Canyon. 1,600–2,800 ft. Mexico: to be expected in Coahuila and Chihuahua; to ca. 30 mi S of Big Bend National Park, in the same type of hot and exposed limestone habitats as in TX. Map 19.

Plate 72. *Opuntia azurea* var. *aureispina* (golden-spined prickly pear) near Mariscal Mt, type locality, Big Bend National Park, Brewster Co., TX.

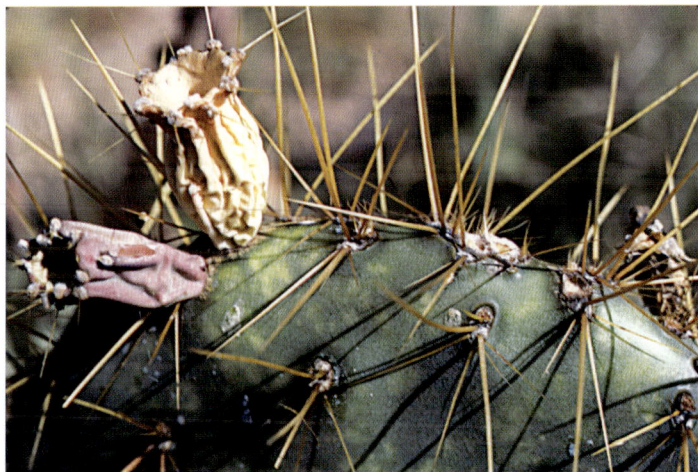

Plate 73. *Opuntia azurea* var. *aureispina* (golden-spined prickly pear) near Mariscal Mt, type locality, Big Bend National Park, Brewster Co., TX; mature fruits.

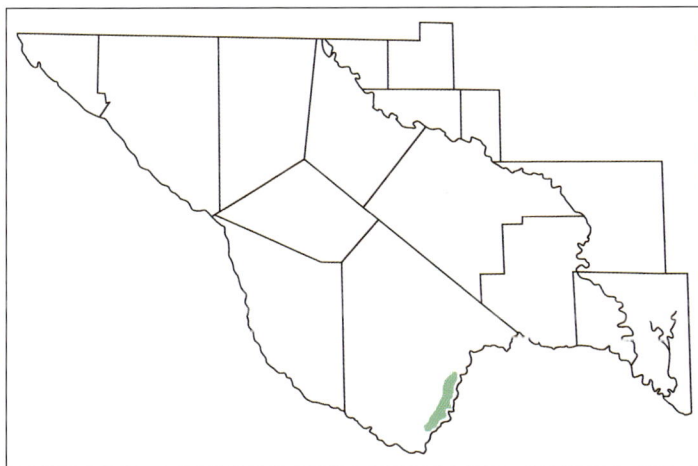

Map 19. *Opuntia azurea* var. *aureispina* (golden-spined prickly pear).

Vegetative Characters. Upright shrubs 0.3–1 m high from a short trunk. Orbicular to broadly obovate pads 8–16 cm long, 8–15 cm wide, light blue-green to yellow-green, glaucous, with spines in areoles from apex to the base of the pad. Spines yellow in many plants, but in others light brown, orange-brown, to nearly black with only yellow tips. Although spines are present in areoles over the entire pad, upper portions of pads appear densely spined because the spines are more numerous and longer in the upper areoles. In upper areoles on mature pads, 4–12 spines to 3.5–6 cm long and 0.7–1.3 mm wide; only 1–3 spines in younger areoles and areoles on sides of the stems.

Flowers. Flowering Mar–May. Yellow flowers with red or orange centers, like those of other members of the *O. azurea* complex. Flowers 5.5–7 cm long and 5–7 cm in diameter. Filaments pale green or yellow with pale green bases, to 1.5 cm long, and light yellow anthers ca. 1.5 mm long. Style ca. 2 cm long, narrowly urceolate, usually 5–7 mm wide, yellowish or pink at the base, supporting 6–8 pale green or pale yellow stigma lobes 3–8 mm long.

Fruits. Fruits at first fleshy and pale green to slightly reddish-tinged, but at maturity dry without pulp or juice, shriveling and turning tan, yellowish-tan, to greenish-tan. Some shriveled fruits are purplish, but these do not appear to be typical. Fruits at first are short-cylindroid or ovoid, 2.5–4 cm long, 1.7–2.5 cm in diameter, but later are constricted below the hard apical rim, which does not shrivel much during drying. Small areoles 13–24, with 0–4 yellowish spines to 1–2 cm or more long. Spines usually in areoles toward the fruit apex and at the upper rim. Rapidly drying spiny fruits help distinguish this taxon from its closest allies with fleshy, spineless fruits. However, fruits of var. *parva* and var. *diplopurpurea* may shrink. Tan to light brown seeds flattened, large, 4–6 mm in diameter, ca. 1.5 mm thick, irregular in shape, with a wide rim/raphe, this accounting for 2 mm or more of the diameter of the seed.

Other Distinctions. At the site north of Solis some plants designated as var. *aureispina* have yellow spines, and some have light brown to orange-brown spines, or even darker spines. We suspect that some spine-color variation is intrinsic within var. *aureispina*, but var. *aureispina* may hybridize with vars. *azurea, diplopurpurea,* and/or *parva;* all of these varieties occur in the general area, and they are mostly dark-spined. The hybrids between var. *aureispina* and var. *diplopurpurea/parva* were described as *O.* X *rooneyi* M. P. Griff.

Full Name and Synonyms. Opuntia azurea var. *aureispina* (S. Brack & K. D. Heil) A. M. Powell & J. F. Weedin. *O. macrocentra* Engelm. var. *aureispina* S. Brack & K. D. Heil; *O. aureispina* (S. Brack & K. D. Heil) Pinkava & B. D. Parfitt.

Other Common Name. Golden-spined opuntia.

Opuntia azurea var. discolor
Big Hill Prickly Pear
PLATES 74, 75

The type locality is at Big Hill, Presidio County, Texas. The varietal Latin epithet, *discolor*, means "of different colors" and refers to the different spine colors and lack of purple pigment in the stems and spines.

Distribution. Known distribution restricted to a small area on the slopes of

Plate 74. *Opuntia azurea* var. *discolor* (Big Hill prickly pear); type locality, Big Hill, 12–13 mi W of Lajitas, Presidio Co., TX.

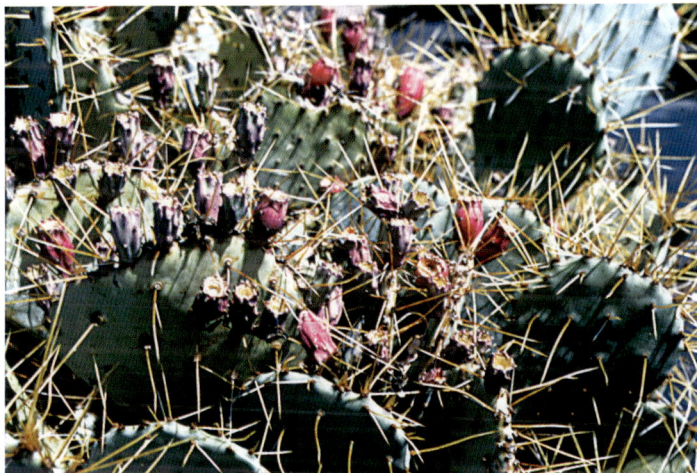

Plate 75. *Opuntia azurea* var. *discolor* (Big Hill prickly pear); type locality, Big Hill, 12–13 mi W of Lajitas, Presidio Co., TX; mature fruits.

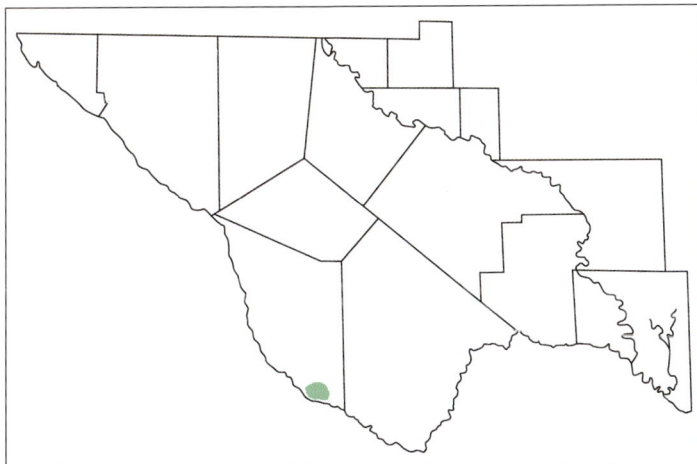

Map 20. *Opuntia azurea* var. *discolor* (Big Hill prickly pear).

"Big Hill," a desertic high ridge approxi- mately 12–13 mi NW of Lajitas that is traversed by the paved road between Lajitas and Redford (the Camino Real, or Hwy 170), and a small population 3.3 mi E of Big Hill, S of Colorado Canyon along the Rio Grande, Presidio Co. 2,500–3,000 ft. Mexico: visible on desertic hills directly across the Rio Grande in Chihuahua. Map 20.

Vegetative Characters. Plants upright and spreading, with pads broadly obovate to suborbicular, to 20 cm long, pale green to pale blue-green and not much, if at all, colored with betacyanins. Spines borne in areoles over upper three-fourths or more of the pad, strongly projecting, some irregu- larly curved or twisted, usually 3–4 per areole, to 10.5 cm long, 1–1.5 mm wide at base, yellow, white, or golden to orange or reddish. Areoles farther apart on the stem, 4–6.5 cm, than in var. *diplopurpurea*. Veg- etative features alone distinguish var. *dis- color* from the related purple prickly pear, var. *diplopurpurea*. Without fruits and seeds, these plants superficially resemble *O. camanchica* as much as *O. azurea* or *O. macrocentra*.

Flowers. Flowering Apr. Yellow flowers with red centers, identical to those of *O. azurea* var. *diplopurpurea* and other mem- bers of *O. azurea* complex. Stamens with pale green filaments; anthers yellow, style cream-colored, stigma lobes, on different plants, pale green to cream-colored.

Fruits. Fruits pale red to pinkish and tend to dry and shrink at maturity. Fruit shape and size, and seeds, like those of var. *diplopurpurea*.

Full Name. *Opuntia azurea* var. *dis- color* J. F. Weedin.

Opuntia tortispina
Twisted Spine Plains Prickly Pear
PLATES 76, 77

Opuntia tortispina is a low, sprawling, often slightly wrinkled prickly pear that is common in the grassy plains from the Davis Mountains northward and west- ward. *Opuntia tortispina* is very similar to *O. cymochila* Engelm., and slightly less similar to *O. phaeacantha* or *O. macrorhiza*. *Opuntia tortispina* has been confused with all of these species in some previous publications. The type locality is "On the Camanchica Plains near the Cana- dian River." The specific epithet refers to twisted spines, after the Latin *tortus*, "twisting," and *spina*, "spine."

Distribution. Igneous or limestone alluvium of mountain basins, in mountain woodlands, or less often in sand or sandy loam, rarely in desert mountains. Present in much of the Trans-Pecos, but mostly in grasslands of mountain basins, Davis Mts W to El Paso Co., not known in Terrell Co. 3,400–6,500 ft. From Trans-Pecos TX, the Panhandle, and NM, to CO; apparently also in N AZ, UT, NV, and S CA and rare in SE AZ. Mexico: in NE Sonora, N Chi- huahua, and NW Coahuila. Map 21.

Vegetative Characters. Habit low, creeping, prostrate, sprawling, with some branches 2–3 pads high; whole plants usually less than 30–40 cm high. At some sites, plants may be sympatric with *O. macrocentra, O. azurea* var. *diplopur- purea, O. phaeacantha*, or *O. polyacantha*, but in central mountain basin grasslands, relatively pure stands are common. Pads obovate to long-obovate, 9–19 cm long, 7–15 cm in diameter; spinose areoles cover all or upper three-fourths of pad, usually

Plate 76. *Opuntia tortispina* (twisted spine plains prickly pear) near Marfa, Presidio Co., TX.

Plate 77. *Opuntia tortispina* (twisted spine plains prickly pear) near Fort Davis, Jeff Davis Co., TX; mature fruit.

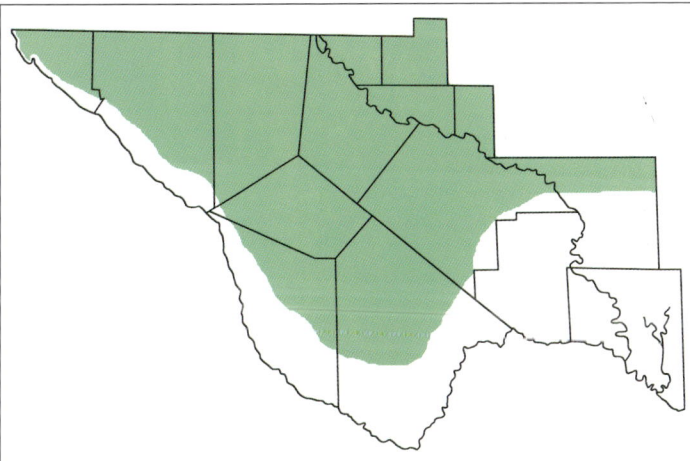

Map 21. *Opuntia tortispina* (twisted spine plains prickly pear).

with 4–5 white central spines per areole. Green stems may exhibit some purplish pigmentation, especially around the areoles, as in *O. phaeacantha*, but much less purplish than *O. macrocentra. Opuntia tortispina* has more closely set areoles than does *O. phaeacantha*, and pads tend to be slightly wrinkled after their first winter, as in *O. macrorhiza, O. pottsii*, and *O. polyacantha* var. *trichophora* (all of which are smaller plants).

Flowers. Flowering May–Jun. Yellow flowers 4.5–7 cm long and 4–6 cm in diameter. In most populations, flowers have pale to prominent red or rusty-red centers, but in some plants inner tepals are yellow to the base. This flower-color variation does not appear to have geographic significance. Plants with "peach color" flowers occur sporadically among yellow-flowered plants in northwest Brewster County. Filaments 1–1.5 cm long, reddish distally and greenish proximally, or reddish from the distal end to the base. Pale yellow anthers 2–2.5 mm long. Cream-colored or whitish style slender-urceolate and 1.5–2.3 cm long. Stigma lobes 5–7, ca. 3 mm long and green.

Fruits. Fruits obovate to elliptic, to 3.5–4.5 cm long, 2–2.8 cm in diameter, dull reddish, purple-red, rose-purple, or pale red on tan. Fruits somewhat juicy and sweet with light red to pale green pulp and fleshy rind thickened toward the top of the fruit. Seeds 4–6 mm in diameter, flattened but irregular in shape, ca. 2 mm thick, with beakless aril-rim ca. 1 mm wide.

Other Distinctions. In the Trans-Pecos *O. tortispina* stands visually intermediate between the often larger *O. phaeacantha* and *O. pottsii. Opuntia tortispina* is identified by its low habit with relatively small pads, usually white spines, yellow flowers with weakly defined reddish centers, and smallish red fruits.

Full Name and Synonym. Opuntia tortispina Engelm. & Bigelow. *O. tenuispina* Engelm. & Bigelow.

Other Common Name. Plains prickly pear.

Opuntia macrorhiza
Plains Prickly Pear
PLATES 78, 79

Although *O. macrorhiza* technically is not known to occur in the Trans-Pecos, its near approach to the western boundary of Ward County and adjacent counties suggests that it might also exist west of the Pecos River in localized, probably sandy habitats. The type locality for *O. macrorhiza* is on the upper Guadalupe River in south-central Texas, with a lectotype designation from between the Pedernales and Guadalupe rivers. The specific epithet refers to the fleshy roots, after the Greek *makros*, "large," and *rhiza*, "root."

Distribution. Sand and limestone. Almost throughout TX except for the Trans-Pecos, W nearly to the Pecos River, Ward, Winkler, Crane, and Ector counties. 2,500–3,000 ft. Widespread E of the Rocky Mts, but its true range is obscured by misidentifications; rare in the arid Southwest. Map 22.

Vegetative Characters. Low, sprawling plants, usually less than 30–40 cm high, closely resembling a northern species, *O. cymochila*. Plants are in the size range of Davis Mountains *O. polyacantha*, slightly smaller than *O. cymochila* and larger than the tiny, bluish-gray clumps of *O. pottsii*. Roots often form tuberlike swellings up to 2.5 cm thick, particularly when rooted in sandy substrates. Pads, wrinkled after the first winter, tend to root where their edges

Plate 78. *Opuntia macrorhiza* (plains prickly pear) near Monahans, Ward Co., TX; cultivated.

Plate 79. *Opuntia macrorhiza* (plains prickly pear) near Monahans, Ward Co., TX; nearly mature fruits; cultivated.

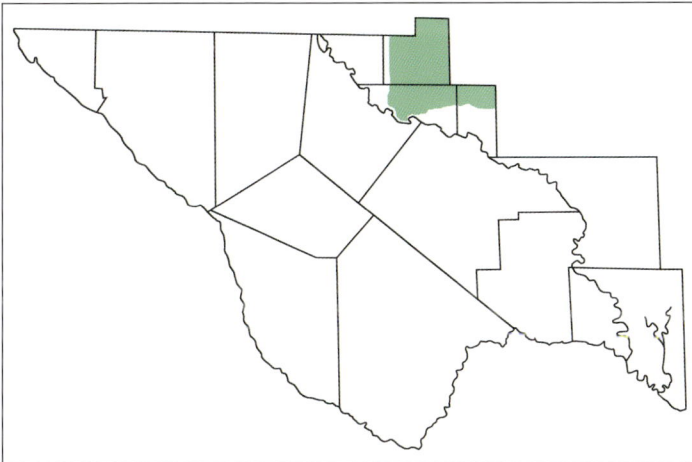

Map 22. *Opuntia macrorhiza* (plains prickly pear).

touch the ground. Individual pads are only 7.5–13 cm long and 8–12 cm wide. Usually 1–3 main spines in areoles on the upper part of the pad. Lowermost spine usually flattened and deflexed, the other one or two usually subterete, projecting outward or upward. Typically, spines chalky-white to whitish, often with their bases yellow, orange, reddish, or brown.

Flowers. Flowering Apr–Jun. Flowers have fewer tepals than those of O. *cymochila,* and are typically yellow with sharply defined orange or red centers, whereas flowers of O. *cymochila* are all yellow or rarely with red centers. An additional prominent floral difference between the two species is cream-colored or pale green stigma lobes in O. *macrorhiza* and deep green stigma lobes in O. *cymochila.*

Flowers of O. *macrorhiza* vary in size; measurements include 6 cm long and 5–8 cm in diameter. Filaments cream distally and greenish basally, anthers yellow. Cream-colored style closely matches stigma. Stigma barely elevated above anthers, at least in some flowers.

Fruits. Reported variable throughout the range of the species, with some greenish, orange, to purplish-red and sour. In Ward County, dull reddish-purple, obovoid, 2.5–3 cm long, 1.5–2 cm in diameter, with few small areoles. Not many glochids and rarely any spines present in the areoles. Clear juice with a taste like unsweetened apple, according to one of our tests. Fruits of the confusingly similar O. *cymochila* are dull purplish-red to brownish, ovoid, more slender below the apex; areoles usually are prominent, with abundant glochids and often slender spines. Fruit pulp of O. *cymochila* juicy and purple, and fruits are said to be among the sweetest of any prickly pear

fruits in the United States. Seeds of O. *macrorhiza* light brownish to tan, irregularly shaped, 3–3.5 mm in diameter (the seeds of O. *cymochila* 5–6 mm in diameter), 1.5–2 mm thick, notched at the hilum, with a narrow margin to 0.5 mm wide (O. *cymochila* seeds have a broad margin).

Full Name and Synonyms. Opuntia *macrorhiza* Engelm. O. *mesacantha* Raf. var. *macrorhiza* (Engelm.) J. M. Coult.; O. *compressa* J. F. Macbr. var. *macrorhiza* (Engelm.) L. D. Benson.

Opuntia pottsii
Potts' Prickly Pear
PLATES 80, 81

Opuntia pottsii is one of the smallest prickly pears in the Trans-Pecos, along with O. *polyacantha.* Opuntia *pottsii* often goes unnoticed among grasses in the basins of the Davis Mountains until the plants produce their red flowers. The type locality is near Chihuahua City, Mexico, which is at an elevation of about 4,000 feet. The species was named after John Potts, who managed the mint in Chihuahua, Mexico, collected the original material of the species, and between the years 1842 and 1850 sent many cactus collections to F. Scheer at the Royal Botanic Gardens, Kew, England.

Distribution. Alluvium in basin grasslands, sand, gypsum, or limestone away from the mountains. Northern Presidio and Brewster counties, reported in the Chisos Mts of S Brewster Co., Jeff Davis Co., basins of the Davis Mts; Culberson Co.; Reeves Co.; probably Pecos and Terrell counties; Winkler Co. and other counties adjacent to the Pecos River NE of the Trans-Pecos. 2,800–5,500 ft. Reported in Maverick Co.; southern Panhandle S to

Plate 80. *Opuntia pottsii* (Potts' prickly pear) near Marfa, Presidio Co., TX.

Plate 81. *Opuntia pottsii* (Potts' prickly pear) near Alpine, Brewster Co., TX; maturing fruits; cultivated.

Ector Co., and sand hills. Across southern NM and into SE AZ; central and NW AZ into SW Utah. Mexico: NE Sonora and Chihuahua. Map 23.

Vegetative Characters. Typically small, upright, with only 6–10 pads, usually not more than 30 cm high. Fleshy roots exude milky sap when injured. Typically, pads obovate, to 10–13 cm long, 8–10 cm in diameter, glaucous blue-green when

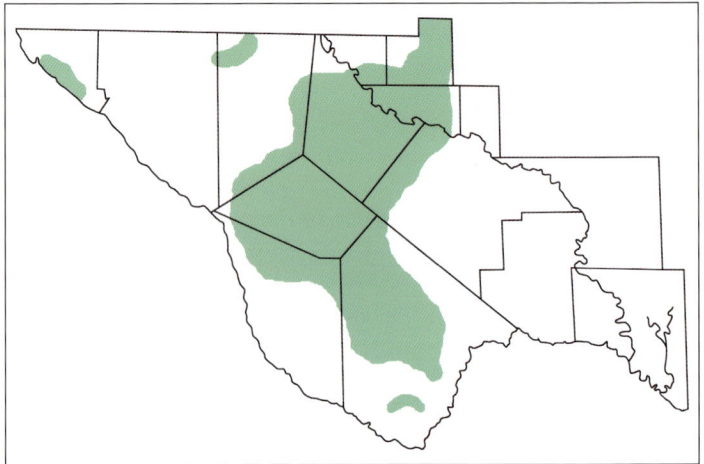

Map 23. *Opuntia pottsii* (Potts' prickly pear).

healthy. Whitish to gray spines 1–3 per areole, on upper part of the pad. Longest spines 1.4–6.4 cm long, ca. 0.2 mm thick, acicular or somewhat flattened.

Flowers. Flowering May. The only Trans-Pecos prickly pear with totally red flowers. Flowers 5–9 cm long and 4–6 cm in diameter. In some, the slender pericarpel alone is often ca. 5 cm long. Filaments 1.2–1.5 cm long, pale yellow, greenish-yellow to purplish. Anthers bright yellow, ca. 2 mm long. Style pinkish, 1.7–2.3 cm long, 2–3 mm wide near the base. Stigma lobes 5–8, cream-colored, 3–4 mm long, stout.

Fruits. Fruits narrowly obovoid to clavate, 2.5–5 cm long, ca. 2 cm in diameter distally, with a slender base and deep umbilicus; typically glaucous, pinkish-purple (pigments confined to the skin). In some populations in Reeves County, and perhaps elsewhere, ripened fruits dull to light red or reddish-brown. Fruits juicy with pale greenish pulp and not very sweet. Seeds 3.5–5.5 mm in diameter, thick-discoid, irregular in outline, wide-margined, beaked at the hilum.

Full Name and Synonyms. Opuntia *pottsii* Salm-Dyck. *O. ballii* Rose; *O. plumbea* Rose; *O. loomsii* Peebles; *O. macrorhiza* Engelm. var. *pottsii* (Salm-Dyck) L. D. Benson.

Opuntia mackensenii
Mackensen's Prickly Pear, Short-Spined Purplish Prickly Pear

Recent workers have not recognized O. *mackensenii*, except possibly D. Ferguson (unpub.) and A. D. Zimmerman (pers. comm.). Benson (1982) included *O. mackensenii* (among diverse other species) in his concept of *O. macrorhiza* var. *macrorhiza*, with the statement that plants described as *O. mackensenii* were intermediate between *O. macrorhiza* and *O. phaeacantha* var. *major*. Weniger (1984) did not mention *O. mackensenii*. The type locality of *O. mackensenii* is "near Kerrville," Kerr County, Texas, where it was originally collected in 1909 by Bernard Mackensen, a collector and author of *Opuntia* species in the vicinity of San Antonio, Texas. The species was named for Mackensen in 1911 by J. N. Rose.

Full Name. Opuntia mackensenii Rose.

Key to the Varieties of *Opuntia mackensenii*

1. Spines usually whitish; eastern distribution, relatively moist habitats
O. mackensenii var. *mackensenii*, p. 89
1. Spines usually dark purple or reddish-brown to nearly black, often with whitish tips; western distribution, relatively dry habitats
O. mackensenii var. *minor*, p. 91

Opuntia mackensenii var. mackensenii
Mackensen's Prickly Pear
PLATES 82, 83

The type locality for the recent synonym *O. edwardsii* is Pedernales Falls State Park, Blanco County, Texas. Despite ending in *-ii* instead of *-ensis,* the epithet *edwardsii* was taken after the Edwards Plateau, an area of major distribution for *O. mackensenii* var. *mackensenii.*

Distribution. Alluvial or rocky substrates, mostly limestone, among grasses in shrubland, woodland, or grassland habitats. Terrell Co., from W of Sanderson E through Val Verde Co. 1,000–3,000 ft. Crockett Co. E throughout the Edwards Plateau, N through rolling plains and high plains to Randall Co. in TX Panhandle. Mexico: presumably in adjacent Coahuila. Map 24.

Vegetative Characters. As in most of

Plate 82. *Opuntia mackensenii* var. *mackensenii* (Mackensen's prickly pear), Terrell Co., TX; cultivated.

Plate 83. *Opuntia mackensenii* var. *mackensenii* (Mackensen's prickly pear) near Longfellow, Terrell Co., TX; mature fruit; cultivated.

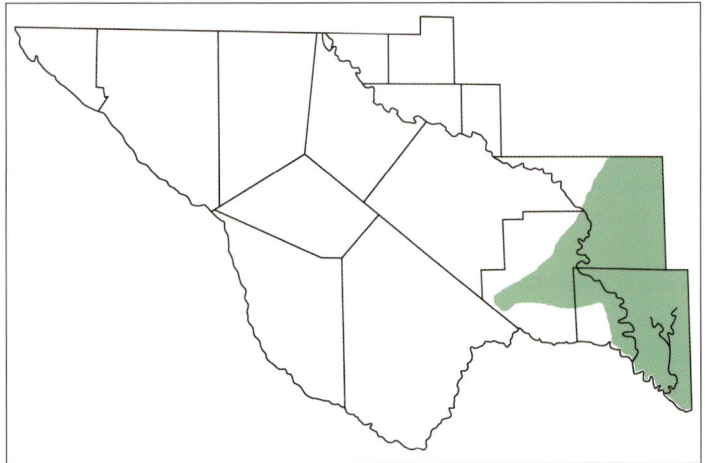

Map 24. *Opuntia mackensenii* var. *mackensenii* (Mackensen's prickly pear).

its allies, low, sprawling subshrubs 20–45 cm high with slightly ascending branches. Sometimes roots include tuberous thickenings. In habit and other morphology var. *mackensenii* closely resembles O. *phaeacantha* var. *phaeacantha*, which has a more westerly distribution and slightly thinner pads. In var. *mackensenii* spines are produced over three-fourths to nearly the entire pad, with 2–5 usually white spines, 1–5.5 cm long, in the distal areoles. In lower part of areole usually one larger deflexed spine and one or more shorter deflexed spines. In upper part of areole spines may be absent, or 1–2 projecting spines.

Flowers. Flowering spring. Flowers 7–8 cm in diameter, basically yellow or (in our experience) yellow with red, pale red, or reddish-brown centers. Filaments colorless, anthers light yellow. Style whitish to cream with pale green to cream stigma lobes. Flowers on some plants have longer styles, exceeding length of stamens by 5–7 mm, but on other plants stigma barely extends above anthers.

Fruits. Fruits purple, rose to pale red, obovate, fleshy, and 3–6 cm long. Fruit pulp, as determined from one specimen in Terrell County, dull to bright green with seeds in a small amount of clear juice, matching the original description. In profile, fruits truncate apically, areoles spineless. Seeds discoid, suborbicular in outline, 4–6 mm in diameter ("5 to 6 mm" in the original diagnosis).

Full Name and Synonym. Opuntia *mackensenii* var. *mackensenii*. O. *edwardsii* V. E. Grant & K. A. Grant.

Other Common Name. Edwards Plateau opuntia.

Opuntia mackensenii var. minor
Short-Spined Purplish Prickly Pear
PLATES 84, 85

Opuntia mackensenii var. *minor* was described *as* O. *macrocentra* var. *minor* by Anthony (1956) as a product of her study of the Opuntiae of the Big Bend region of Texas. The type locality is 1.4 miles southeast of Ruidosa in southern Presidio County, Texas. The varietal epithet is after the Latin *minor*, "smaller," supposedly a reference to the smaller characters of var. *minor* as compared by Anthony to O. *macrocentra* var. *macrocentra*.

Distribution. Sand and sandy loam, *Larrea* and *Prosopis* flats, Presidio and Brewster counties; grasslands and slopes of igneous hills, S Davis Mts, N Brewster Co.; limestone hills and mesas, Terrell and Val Verde counties. 2,500–4,600 ft. Mexico: presumably in adjacent Chihuahua and Coahuila. Map 25.

Vegetative Characters. Plants sprawling and many-branched, often with low, creeping habit, ca. 25–60 cm high, with some branches ascending. Stems glaucous and blue-green, with reddish-purple pigmentation increasing in plants (at least those near the type locality) under environmental stress. Pads obovate, 12–16 cm long, 8–12.5 cm wide. Spines 3–4.5 cm long, dark purple to dark reddish-brown, some with whitish tips. Usually 4–9 spines in areoles, and spiny areoles present on upper one-half to three-fourths of the pad.

Plate 84. *Opuntia mackensenii* var. *minor* (short-spined purplish prickly pear); type locality, near Ruidosa, Presidio Co., TX.

Plate 85. *Opuntia mackensenii* var. *minor* (short-spined purplish prickly pear) near Dryden, Terrell Co., TX; mature fruits.

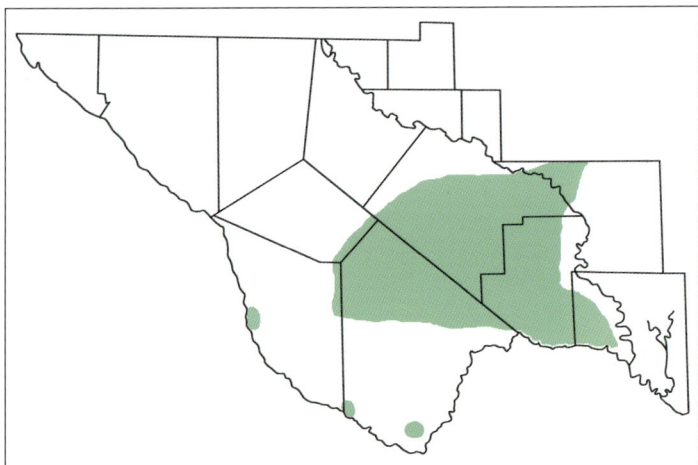

Map 25. *Opuntia mackensenii* var. *minor* (short-spined purplish prickly pear).

Flowers. Flowering Apr–May. Yellow flowers with red centers, or at least reddish blush, similar to those of *O. macrocentra* and *O. azurea*. Flowers up to 7.5 cm long and 7.5 cm wide when fully open. Yellow inner tepals with reddish midstripes (extending to the tepal tips), not merely red bases. Filaments and anthers cream-yellow. Style and stigma lobes cream-colored.

Fruits. Fruits red-brown to dull red and fleshy at maturity. Base of fruit often remaining green long after the distal portion turns reddish. Fruits short ovate-truncate to oblong-elliptic, 2.5–3 cm long, 1.5–1.7 cm in diameter, with shallow umbilicus and without constriction below upper rim. Rind and pulp green; juice moderate in amount, sweet, clear. Seeds tan, flattened, 4–5 mm in diameter, including a prominent margin 0.5–0.6 mm wide.

Other Distinctions. Resembles *O. macrocentra* and *O. azurea* in habit, pad shape, purplish stem pigments, spine configuration (but spines shorter), flower morphology. Plants most closely matching original description of var. *minor* occur from near Candelaria to east of Ruidosa in southern Presidio County and near Santa Elena Canyon in southern Brewster County. Other plants (the "grassland tetraploids") in Pecos County, Presidio County, northern Brewster County, near Dryden in Terrell County, and near Langtry in Val Verde County closely resemble var. *minor* in spine and flower morphology, although in general the nearly black to dark reddish-purple spines are darker proximally than are those of var. *minor* and they are more typically white-tipped. See Powell and Weedin (2004) for additional discussion of var. *minor*.

Full Name and Synonym. Opuntia mackensenii var. *minor* (M. S. Anthony), comb. nov., forthcoming. *O. macrocentra* Engelm. var. *minor* M. S. Anthony.

Opuntia chisosensis
Chisos Prickly Pear
PLATES 86, 87

Opuntia chisosensis is a distinct species with an interesting taxonomic history (see Powell and Weedin, 2004). The type locality is the Basin, Chisos Mountains, Big Bend National Park, Brewster County, Texas. The species is named after the Chisos Mountains, using the Latin suffix *-ensis*, meaning "belonging to."

Distribution. Oak-juniper-pinyon woodland and grassy meadows, middle to upper mountain slopes and canyons. Brewster Co., Chisos Mts. 5,200–7,200 ft. Mexico: Coahuila, Maderas del Carmen (igneous, wooded portion of the Sierra del Carmen); to be expected elsewhere in Mexico. Map 26.

Vegetative Characters. Readily identified in the Chisos Mountains as a medium-

Plate 86. *Opuntia chisosensis* (Chisos prickly pear), Chisos Mts, Big Bend National Park, Brewster Co., TX; cultivated.

Plate 87. *Opuntia chisosensis* (Chisos prickly pear), Chisos Mts, Big Bend National Park, Brewster Co., TX; mature fruits; cultivated.

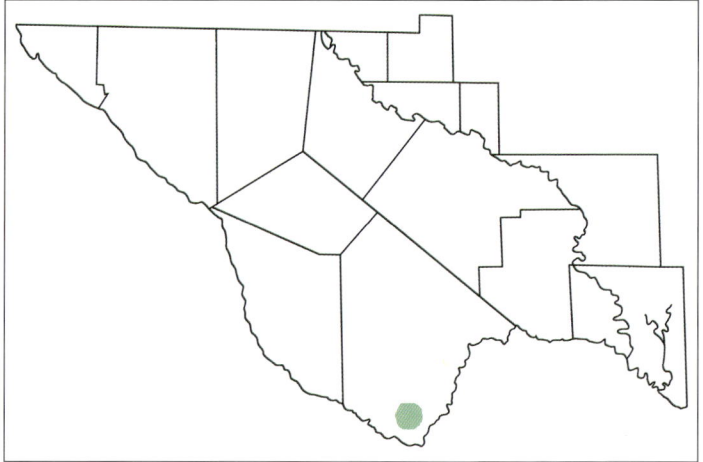

Map 26. *Opuntia chisosensis* (Chisos prickly pear).

sized prickly pear with bright yellow spines. A few have dark spines instead, especially on lower pads, but otherwise are identical. Plants, common in wooded areas, are upright shrubs to ca. 1 m tall, with branches spreading and ascending from a thick base but not a trunk. Orbicular to short-obovate pads 16–29 cm long, 13–22 cm wide, pale gray-green or blue-green and glaucous to medium- or yellow-green with spines in the upper two-thirds of the areole. Spines yellow on whole plants or on upper stems, but spines on lower stems possibly reddish to nearly black, apparently a function of age. Central spines 1–5 per areole, mostly deflexed on the face of the pad, to upright and spreading at all angles on the upper margin of the pad. Usually one larger spine, to 6.7 cm long, 1.5 mm wide, in each areole. Other largest spines about 3–6 cm long.

Flowers. Flowering May. Flowers about 5–6.5 cm long and wide. Tepals are an unusual yellow, described as pale yellowish-buff. Pale yellow fades to pale salmon at the end of the day. Flowers sometimes open for two or rarely three days. Filaments pale green, anthers yellow. Style yellow, stigma lobes green. Pericarpel areoles small and distant.

Fruits. Fruits are relatively small compared to those of most other prickly pears in the Trans-Pecos and conspicuous in their smaller size compared to those of other prickly pears in the upper Chisos Mountains. Small fruits are variable in size and shape, usually 2–3.5 cm long, 2–3.5 cm in diameter, globose, depressed globose and truncate at both ends except for an apical umbilicus, bell-shaped, cup-shaped, short oblong, and obovate. Umbilicus saucerlike to 4–7 mm deep and 0.7–1 cm wide. Areoles distant, bearing only a few short glochids; one or several areoles at the upper rim bear one to several slender yellow spines or bristles ca. 0.5–1.2 cm long. Fruits juicy, red to beet-red, or dark red to reddish-purple, often glaucous. Rind and pulp beet-red; juice sweet and clear, stained with red pigment from the rest of the fruit.

Seeds yellowish, compressed or flattened, irregular in shape, 3–4.5 mm long, 3–4 mm wide, 1.3–1.5 mm thick. Aril-margin approaches 1 mm wide, appears to be beaked on some seeds and not beaked on others. Seeds notched on one side at the hilum.

Full Name and Synonym. Opuntia chisosensis (M. S. Anthony) D. J. Ferguson. *O. lindheimeri* Engelm. var. *chisosensis* M. S. Anthony.

Opuntia spinosibacca
Spiny-Fruited Prickly Pear
PLATES 88, 89

Opuntia spinosibacca is one of several cactus species with its distribution limited to the hot, dry, limestone substrates in the southern Big Bend region. Good places to observe the species are at Hot Springs and near the Boquillas Tunnel in Big Bend National Park. The type locality is on the slopes of a limestone hill just west of the rangers' quarters, Boquillas, Big Bend National Park, Brewster County, Texas. The specific epithet is descriptive of the

Plate 88. *Opuntia spinosibacca* (spiny-fruited prickly pear) near tunnel, Big Bend National Park, Brewster Co., TX.

Plate 89. *Opuntia spinosibacca* (spiny-fruited prickly pear), Big Bend National Park, Brewster Co., TX; mature fruits; cultivated.

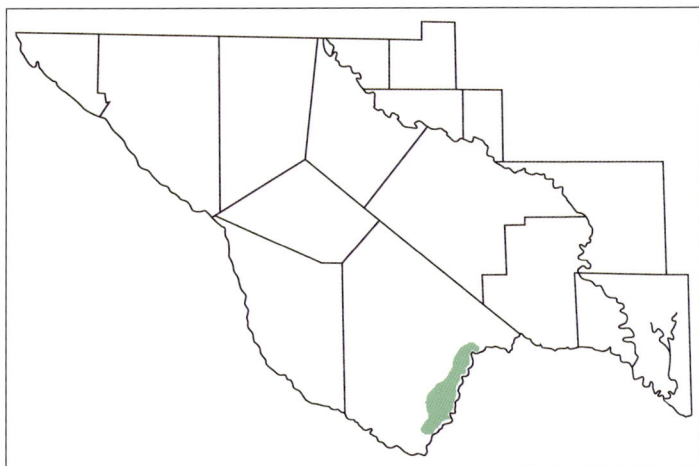

Map 27. *Opuntia spinosibacca* (spiny-fruited prickly pear)

spiny fruits, after the Latin *spina*, "spine," and *bacca*, "a small round fruit."

Distribution. Limestone hills, slopes, and canyons. Brewster Co., W of Hot Springs E along and near Rio Grande to near Reagan Canyon. 1,600–2,300 ft. Mexico: not documented, but probably in limestone near the Rio Grande, vicinity of Boquillas. Map 27.

Vegetative Characters. Plants are basically upright, compact shrubs to 1–1.5 m tall, but some spreading plants are found near Heath Canyon and other sites.

"Spreading growth forms" closely resemble *O. camanchica*, unless characteristic spiny flowers and fruits of *O. spinosibacca* are present. Pads 10–20 cm long, 7.5–15 cm wide, light green to yellowish-green with spines in areoles across most of the pad except near the base. Expect a purple blotch near each areole. Areoles elevated on low but conspicuous tubercles, unusual among upright prickly pears. Areoles number relatively few per pad; pads smaller than might be expected from the size of the shrub. Spines usually reddish-brown and number 4–8 in upper areoles, where they are up to 2–7 cm long, 1.5–2 mm wide, divergent, usually flattened, sometimes angular, often twisted and curved.

Flowers. Flowering Mar–May. Flowers bright to golden-yellow or orange-yellow, with red centers, 5–7.5 cm long and 5–7 cm in diameter. Red pigmentation of inner tepals often extends distally up the midregion to the apex, as in *O. macrocentra* and *O. mackensenii* var. *minor.* With age, red centers may fade to washed-out red-pink. Filaments pale green to cream-colored, 0.6–1.4 cm long. Anthers pale yellow and 1.5–2 mm long. Whitish to pinkish style slender-urceolate, 1.5–2.2 cm long, supporting 7–9 pale green to cream-yellow stigma lobes to ca. 3 mm long.

Fruits. Spiny, fleshy, greenish-yellow at first but at maturity dry, shrunken, tan to yellowish or reddish. On some plants fruits remain fleshy, somewhat juicy, and mottled red, but some plants with fleshy fruits may be misidentified as the lowland (tall), red-spined form of *O. dulcis.* Fruits in *O. spinosibacca* slightly tuberculate, 2.5–4.5 cm long, 1.3–2.5 cm in diameter, basically ovoid with a flared apical rim

and with a prominent umbilicus. Spines in scattered areoles, but particularly evident at and near the rim, where usually reddish-brown spines to 2.5 cm long project mostly horizontally around the entire fruit apex, or nearly so. Yellowish seeds flattened, irregularly shaped, 4–6.5 mm in diameter, 1.5–2 mm thick, notched at the hilum, with a prominent aril-margin 1–1.5 mm wide.

Full Name and Synonyms. Opuntia *spinosibacca* M. S. Anthony. *O.* X *spinosibacca* M. S. Anthony; *O. phaeacantha* Engelm. var. *spinosibacca* (M. S. Anthony) L. D. Benson.

Other Common Name. Big Bend prickly pear.

Opuntia phaeacantha var. phaeacantha
Brown-Spined Prickly Pear
PLATES 90, 91

Opuntia phaeacantha and its allies, collectively called "the hexaploid juicy-fruited northern prickly pear cacti," comprise one of the most poorly understood groups of prickly pears in the Trans-Pecos (see Powell and Weedin, 2004). Our concept of the *O. phaeacantha* group in the Trans-Pecos

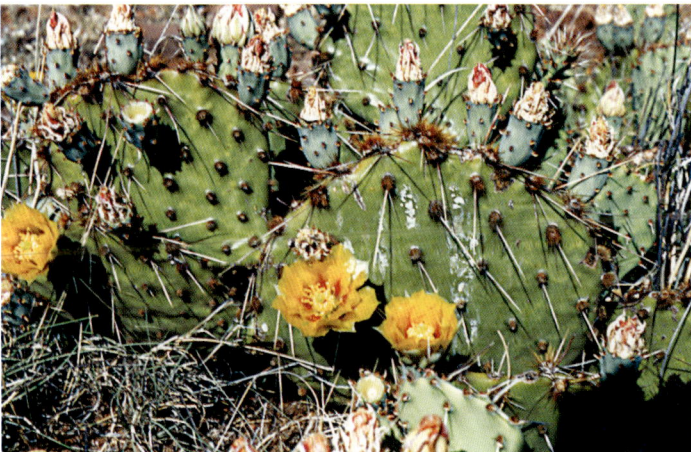

Plate 90. *Opuntia phaeacantha* var. *phaeacantha* (brown-spined prickly pear) near Alpine, Brewster Co., TX.

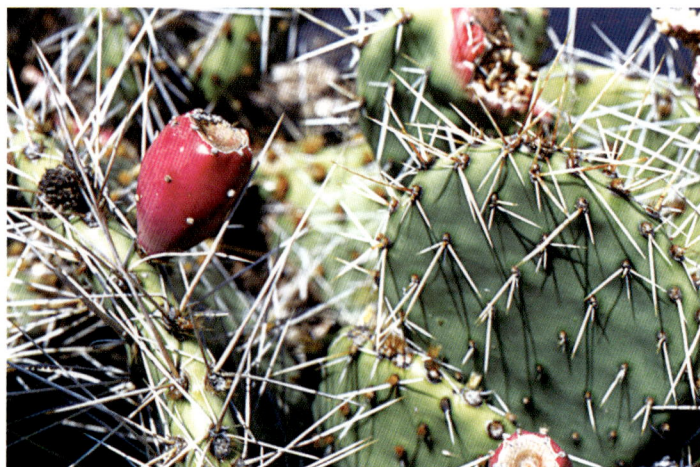

Plate 91. *Opuntia phaeacantha* var. *phaeacantha* (brown-spined prickly pear), Wildrose Pass, Jeff Davis Co., TX; fruits.

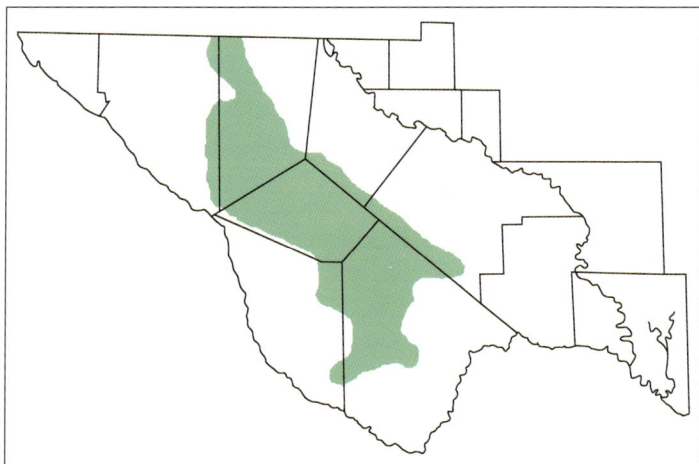

Map 28. *Opuntia phaeacantha* var. *phaeacantha* (brown-spined prickly pear).

incorporates generally unpublished information from D. Ferguson (pers. comm.) as well as our own observations. The type and lectotype localities of var. *phaeacantha* are near Santa Fe, New Mexico, near the Rio Grande. The epithet *phaeacantha* is after the Greek *phae*, "dark," and *akantha*, "spine," a misleading descriptor because the spines of O. *phaeacantha* var. *phaeacantha* usually are light-colored, usually whitish, at least in our region.

Distribution. Infrequent in grasslands and woodlands of the central mountains, mid- to higher elevations; less frequent to rare in rocky and woodland areas of more desertic mountains to the W, S, and E, usually above 4,000 ft; very rare in the desert at elevations below 4,000 ft; El Paso, Hudspeth, Culberson, Jeff Davis, Presidio, and Brewster counties. 4,000–7,500 ft. In NM N to N CO and N UT, W to AZ and CA. Mexico: N Sonora, Chihuahua, and Coahuila. Map 28.

Vegetative Characters. One of the best distinguishing features is the plant's winter habit of weak stems and pads that sag or even lie flat on the ground, as in *O. macrorhiza* and *O. tortispina,* probably because the wood is relatively poorly developed. A sometimes subtle difference from *O. tortispina* (and *O. macrorhiza* and *O. pottsii*) is that there are no sharply defined transverse wrinkles in winter conditions and no corresponding visible markings on pads that have passed through the winter.

Turgid, smooth-skinned summer habit more nearly erect but still sprawling, to ca. 30–60 cm high (taller than *O. tortispina*). Pads obovate to broadly elliptic or suborbicular, 10–22 cm long, 9–18 cm wide, averaging larger than those of *O. tortispina.* Spines (fewer than in *O. tortispina*) produced over upper one-half to three-fourths of the pad. Usual spine pattern in subapical areoles includes two (sometimes three) spines, one 2–3 cm long, the other 3.5–5 cm long. Spines whitish to tan throughout, or brownish or reddish proximally.

Flowers. Flowering May–Jun. Flowers yellow with pale to prominent reddish centers, 5–7 cm long and wide. Filaments and anthers pale yellow. Style cream-colored to pinkish, and short stigma lobes yellowish to pale green or perhaps deep green.

Fruits. Reddish fruits 2.4–5.5 cm long, 1.9–2.8 cm wide, turbinate or broadly clavate, with a shallow or cuplike umbilicus. Pulp bright red to pinkish with sweet taste. Seeds 4–5 mm in diameter, tan, discoid-reniform or irregularly orbicular, with a hilar notch, the margin ca. 0.5 mm wide.

Other Distinctions. Plants often found in same vicinity, if not side by side, with *O. tortispina, O. engelmannii, O. macrocentra* and/or *O. azurea,* sensu lato, *O. polyacantha, O. mackensenii* var. *minor,* and *O. pottsii.* Variety *phaeacantha* more easily distinguished from other associated prickly pears than from *O. tortispina.* In central Trans-Pecos mountains, var. *phaeacantha* more likely to be found on rocky hillsides and in woodlands than in alluvial grasslands that are the typical habit of *O. tortispina* and *O. pottsii.*

Full Name and Synonym. Opuntia phaeacantha Engelm. var. *phaeacantha. O. phaeacantha* Engelm. var. *major* Engelm., sensu Benson (in part).

Opuntia camanchica
Comanche Prickly Pear
PLATES 92, 93

Opuntia camanchica is one of the most common prickly pear species in the Trans-Pecos. Both Benson (1982) and Weniger (1984) interpreted *O. camanchica* as a variety of *O. phaeacantha;* they did not agree upon the morphological and geographic criteria considered diagnostic for var. *camanchica.* Neither author considered *camanchica* to be a Trans-Pecos entity. Our concept of *O. camanchica* in the Trans-Pecos (see Powell and Weedin, 2004) includes mostly specimens/sites cited by Benson as *O. phaeacantha* var. *major.* The type locality for *O. camanchica* is on the Llano Estacado near the Canadian River, the Comanche Plains country. Presumably the specific epithet is a geographic reference to the Comanche Plains.

Distribution. Mostly desert or semidesert habitats throughout most of the Trans-Pecos. 2,000–4,500 ft. Throughout most of western half of TX, infrequent in deep S TX; OK (W portion including Panhandle), CO (SE corner), NM, AZ, SW UT, S NV, and SE CA. Mexico: Coahuila S to Durango. Map 29.

Vegetative Characters. Similar in habit to *O. phaeacantha* var. *phaeacantha,* but with more substantial supportive tissues in

Plate 92. *Opuntia camanchica* (Comanche prickly pear), Brewster Co., TX; cultivated.

Plate 93. *Opuntia camanchica* (Comanche prickly pear), Brewster Co., TX; cultivated.

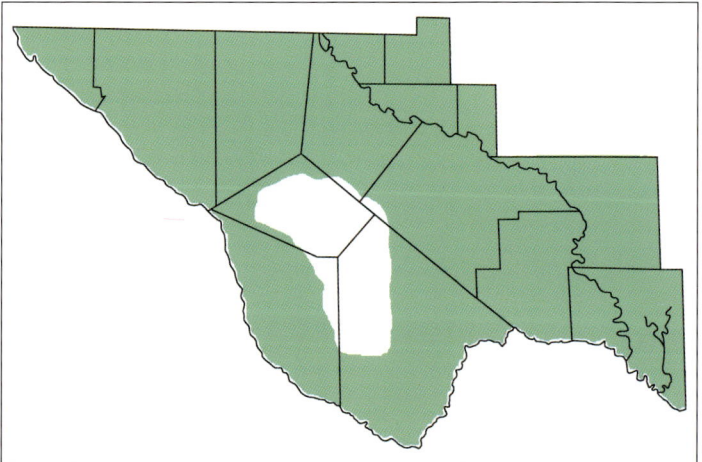

Map 29. *Opuntia camanchica* (Comanche prickly pear)

the stems, that is, woodier, and so under good conditions taller and remaining relatively erect in winter. Plants low and spreading to mostly erect, hemispheric, or lower shrubs usually less than 70 cm high. Pads usually broadly obovate to suborbicular, 15–25 cm long, 15–20 cm wide. Much variation in spine number and color, including red and black, whereas *O. phaeacantha* var. *phaeacantha* is basically brown and white. Plants of *O. camanchica* nearly spineless or with four or more main spines in each areole. When present (most Trans-Pecos populations), 3–7 larger spines 3–6.5 cm long, typically stout, divergent in several directions, often curved.

Flowers. Flowering Apr–Jun. Flowers yellow with red to pale reddish centers, 5–7 cm long and wide. Filaments 1.5–1.7 cm long, cream-colored; yellow anthers ca. 2 mm long. Style cream-colored, ca. 2 cm long; stigma lobes cream-colored to pale green or deep green.

Fruits. Reddish to cherry-red, 3.7–5 cm long, 2.3–3.3 cm wide, usually with a deep umbilicus. Fruits spineless or with 1–2 spines in areoles near the rim. The relatively few areoles cottony-white with brownish glochids ca. 1 mm long. Fruit pulp green, with clear juice, but juice not abundant and not very sweet or flavorful but with light, sweet scent. Only the outer layer of the pericarpel red, with most of the fruit rind colorless. Although red fruits with greenish pulp seem typical, many plants in the Trans-Pecos with spine characters matching those of *O. camanchica* produce red fruits with reddish pulp and juice. Seeds of *O. camanchica* 4–5 mm in diameter, tan to light brown, irregularly discoid with a broad hilar notch, and a prominent aril 0.5 mm or more wide.

Full Name and Synonyms. Opuntia camanchica Engelm. & Bigelow. *O. phaeacantha* Engelm. var. *brunnea* Engelm.; *O. phaeacantha* Engelm. subsp. *camanchica* (Engelm.) Borg; *O. phaeacantha* Engelm. var. *camanchica* (Engelm. & Bigelow) L. D. Benson; *O. phaeacantha* Engelm. var. *major* Engelm., in part, sensu Benson.

Opuntia dulcis
Sweet Prickly Pear
PLATES 94–96

Although *O. dulcis* appears to be relatively common in the Trans-Pecos, it has not been interpreted consistently as a distinct taxon. The syntype locality was "near the middle course of the Rio Grande, near Presidio del Norte, etc.," and "frequently observed near Presidio del Norte and Eagle Pass." The lectotype is a Wright collection with the ambiguous data "El Paso? West Texas? Probably Presidio del Norte," Wright in 1852, MO. The specific epithet is after the Latin *dulcis*, "sweet," a reference to the very sweet fruits as noted on the label of the type specimen.

Distribution. Common in desert habitats, often along and near the Rio Grande. El Paso Co., SE at least to Terrell Co. Restricted mostly to the Trans-Pecos. 2,200–3,500 ft. To be expected in adjacent NM and Mexico. Map 30.

Vegetative Characters. Overlaps the geographic range of *O. camanchica*, but the contact zone is undocumented, at least in the Trans-Pecos. Like most prickly pears, *O. dulcis* is most easily identified by growth habit, which is similar to that of *O. phaeacantha* but larger and more upright than that of *O. camanchica* (that is, almost as large as *O. engelmannii*). Pads not distinctive, being obovate to suborbicular, 15–

Plate 94. *Opuntia dulcis* (sweet prickly pear) near Boquillas, Big Bend National Park, Brewster Co., TX.

Plate 95. *Opuntia dulcis* (sweet prickly pear) near Boquillas, Big Bend National Park, Brewster Co., TX; cultivated.

18 cm long, 10–15 cm wide. Spines 1.27 cm long, usually fewer in each areole, typically more slender than in *O. camanchica*. Usually two spines per areole in *O. dulcis*. Spines in many distinct populations of *O. dulcis* light brown proximally and nearly white on the distal portion, but spines in some populations are almost any color.

Flowers. Flowering Apr–May. We have observed the flowers of relatively few plants identified as *O. dulcis*. Those flowers yellow with conspicuous or incon-

Plate 96. *Opuntia dulcis*
(sweet prickly pear)
near Boquillas, Big Bend
National Park, Brewster
Co., TX; mature fruits;
cultivated.

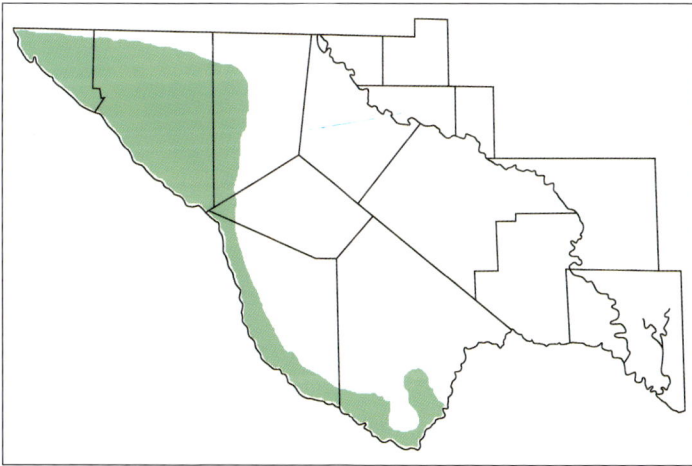

Map 30. *Opuntia dulcis*
(sweet prickly pear)

spicuous red centers. On average they are slightly larger (ca. 5–7 cm), wider open, and with more tepals than those of *O. camanchica*. Yellow tepals suffused, not sharply marked with red basally, or with merely an inconspicuous reddish flush in the midregion. Filaments pale green to cream-colored or colorless, ca. 1.5 cm long; yellow anthers ca. 2 mm long. Colorless or rosy style 2–2.5 cm long, with stigma lobes usually light green.

Fruits. Ripe fruits of *O. dulcis* red to

purplish, obovate to obconic, 5–5.5 cm long, 2.5–3 cm in diameter, with a shallow or deep umbilicus. Fruit surface smooth with few areoles bearing a small number of glochids and no spines. Fruit rind purple; pulp either pink, purple, red, or greenish. Apparently mature fruits always juicy, with deep purple, pink, to clear sweet juice. Seeds tan, irregularly discoid, 3.5–4.5 mm in diameter, with a narrow hilar notch and a prominent aril-margin 0.7–1 mm wide.

Full Name and Synonyms. Opuntia dulcis Engelm. *O. lindheimeri* Engelm. var. *dulcis* (Engelm.) J. M. Coult.; *O. engelmannii* Salm-Dyck var. *dulcis* (Engelm.) J. M. Coult. ex K. Schum.

Other Common Name. Sweet opuntia.

Opuntia engelmannii
Engelmann's Prickly Pear, Texas Prickly Pear

After two decades of nomenclatural and taxonomic confusion (see Benson, 1982; Weniger, 1984) concerning *O. engelmannii* (*O. phaeacantha* var. *discata*) and *O. lindheimeri*, Parfitt and Pinkava (1988) reinstated the correct application of the name *O. engelmannii*. Secondarily, they revised some of the several closely related taxa within *O. engelmannii*, sensu lato. Parfitt and Pinkava recognized six taxa as varieties of *O. engelmannii*, three of which, var. *engelmannii*, var. *lindheimeri*, and var. *linguiformis*, are known to occur in the Trans-Pecos. Two of these taxa are treated in the following discussions as natural populations, and var. *linguiformis*, a cultivar with no known natural populations, is reviewed at the end of the *Opuntia* section. See Powell and Weedin (2004) for further systematic discussion of *O. engelmannii* and its allies. The specific epithet honors George Engelmann, one of the early authorities in Cactaceae, and a prolific worker who described numerous southwestern cacti, including many of those in the Trans-Pecos region.

Full Name. Opuntia engelmannii Salm-Dyck ex Engelm.

Key to the Native Trans-Pecos Varieties of *Opuntia engelmannii*

1. Spines mostly white; mostly western

   O. engelmannii var. *engelmannii*, p. 104

1. Spines all yellow; mostly eastern

   O. engelmannii var. *lindheimeri*, p. 107

Opuntia engelmannii var. engelmannii
Engelmann's Prickly Pear

PLATES 97, 98

This is the taxon called *O. phaeacantha* var. *discata* by Benson (1982) and *O. engelmannii* var. *engelmannii* by Weniger (1984). The type locality of var. *engelmannii* is north of Chihuahua City in Chihuahua, Mexico.

Distribution. Probably the most common prickly pear species in the Trans-

Plate 97. *Opuntia engelmannii* var. *engelmannii* (Engelmann's prickly pear) near Terlingua, Brewster Co., TX.

Plate 98. *Opuntia engelmannii* var. *engelmannii* (Engelmann's prickly pear), Alpine, Brewster Co., TX.

Pecos, and certainly the most conspicuous; desert habitats to mountain grasslands and woodlands. 1,000–6,000 ft. In TX rare and poorly documented E of the Pecos River, SE reportedly to Bexar and Travis counties, scattered NE to the S plains and probably into OK. West through central and S NM to NW AZ, and S CA. Mexico: northernmost Sonora, Chihuahua, and Coahuila. Map 31.

Vegetative Characters. Plants usually 90–140 cm high, with upright, spreading branches. At some localities, particularly in deep soil along and near the Rio Grande, they may be taller than 1.4 m. Shape and size of pads variable, but usually obovate, broadly elliptic, or orbicular, 20–30 cm long, 15–25 cm in diameter. Under optimal growth conditions, pads of some plants large, thick, heavy. Larger plants are the largest wild prickly pears in the Trans-Pecos. In var. *engelmannii*,

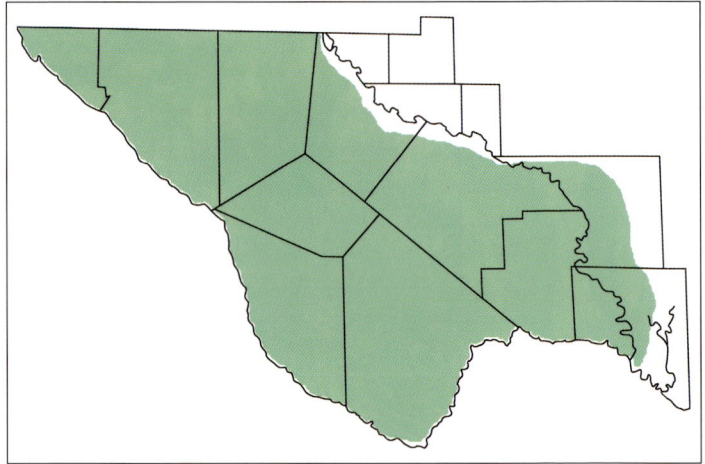

Map 31. *Opuntia engelmannii* var. *engelmannii* (Engelmann's prickly pear).

characteristically spines are distributed over most of the pad, usually in all but the lower few areoles. There is a great deal of spine variation in var. *engelmannii*, but the typical spine pattern is distinctive, including more or less three flattened bone-white or chalky-white spines deflexed in a "bird's foot." Spineless and weakly spined plants are also found in the Trans-Pecos. Glochids of var. *engelmannii* and its immediate allies also are distinctive: glochids of mid-stem areoles are distributed around much of the areole periphery or throughout the areole, spaced apart (not touching each other and basically not densely crowded), conspicuous, stout, 3–5 mm long, of unequal lengths. In other taxa, including commonly sympatric *O. camanchica*, glochids tend to be more organized in dense apical tufts or crescents where glochids are mostly of equal or subequal lengths.

Flowers. Flowering Apr–Jul. Flowers of var. *engelmannii* uniformly clear yellow, without red centers. Orange venation may be evident upon close examination of inner tepals. At the end of a day, flowers may change to pale orange. Relatively large flowers, 7–8 cm in length and diameter, widely funnelform in the afternoon sun. Filaments cream-colored to pale green, 1–2 cm long. Yellow anthers 2–3 mm long. Style cream-colored; stigma lobes green.

Fruits. Fruits deep purple, reddish-purple, or dark beet-red; older fruits almost black from a distance. Relatively large fruits typically barrel-shaped or may be oval or obovate, 5.5–8 cm long, 3.3–5 cm in diameter, with shallow umbilicus. Fruit pulp deep red, very juicy, very sweet to taste and smell. Juice beet-red, tending to stain any surface. Usually fruits spineless, but slender spines 5–8 mm long are found in some upper areoles on fruits of some plants. Areoles dominated by white wool and glochids. Seeds of var. *engelmannii* and allies smaller and more numerous than those of most other Trans-Pecos prickly pears. Seeds tan, irregularly discoid, 3–4 mm in diameter, with a narrow hilar notch and a narrow beakless aril-margin.

Full Name and Synonyms. *Opuntia*

engelmannii var. *engelmannii*. O. *engel-mannii* Salm-Dyck; O. *discata* Griffiths; O. *engelmannii* Salm-Dyck var. *discata* (Griffiths) A. Nelson; O. *phaeacantha* Engelm. var. *discata* (Griffiths) L. D. Benson & Walk.

Other Common Names. Engelmann's opuntia; purple-fruited prickly pear.

Opuntia engelmannii var. lindheimeri
Texas Prickly Pear
PLATES 99, 100

Traditionally, var. *lindheimeri* has been treated either as a distinct species with several varieties (e.g., Benson, 1982) or as a variety of O. *engelmannii* (Weniger, 1984; Parfitt and Pinkava, 1988). The cultivar O. *engelmannii* var. *linguiformis* is not known from any native populations. See Powell and Weedin (2004) for biosystematic discussion concerning var. *lindheimeri*. The type locality of O. *lindheimeri* is near New Braunfels in Comal County, south-central Texas. The epithet honors Ferdinand Lindheimer, who obtained significant plant collections from southern Texas from 1843 to 1852.

Distribution. Rare in relatively deep soil mostly along and near the Rio Grande, Brewster Co., in Big Bend National Park; more common in the southeastern Trans-Pecos, near the Pecos River. 1,000–1,800 ft. South TX, NE to Central TX, E to SW LA, N to S OK. Mexico: N Tamaulipas E to Durango and Chihuahua. Map 32.

Vegetative Characters. Typically relatively large, sprawling shrubs 0.5–2 m high, in South Texas to 5 m or more in diameter, with obovate to orbiculate pads 15–30 cm long, 12–25 cm wide, ca. 2 cm thick. Spines usually 1–5 per areole, clear yellow, sometimes whitish-yellow, or with reddish-brown or blackish bases. In each areole typically 3–4 spines, sometimes more, with 1–3 longer spines to 3–7.5 cm long, angled downward or porrect but not arranged in a "bird's-foot" pattern. In Trans-Pecos populations of var. *lind-heimeri*, 1–4 spines per areole are usual. Some spines on the lower half of the pad are as long, or nearly as long, as spines on

Plate 99. *Opuntia engelmannii* var. *lindheimeri* (Texas prickly pear) from Jim Hogg Co., TX; cultivated.

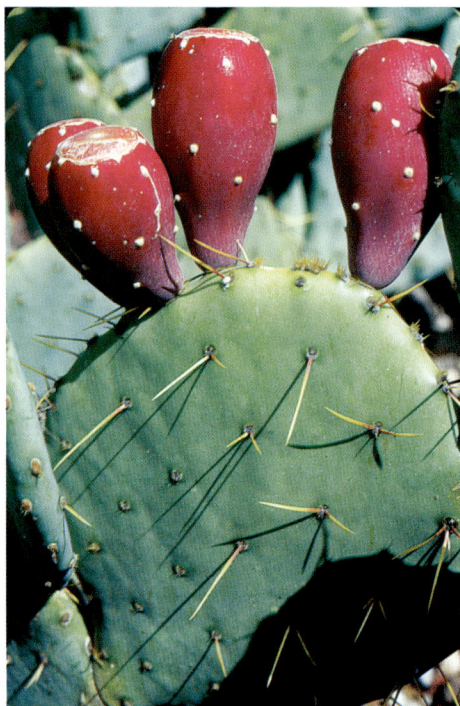

Plate 100. *Opuntia engelmannii* var. *lindheimeri* (Texas prickly pear) near Dryden, Terrell Co., TX; mature fruits; cultivated.

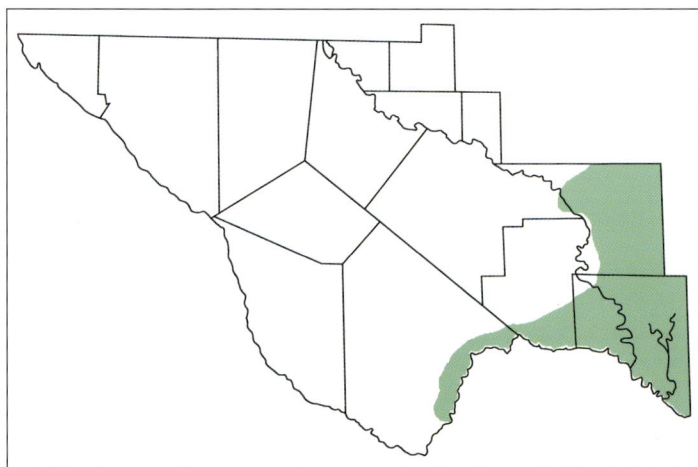

Map 32. *Opuntia engelmannii* var. *lindheimeri* (Texas prickly pear).

the upper half of the pad. Spines usually more slender than those of var. *engelmannii*. Glochid pattern the same in var. *lindheimeri* and var. *engelmannii*.

Flowers. Flowering Apr–Jun. In the Trans-Pecos, flowers rich yellow without red centers, with tepals of late first-day or second-day flowers orange or reddish. In South Texas, flowers yellow, orange-red, orange-yellow, various shades of yellow and orange-red, and rarely red. Possibly most orange or reddish flowers have opened for more than one day. Flowers relatively large, 5–8 cm long and 5–8 cm in diameter. Filaments ca. 1.5 cm long, cream-colored, as are anthers, ca. 2 mm long. Stamens notably sensitive. Style greenish-yellow to whitish, 1.7–2 cm long, with bulbous base. Stigma lobes 6–8, heavy, ca. 5 mm long, usually dark green.

Fruits. Fruits purple to reddish-purple, often pyriform, 3–7 cm long, 2.5–4 cm in diameter, spineless, with a shallow umbilicus. Fruit pulp and juice beet-red, very juicy, sweet, edible. Seeds tan, irregularly discoid, 3–4 mm in diameter, 1–2 mm thick, with a narrow aril. Seeds from some populations in the Trans-Pecos and South Texas ca. 3 mm in diameter, slightly but consistently smaller than those from most populations of var. *engelmannii.*

Other Distinctions. Variety *lindheimeri,* although abundant eastward, is rare in the Trans-Pecos except for the southeastern Big Bend along the Rio Grande and associated drainages in deep soil. Probably the westernmost robust population is near Rio Grande Village in Big Bend National Park. Plants are reported scattered along and near the Lower Scenic River to Val Verde County, where they become more common east across the Edwards Plateau and into South Texas. In Texas, plants of var. *lindheimeri* are most common southeast of the Balcones Escarpment in the deep soils of the coastal plain. In the southeastern Trans-Pecos and on the Edwards Plateau, var. *lindheimeri* is sympatric with var. *engelmannii* and *O. camanchica.* When typical spine configurations are exhibited, all three of these taxa are easily distinguished.

Full Name and Synonym. Opuntia engelmannii var. *lindheimeri* (Engelm.) B. D. Parfitt & Pinkava. *O. lindheimeri* Engelm.

Opuntia polyacantha
Plains Prickly Pear

This is a diploid-polyploid complex with chromosome numbers ranging from diploid (as in the Trans-Pecos) to octoploid and higher in some hybrids, distinguished by dry, often spiny fruits; large, flat seeds with a wide aril-margin; and green stigmas. The taxonomy of the *O. polyacantha* complex, mostly as presented by Parfitt (1991), is reviewed in Powell and Weedin (2004). The type locality of *O. polyacantha* is near Fort Vanderburgh on the Missouri River in Mercer County, North Dakota. Consequently, *O. polyacantha* var. *polyacantha,* the nomenclaturally typical part of the species, refers to the northeastern (presumably tetraploid) plants. The specific epithet is after the Greek *poly,* "many," and *akantha,* "spine," in reference to the many-spined stem segments.

Full Name. Opuntia polyacantha Haw.

Key to the Varieties of *Opuntia polyacantha*

1. Plants 10–25 cm high; pads all flat, usually obovate or orbiculate, 7–13 cm long, 5.5–11 cm wide, ca. 1 cm thick; pericarpel elongated but not "slender," roughly 1.5 times longer than thick; habitat mountain woodlands, grasslands, desert sand, or gypsum

    O. polyacantha var. *trichophora*, p. 110
1. Plants lower shrubs to 5–15 cm high; pads, or some of them, weakly compressed or terete, superficially resembling joints of a cholla cactus, 4–7 cm long, 2–3 cm wide, often to 2 cm thick; pericarpel slender, sometimes 2–3 times as long as thick; habitat deep sand

    O. polyacantha var. *arenaria*, p. 112

Opuntia polyacantha var. trichophora
Southern Plains Prickly Pear
PLATES 101, 102

Opuntia polyacantha var. *polyacantha*, whether in the broad sense of Parfitt (1991) or in the narrowest possible sense of only the tetraploid Great Plains and Rocky Mountains part of *polyacantha*, is documented in herbaria as the second most widespread taxon in the O. *polyacantha* complex, after O. *fragilis*. They share basically the same range except that O. *fragilis* extends farther west, northwest, northeast, and farther into Canada and higher in the mountains than does O. *polyacantha*. *Opuntia fragilis* is unknown in the Trans-Pecos but does occur rarely in the Texas Panhandle. *Opuntia polyacantha* var. *polyacantha* is known for Briscoe, Armstrong, and Potter counties of the Texas Panhandle (Parfitt).

Distribution. Clay, gravel, or sandy soils, oak-juniper-pinyon woodlands, grasslands, less often desert sands or gypsum. Expected in E extreme of El Paso Co., Hudspeth Co., mostly Sierra Diablo Plateau in sandy grassland and desert,
Culberson Co., Apache Mts N to Guadalupe Mts. Presidio Co., Chinati Peak. Jeff Davis Co., Davis Mts. Brewster Co., NW portion in Davis Mts. 3,600–8,382 ft. South-central NM, from the Guadalupe Mts. Mexico: Coahuila, Sierra del Carmen; to be expected in Chihuahua adjacent to the Trans-Pecos. Map 33.

Vegetative Characters. Apparently nonrhizomatous, low plants in mountain woodlands and grasslands with closely spaced areoles and usually 6–17 rather slender spines per aerole. Longest spines, usually 1–3 per areole, may be 4–7 cm, while other spines are often appressed and less than 2 cm long. Longest spines whitish to gray, less often reddish to brown or nearly black. Prostrate stems often produce roots where the pads touch the ground, especially in sandy substrates, but rarely "root-sprout" from rhizomes, as does var. *arenaria*. Usually plants of var. *trichophora* are readily distinguishable from other prickly pears in the Trans-Pecos, except perhaps var. *arenaria;* small pads of var. *trichophora* are partially obscured by thin spines produced in all but the lowermost areoles.

Plate 101. *Opuntia polyacantha* var. *trichophora* (southern plains prickly pear), Hueco Tanks, El Paso Co., TX (photo by Richard D. Worthington).

Plate 102. *Opuntia polyacantha* var. *trichophora* (southern plains prickly pear) from Davis Mts, Jeff Davis Co., TX; mature fruits; cultivated.

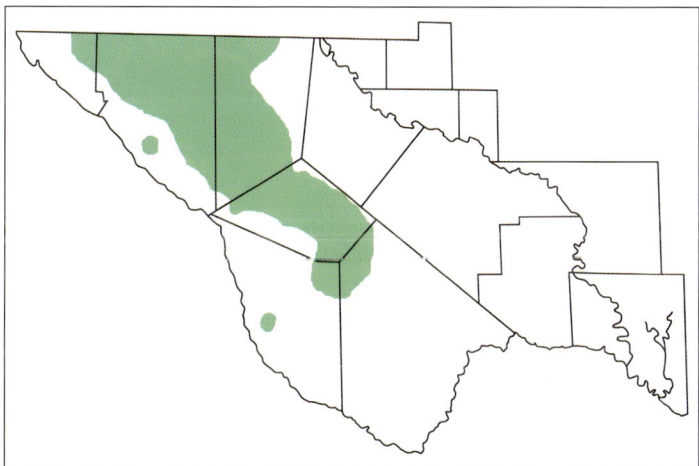

Map 33. *Opuntia polyacantha* var. *trichophora* (southern plains prickly pear)

Flowers. Flowering May–Jun. Flowers yellow in all Trans-Pecos populations of var. *trichophora,* so far as known, although they may darken slightly before wilting (late in first day after anthesis). In New Mexico and farther north, *O. polyacantha* may have yellow flowers with red centers, or flowers may be pale yellow to magenta, rarely white, or combinations of these colors. Flowers usually 4–7 cm long, 4–5 cm wide. Filaments white or barely green-tinted to cream-colored or pale yellow, ca. 1 cm long; anthers cream-yellow. Style cream-colored to pinkish, ca. 2 cm long, only slightly expanded at the base. About eight stigma lobes ca. 3 mm long, dark green.

Fruits. Fruits dry, relatively small, very spiny at maturity. As fruit matures, green surface turns dull red, then whole fruit, even previously green cortex, dries rapidly, turning tan or cream-colored. Dried mature fruits somewhat obconic, 1.9–2.5 cm long, 0.8–1 cm (to 1.7 cm immediately before drying) in diameter, ultimately with a shriveled or wrinkled tan surface. Fruit surface supports numerous (ca. 12–42) uniformly distributed, whitish, cottony areoles, with 6–15 spines per areole, 0.4–1.2 cm long. Papery, dry fruit surface partially hidden by spines. Umbilicus reaches ca. 8 mm deep. After abscission from the plant, the fruit base in some individuals in the Davis Mountains is covered by dried, hard pericarpel tissue, whereas in other individuals fruits have a circular basal pore at the time of abscission. Flattened seeds of *O. polyacantha* are largest of any Trans-Pecos prickly pear species and relatively few per fruit. Seeds irregularly discoid, tan or cream-colored, 6–6.8 mm in maximum diameter, 1.5–2 mm thick, with a beaked aril-margin 1–2 mm wide. Horizontally oriented seeds either numerous or as few as 8–10 per fruit.

Other Distinctions. Benson (1982) attempted to distinguish *O. polyacantha* var. *trichophora* primarily by the presence of remarkably long, hairlike, white, flexible, curly, or down-curving spines, which almost obscure the pads in some populations. Benson described the spines as resembling the hair of an Angora goat. The pads often exhibit a shaggy appearance, particularly on their lower portions. This unusual spine morphology is sporadic almost throughout the range of *O. polyacantha,* even in some tetraploid plants of Wyoming, and should not be recognized taxonomically. See Powell and Weedin (2004) for additional discussion of *O. polyacantha* in the Trans-Pecos.

Full Name and Synonyms. Opuntia polyacantha var. *trichophora* (Engelm. & Bigelow) J. M. Coult. *O. missouriensis* DC. var. *trichophora* Engelm. & Bigelow; *O. polyacantha* Haw. var. *rufispina* (Engelm. & Bigelow) L. D. Benson, in part; *O. trichopohora* (Engelm.) Britton & Rose.

Other Common Names. Starvation prickly pear; the brown-spined morphs, erroneously identified as *rufispina,* have been called red-spined prickly pear; the "hairy" morphs, easily and accurately identified as *trichophora,* have been referred to as hair-spined prickly pear and bristlehair prickly pear cactus.

Opuntia polyacantha var. arenaria
Sand Prickly Pear
PLATES 103, 104

Prior to the research by Parfitt (1991) this taxon was treated as the distinct species *O. arenaria.* It is a sand-loving prickly pear that is almost entirely restricted to

Plate 103. *Opuntia polyacantha* var. *arenaria* (sand prickly pear), El Paso Co., TX (photo by Richard D. Worthington).

Plate 104. *Opuntia polyacantha* var. *arenaria* (sand prickly pear), Fabens, El Paso Co., TX; fruit (photo by James F. Weedin).

the deep sands in the Rio Grande valley near El Paso and Juarez. The type locality is "Sandy bottoms of the Rio Grande near El Paso." The varietal epithet is after the Latin *arenarius*, pertaining to sand, in reference to this habitat specialization.

Distribution. Relatively uncommon in deep sand and silt, usually in dune areas. El Paso Co., in the Rio Grande valley from NM boundary at Anthony downstream through Fabens and the S Hueco Mts to SW Hudspeth Co. near McNary.

3,600–4,500 ft. Doña Ana Co., NM, from E of Columbus E to the Rio Grande valley. Mexico: Chihuahua, near Samalayuca S of Juarez, presumably W to near Palomas. Map 34.

Vegetative Characters. Superficially resembles dog chollas. Low, creeping, relatively small mats loosely and irregularly spread to 1–3 m in diameter. Branches growing in shifting sands are often partially buried in substrate and misinterpreted as rhizomes. Glochid-filled areoles,

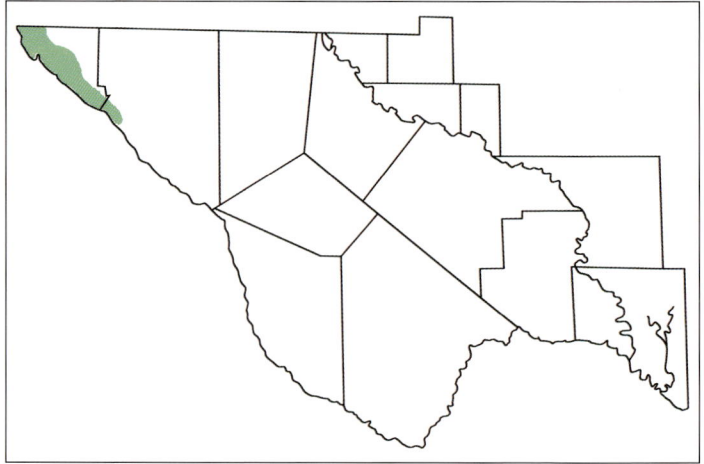

Map 34. *Opuntia polyacantha* var. *arenaria* (sand prickly pear).

pads, and ultimately whole clumps are produced ("root-sprouting") as adventitious buds developing directly from horizontal, undersand, rhizomelike roots extending up to 2 m. Small pads are narrower and less flattened than those of any other prickly pear in the Trans-Pecos, 4–7 cm long, 2–3 cm wide, 1–2 cm thick. Largest pads oval, ovate, obovate-elongate, subcylindroid or clavate, or narrowly obovoid, almost terete in cross section. Pad faces have the areoles raised on short tubercles. Spines are produced in all but the lowermost areoles, but the rather full spine cover does not obscure the glaucous or light green stem surface. Spines white, or mostly so, usually with one spine much longer (to 4 cm) than 5–10 others in each areole.

Flowers. Flowering May–Jun? Yellow flowers 4–6.5 cm long, 4–6 cm wide. Filaments white to pink-tinged, 0.6–1 cm long, supporting yellow anthers ca. 1.5 mm long. Style white to pale green, 1.2–2 cm long, 2–2.5 mm in diameter at the bulbous base.

Stigma lobes 5–8, ca. 2 mm long, green.

Fruits. Fruit turns reddish, then tan and dry at maturity. Fruit surface somewhat obscured by 3–6 white spines 6–9 mm long in each of 10–14 areoles. Fruit narrowly obovate-obconic, constricted below apex, 2.5–3 cm long, 0.9–1.2 cm in diameter. Umbilicus relatively deep. Seeds tan, shiny, irregularly discoid, large, like those of var. *trichophora.*

Full Name and Synonym. Opuntia polyacantha var. *arenaria* (Engelm.) B. D. Parfitt. *O. arenaria* Engelm.

Other Common Names. El Paso prickly pear; sand-loving opuntia.

Other Opuntia Species

The seven prickly pear taxa briefly treated below either are cultivars or escapees in the Trans-Pecos that have not become naturalized in significant wild populations, or they are natives of eastern Texas.

Opuntia basilaris var. basilaris
Beavertail Cactus
PLATES 105, 106

The specific epithet is after the Latin *basilis*, "basal," and *aris*, "belonging to," in reference to the new growth that develops only from the plant base. The spineless pads are not harmless because, as in *O. rufida* and *O. microdasys*, the areoles are loaded with pesky glochids.

Distribution. Occasionally seen as an escapee in El Paso. 3,600–3,800 ft. Variety *basilaris* is native to southern CA and adjacent AZ, NV, and southern UT.

Vegetative Characters. Plants low-growing, clumps 15–30 cm high, 0.3–1.5 m across. Pads green to blue-green or irregularly purplish, minutely canescent or velvety as in *O. rufida* and *O. microdasys*, usually obovate, 5–18 cm long, 4–9 cm wide; areoles usually 1–1.5 cm apart, circular, 1.5–3 mm across. Spines absent; glochids brown to tannish, ca. 3 mm long.

Plate 105. *Opuntia basilaris* var. *basilaris* (beavertail cactus), Desert Botanical Garden, Tempe, AZ; in cultivation (photo by David J. Ferguson).

Plate 106. *Opuntia basilaris* var. *basilaris* (beavertail cactus), Tucson Botanic Garden; fruits, in cultivation (photo by James F. Weedin).

Flowers. Flowering in spring. Flowers purplish-pink to reddish, 5–7.5 cm long and wide; filaments reddish, anthers yellow; style pale pink to reddish, stigma white.

Fruits. Fruits green at first, turning tan or gray, dry at maturity, spineless, narrowly urceolate, 2.5–3 cm long, 1.5–2.3 cm in diameter; seeds irregularly discoid, ca. 6 mm in largest diameter, 2–3 mm thick, bone-white or grayish. Fruits persist for several months.

Full Name. Opuntia basilaris Engelm. & Bigelow var. *basilaris.*

Opuntia ellisiana
Ellis' Prickly Pear
PLATES 107, 108

Opuntia ellisiana is commonly planted as an ornamental in Alpine and some other towns of the Trans-Pecos. The type specimen was from a garden in Corpus Christi, Texas. In Alpine, *O. ellisiana* is heavily browsed by mule deer that come into town during dry spells. The species was named after Prof. J. Coswell Ellis, who collected the plants for Griffiths.

Distribution. In Trans-Pecos horticulture, 1,800–5,000 ft. Distribution in the Trans-Pecos is not fully known. Widely grown in the S half of TX.

Vegetative Characters. Plants 1–2 m high, usually twice as wide as tall (contrasting with *O. ficus-indica,* which is erect). Pads typically obovate, 15–23 cm long, 10–15 cm across; areoles 2.5–3.5 cm apart, small, circular, 2–3 mm across, slightly elevated on low tubercles; glochids few, ca. 1 mm long. Spines of cultivar *ellisiana* usually absent.

Flowers. Flowering Jun. Flowers plain yellow (no red centers); filaments and anthers cream-yellow, style cream-colored, stigma lobes green. Apparently the only prickly pear in the Trans-Pecos with stamens nonsensitive or barely so.

Fruits. Maturing in late September through November in Alpine. Fruits pyriform or turbinate, turning pink to rose (partially to mostly), then reddish-purple, ca. 3 cm long, 2.5 cm in diameter, with a shallow umbilicus, fleshy, pulp reddish-purple, moderately juicy, juice purple; areoles few, mostly distal, with a few glo-

Plate 107. *Opuntia ellisiana* (Ellis' prickly pear), Alpine, Brewster Co., TX; cultivated.

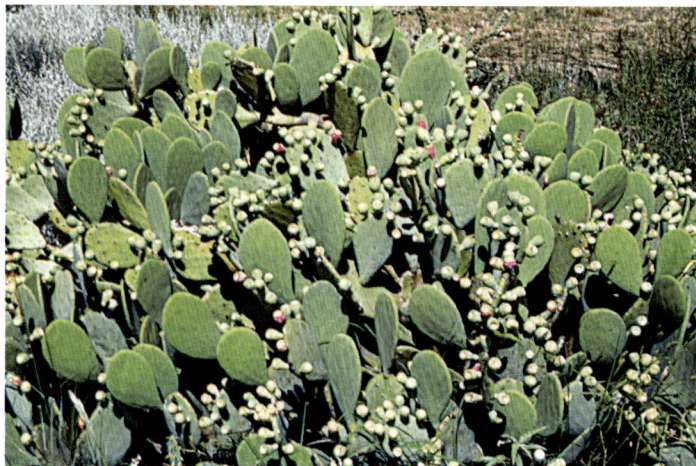

Plate 108. *Opuntia ellisiana* (Ellis' prickly pear); immature fruits.

chids and sometimes bristles; seeds discoid, dark gray to tan, 2.5–3.6 mm in diameter, aril-margin 0.3–0.6 mm wide, tan.

Full Name and Synonym. *Opuntia ellisiana* Griffiths. *O. lindheimeri* Engelm. var. *ellisiana* (Griffiths) K. Hammer.

Opuntia engelmannii var. linguiformis
Cow-Tongue Prickly Pear
PLATES 109, 110

Opuntia engelmannii var. *linguiformis* supposedly is a "mutant form" of *O. engelmannii* var. *lindheimeri*. The colony of "cow tongue" in Alpine has been reproducing "true" from seed since the 1980s. The varietal epithet is after the Latin *lingua*, "tongue," and *forma*, "shape," a reference to the tongue-shaped stems. The characteristic stem shape also inspired the English name and the Mexican name, *lengua de vaca*.

Distribution. Reported growing wild only in Bexar Co. near San Antonio. In the Trans-Pecos, widely propagated vegetatively and dispersed as an ornamental; particularly common in Alpine, where it easily survives the winters. In Big Bend National Park scattered colonies near the Rio Grande are all presumed to be historical/archeological remnants persisting from otherwise inconspicuous ruins of human habitation.

Vegetative Characters. Plants erect, 1–2 m high, often with spreading branches. Pads typically lanceolate to ca. 1 m long, usually shorter, varying from narrowly ovate to linear, and often subcrescentiform, 10–15 cm across at the widest point near the base of the pad. All vegetative and floral characters are those of *O. engelmannii* var. *lindheimeri*, sensu stricto, except for the elongated stems that result from continued growth of the apical meristem.

Flowers. Flowering May–Jun. Flowers in Trans-Pecos plants are yellow, changing in late afternoon (or if reopening a second day) to yellow-orange, orange, orange-red. Flowers reportedly orange or red in Bexar County.

Fruits. In Alpine, fruits usually mature in mid- to late August but may not ripen until mid- to late September.

Full Name and Synonyms. *Opuntia*

Plate 109. *Opuntia engelmannii* var. *linguiformis* (cow-tongue prickly pear), Alpine, Brewster Co., TX; cultivated.

Plate 110. *Opuntia engelmannii* var. *linguiformis* (cow-tongue prickly pear), Alpine, Brewster Co., TX; fruits; cultivated.

engelmannii Salm-Dyck ex Engelm. var. *linguiformis* (Griffiths) B. D. Parfitt & Pinkava. *O. linguiformis* Griffiths; *O. lindheimeri* Engelm. var. *linguiformis* (Griffiths) L. D. Benson; *O. engelmannii* Salm-Dyck var. *linguiformis* (Griffiths) Weniger, nom. nud.; *O. engelmannii* Salm-Dyck cv. *linguiformis* (Griffiths) A. D. Zimmerman (forthcoming).

Opuntia ficus-indica
Indian Fig
PLATES 111, 112

Opuntia ficus-indica is a species that has been widely introduced and naturalized throughout warm areas of the world. Spineless cultivars have been the most widely dispersed, even in prehistoric times, because the pads are edible as well as the fruit. Feral populations of this (and similar) species have been prolific in some areas; in Australia and parts of Africa they are pests worthy of extermination. *Opuntia ficus-indica* has been cultivated in Mexico for about 9,000 years and probably is native in Mexico. Historically, many cultivars were developed there and disseminated elsewhere.

In Texas and northern Mexico these are the tallest prickly pear cacti seen as occasional single plants or in colonies, as relicts or escapees in towns or near isolated dwellings, and mostly as casually maintained hedges or dooryard plants in the suburbs. In the Trans-Pecos, *O. ficus-indica* is practically restricted to rural and suburban gardens along the Rio Grande. It is to be expected at any old or Spanish-speaking family residence from El Paso to Brownsville, but above 2,000 feet altitude the plants are mostly frozen-down and inconspicuous below rooftop level against south-facing walls. The specific epithet is after the Latin *ficus*, "fig tree," and the Greek *indikos*, "Indian," a reference to the figlike fruit that was used by North American Indians.

Distribution. Cultivated from sea level to (in the Trans-Pecos) 1,800–3,600 ft.

Vegetative Characters. Plants large

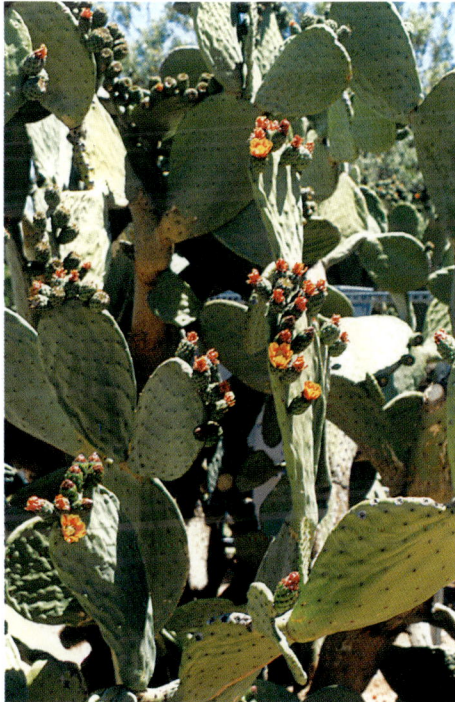

Plate 111. *Opuntia ficus-indica* (Indian fig), Tucson Botanic Garden; in cultivation (photo by James F. Weedin).

Plate 112. *Opuntia ficus-indica* (Indian fig), Tucson Botanic Garden; fruits, in cultivation (photo by James F. Weedin).

shrubs or trees, potentially 3–5 m or more high; trunk large, to 60–120 cm long. Pads green, typically obovate to oblong, the largest pads 28–60 cm long, 15–40 cm in diameter, 2–2.5 cm thick; areoles 2–6 cm apart, elliptic-oblong, 2–5 mm long, 3–4 mm broad. Spines absent, present in a few areoles, or 1–6 present in most of the areoles on each pad; spines usually white, or tan to pale brown, deflexed and spreading, straight, flattened, subulate, to 1.2–2.5 cm long, 0.6–0.9 mm wide at the base; glochids yellow, protruding 1–2 mm, numerous, falling early.

Flowers. Flowering in spring. Flowers 5–7 cm long and wide, yellow to orange-yellow with reddish or greenish centers; filaments and anthers yellow; style and stigma lobes greenish.

Fruits. Fruits potentially of several different colors, including red to purplish and yellow to orange, fleshy, juicy, edible, typically oval, 5–10 cm long, 4–9 cm in diameter, with a shallow umbilicus, spineless, or sometimes with spines, persistent for several months. Seeds irregularly discoid, gray or tan, 3–4 mm in largest diameter.

Full Name and Synonym. Opuntia ficus-indica (L.) Miller. *Cactus ficus-indica* L.

Opuntia humifusa
Eastern Prickly Pear

Distribution. Two varieties are recognized for the eastern United States and southeastern Canada (Pinkava, 2003d), only one, var. *humifusa*, reaching eastern TX (Map 35). Variety *humifusa* occurs throughout the eastern United States as far west as SD and NE.

Vegetative Characters. Plants low-growing, often prostrate, usually 1–2 stem segments tall, to 45 cm tall. Stem segments (joints, pads) dark or shiny green, fleshy, cross-wrinkled when stressed, circular to broadly oblong or obovate, 6–17 cm long, 5–12 cm wide. Areoles oval to circular, 2–4 mm across. Spines absent or 1–3 per areole, spreading, terete, whitish to brownish, 2.5–5.5 cm long, one deflexed spine sometimes present; glochids in a dense apical crescent or tuft, yellow to reddish-brown, to 4 mm long.

Flowers. Flowering spring–summer.

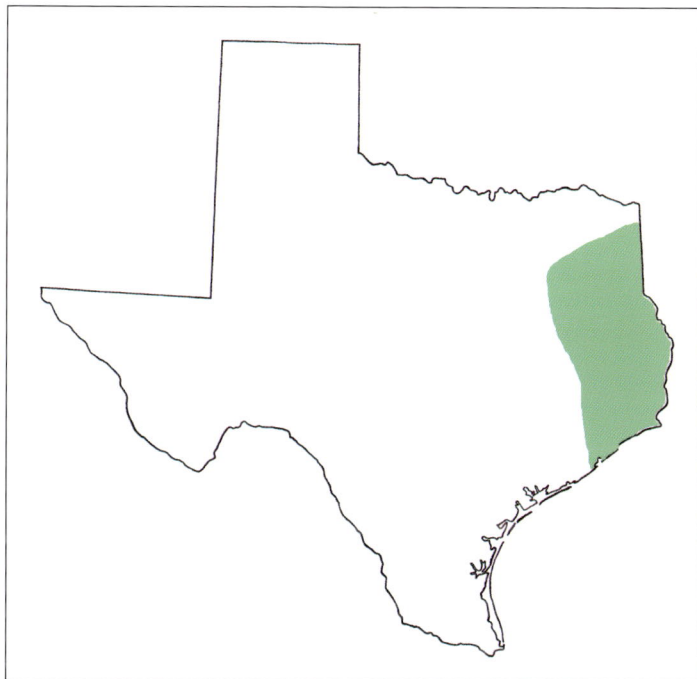

Map 35. *Opuntia humifusa* (eastern prickly pear).

Flowers pale to bright yellow throughout, 2–3 cm in diameter; filaments yellow to orange, anthers pale yellow to cream; style and stigma lobes white.

Fruits. Fruits greenish at first, later peach to brownish-red, elongate, fleshy, 3–5 cm long, 1.2–2 cm wide, spineless, base tapering. Seeds tan, 3.5–4.5 mm across.

Full Name and Synonym. Opuntia humifusa (Raf.) Raf. *O. compressa* var. *austrina* (Small) L. D. Benson.

Opuntia pusilla
Cockspur, Creeping Cactus, Sandbur, Little or Crow-Foot Prickly Pear

Distribution. Near Galveston Bay in TX (Map 36), E to FL and NC, mostly near the coast (Pinkava, 2003d).

Vegetative Characters. Plants prostrate, trailing, creeping, often forming mats to 1 m across. Stem segments (joints, pads) easily detached, green to purplish-red, wrinkled under stress, flattened to nearly cylindric or globular, elliptic to linear, 2.5–6 cm long, 1.2–2.5 cm wide. Areoles nearly circular, 2–3 mm across. Spines strongly barbed, reddish-brown to gray, to 3 cm long, usually 1–2 per areole, sometimes absent or as many as 4; glochids in apical crescent, pale yellow to brown, to 3 mm long.

Flowers. Flowering spring. Flowers yellow throughout, 2–3 cm in diameter; filaments and anthers yellow; style and stigma lobes white.

Fruits. Fruits green at first, later reddish-purple, barrel-shaped, 1.8–3 cm long, 1.2–2 cm wide, spineless. Seeds tan, 4–6 cm across.

Full Name and Synonym. Opuntia

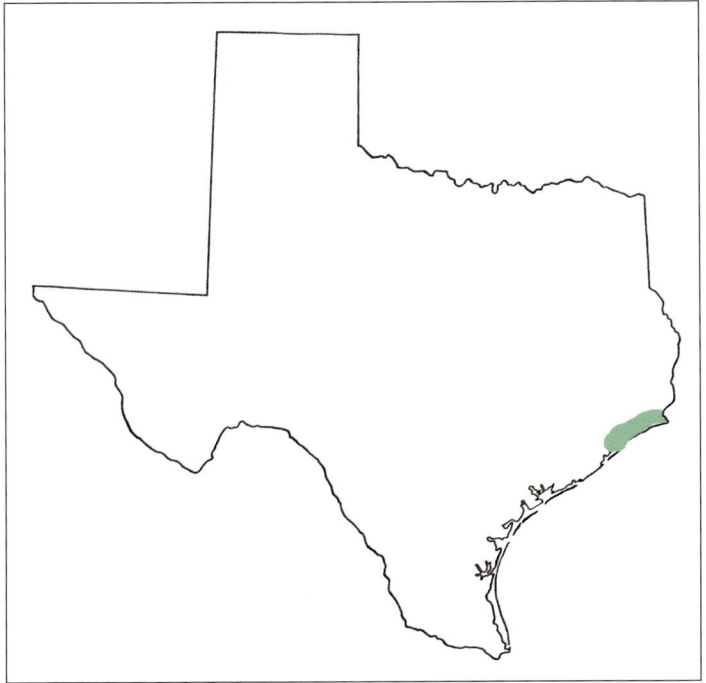

Map 36. *Opuntia pusilla* (cockspur, creeping cactus, sandbur, little or crow-foot prickly pear).

pusilla (Haw.) Haw. *O. drummondii* Graham.

Opuntia stricta
Pest Prickly Pear, Coastal Prickly Pear

Opuntia stricta hybridizes with some other prickly pear species in Texas and Louisiana (Pinkava, 2003d), forming taxonomically confusing populations. Several Texas coastal endemic taxa of *Opuntia*, not necessarily recognized by recent authors, appear to be the basis for erroneous reports of *O. stricta*.

Distribution. Near Galveston Bay in TX (Map 37), east to AL, GA, FL, and SC, usually near the coast. West Indies. Central America. Introduced in South Africa and Australia.

Vegetative Characters. Plants erect or spreading, 60–150 cm or more tall. Stem segments (joints, pads) green, flattened, narrowly elliptic or ovate, 10–30 cm long, 7–20 cm wide, tuberculate. Areoles oval, 3–6.5 mm long, ca. 3.5 mm wide. Spines absent or to 11 per areole, 12–50 cm long, spreading, yellow, turning brown in age, straight or curved; glochids few to many, mostly apical in the areole, yellow to brown, to 4 mm long.

Flowers. Flowering spring–summer. Flowers light yellow throughout, 2–3 cm across; filaments yellow, anthers yellow; style and stigma lobes yellowish.

Fruits. Fruits purplish, nearly barrel-shaped, 4–6 cm long, 2.5–3.5 cm wide, spineless. Seeds tan, 4–5 mm across.

Full Name. *Opuntia stricta* (Haw.) Haw.

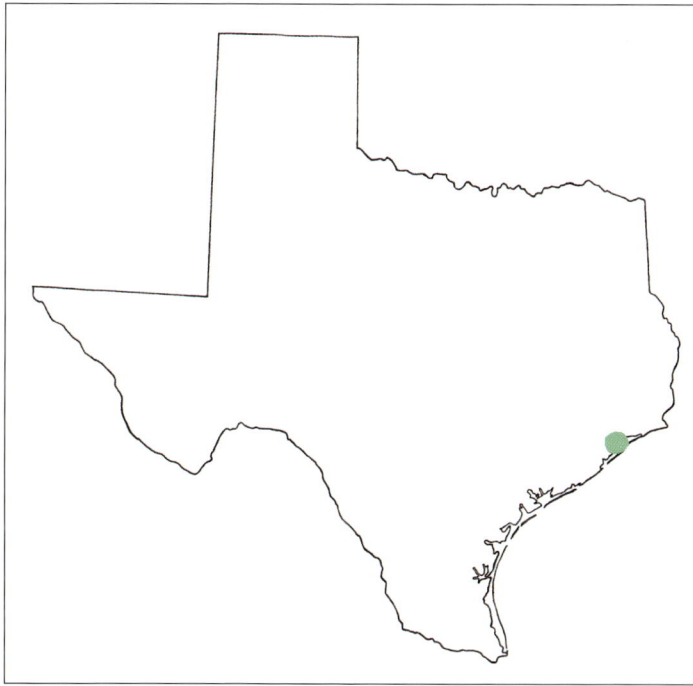

Map 37. *Opuntia stricta* (pest prickly pear, coastal prickly pear).

Peniocereus (A. Berger) Britton & Rose is a genus of about 15 species distributed mostly in Mexico and Central America. Benson (1982) retained *Peniocereus greggii* in *Cereus*, where it was placed originally by Engelmann. *Cereus* is strictly a South American taxon of perhaps 30 species. The genus name *Cereus* is from the Latin *cereus* (adj., "waxen," "a torch"; see *ceraceus*, "like candle wax"; and/or, as noun, applied to the earliest known group of these cacti because they resembled either wax candles or candelabras), here combined with the Latin *penis*, "tail," in reference to the slender stems of *P. greggii*.

Peniocereus greggii var. greggii
Desert Night-Blooming Cereus
PLATES 113, 114

Peniocereus greggii has long been considered by some to be rare in Texas, having been listed as a Category 2 species in the Federal Register. Our observations suggest that *P. greggii* is widespread in the Trans-Pecos and common in certain areas, although very rare in Big Bend National Park. Most plants of *P. greggii* apparently begin under *Larrea, Acacia, Prosopis,* or other shrubs, which serve as nurse plants during their establishment. The slender stems of *P. greggii* resemble the stems of shrubs in which they grow and are thus difficult to locate by casual observation.

Distribution. Desert flats, gravel benches, bajadas, and in degraded grassland and desertscrub in alluvial basins between mountains, documented or expected in every county of the Trans-Pecos. 3,500–5,000 ft. West across southern NM to the E edge of Cochise Co., AZ. Adjacent Mexico S to Zacatecas. Map 38.

Vegetative Characters. Slender, gray-green to dull purple, angular stems unlike those of any other plant species in the Trans-Pecos. Stems 1–2.3 cm in diameter, 15–60 cm long (rarely 2–3 m long if sheltered inside a large shrub), sparsely branched or unbranched, erect or sprawling, strongly ribbed (angled) above; epidermis densely and finely velutinous with microscopic unicellular trichomes, somewhat like the skin of *Opuntia rufida;* ribs 4–6. Plants qualify weakly as "shrubs" in the botanical sense, but not bushy. Areoles elongate-elliptic, 1.5–4.4 mm long, usually 4.5–7 mm apart, on tiny tuberclelike projections along rib crest, with white wool in those on new growth. Areoles exhibit distinctive "arachnoid" spine clusters, best seen under magnification, appearing "plastered" to the narrow

Plate 113. *Peniocereus greggii* var. *greggii* (desert night-blooming cereus); cultivated.

Plate 114. *Peniocereus greggii* var. *greggii* (desert night-blooming cereus); mature fruits; cultivated.

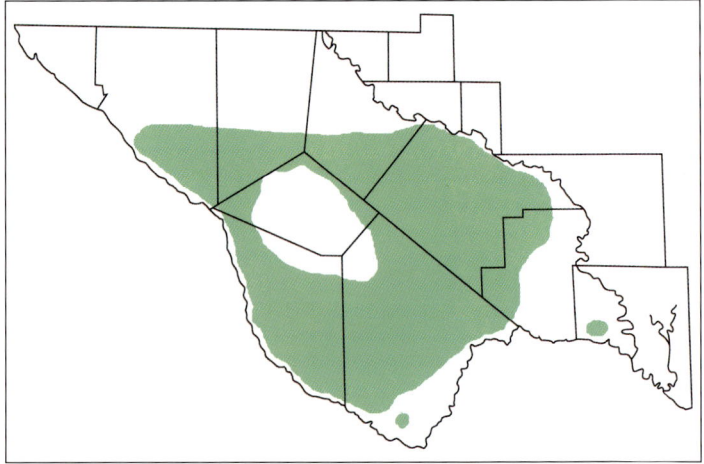

Map 38. *Peniocereus greggii* var. *greggii* (desert night-blooming cereus).

rib crests. Spines small, 10–13 per areole; central spines 1–2, porrect or deflexed, ca. 1 mm long. Radial spines 6–9, blackish, becoming grayish with age, most 0.8–1 mm long, with three lower radials to 3.2 mm long, appressed, straight or slightly curved, basally swollen and stout, acicular, pubescent. In older areoles, dark hairs of elliptic-elongate areoles subtend usually 8–12 black, needlelike, bulbous-based radial spines and 1–2 black central spines. Middle one of the three lower radials usually stout, and lateral ones typically slender and lighter in color. In older areoles, spines may become jet-black. Plants produce fleshy taproots that range from turnip- or carrot-sized in younger plants to much larger in older plants. Taproots 20–65 cm in diameter, weighing up to 125 pounds. Plants in containers may produce turnip-sized taproots in a few years.

Flowers. Flowering Apr–Jun. Flowers strictly nocturnal, white, fragrant, salverform, 5–8 cm in diameter, to 17 or more cm long, borne laterally well below the stem apex from areoles at least one year old. Flowers begin opening about 7:00 p.m. (DST), fully open by dusk, remain open all night for one night, and often close just before, or soon after, sunrise. If temperatures are cool, shaded flowers may remain open until about 9:00 a.m or rarely, under apparently optimal conditions, until after noon. Flowers appear to attract hawkmoths, which serve as primary pollinators, although pollinators of *P. greggii* have not been thoroughly studied. Throat of the open flower is filled with numerous spreading, exserted stamens. Filaments cream-colored, ca. 2.5 cm long; anthers pale yellowish, tan, or creamy-yellow, 1.5–2.5 mm long. Stigma lobes whitish, ca. 10 in number, 10–15 mm long, slender.

Fruits. Ripening in midsummer to early fall, abruptly turning shiny bright red or red-orange at maturity, ultimately turning darker red with age, ellipsoid, 4.5–7.5 cm long, 2.5–3.9 cm in diameter, attenuated apically to form a sterile "beak," weakly tuberculate, each tubercle bearing an areole with short spines. Floral remnant persistent. Seeds black, obovoid, ca. 3 mm long, 2.2–2.3 mm broad, 1.5 mm thick, hilum "basal."

Full Name and Synonym. *Peniocereus greggii* (Engelm.) Britton & Rose var. *greggii. Cereus greggii* Engelm.

Other Common Names. Night-blooming cereus; queen-of-the-night.

Hedgehog Cactus

Echinocereus Engelm. is a genus of ca. 49 or up to 70 species throughout west Texas and ranging from south-central Texas to Oklahoma, north to the Black Hills of South Dakota, west to the Pacific Ocean in Baja California, and south to northern Oaxaca, Mexico. Numerous species occur in the Sonoran and Chihuahuan desert regions, in true desert habitats and in mesic mountains. Most of the species are Mexican endemics; about 23 species occur in the United States, and of those about 17 occur in Texas and about 12 in the Trans-Pecos.

The genus name is derived from *echino-*, a combining form denoting spiny or bearing spines, from the Greek *echinos*, "sea urchin" or "hedgehog," and the previously named cactus genus *Cereus* from the Latin *cereus*, "waxen," presumably in reference to the upright, candlelike habit of many species.

Key to the Trans-Pecos Species

1. Stems of mature plants 1–2 cm tall (aboveground); habitat restricted to Caballos Novaculite substrate in the Marathon Basin, Brewster Co.

E. davisii, p. 166

1. Stems of mature plants 4–30 cm tall; habitats in the Trans-Pecos various (2).

2(1). Flowers 1.5–2.5 cm in width and length; inner tepals green, dull yellow, bronze, or reddish-brown, rarely dull red

E. viridiflorus, p. 168

2. Flowers 4–13 cm in width and length; inner tepals brilliantly colored: red, magenta, pink, yellow, or multicolored (if dull red or brownish, in some *E.* X *neomexicanus*, then always sterile) (3).

3(2). Areoles elliptic to linear (elongate vertically) and close-set (2–6 mm apart); spines all short but often dense, obscuring the stem; stems often solitary (multi-stemmed in older plants) (4).

3. Areoles basically circular, the spacing either close or distant but often far apart; spines relatively long but widely divergent, not obscuring the stem (except as seen from a distance in *E. stramineus*); stems characteristically clumped (likely to form large mounds of numerous stems) except for *E. fendleri,* which is smaller (7).

4(3). Fruit and flower areoles with long, conspicuous white wool and relatively slender spines; bases of inner tepals may be contrastingly marked with dark red-brown; relatively small plants (5).

4. Fruit and flower areoles usually with much shorter, inconspicuous white wool and relatively rigid spines; tepal bases green or white (6).

5(4). Stems soft, relatively conical; central spines usually present and similar to the radials; spines not tightly appressed (plants appear bristly); Big Bend National Park
E. chisoensis var. *chisoensis,* p. 159

5. Stems thick, firm, compact, rounded or short-cylindroid; central spines absent or very short; radial spines tightly appressed (plants appear smooth); not in Big Bend National Park
E. reichenbachii, p. 162

6(4). Spines all whitish, not forming seasonal dark bands down the stem; inner tepals three-colored, pinkish distally, whitish in the middle, greenish proximally; distribution near Sanderson E to Del Rio; diploid
E. pectinatus var. *wenigeri,* p. 148

6. Spines reddish-brown, gray, or whitish, often color-zoned down the stem (a "rainbow" cactus); inner tepals variously colored (yellow, red, pink, magenta, etc.), but never white-centered; distribution widespread in the Trans-Pecos; tetraploid
E. dasyacanthus, p. 144

7(3). Flowers (inner tepals) deep pure red (rarely orange), 2.5–5 cm long and wide, the inner tepals stiff, usually rounded (or mucronulate) apically; flowers remaining constantly open for two or more days and nights
E. coccineus, p. 130

7. Flowers (inner tepals) variously colored: rarely red or orange (mostly in hybrids with *E. coccineus,* rarely in *E. viridiflorus* var. *cylindricus* and var. *russanthus*), rarely white (in *E. stramineus*), 5–13 cm long and wide, rounded, acute, or acuminate apically; flowers either ephemeral or at least closing at night (except in some *E.* X *roetteri*), shorter-lived than those of *E. coccineus* (8).

8(7). Radial spines (often whitish) usually 5–6; central spines usually one, usually light gray to black, or with light gray tips; plants usually with 1–5 stems, to a maximum of ca. 18 stems, these typically 7.5–18 cm long

E. fendleri var. fendleri, p. 150

8. Radial spines usually 7–14; central spines usually 2–4; plants usually many-stemmed, to a maximum of ca. 350, typically 30 cm or more long (9).

9(8). Central spine(s) 1.2–8.7 cm long, usually a main one or two, somewhat flattened in older areoles (10).

9. Central spines usually less than 1.2 cm long, the 2–5 of them all much alike, needlelike, dark gray, reddish to whitish; polyphyletic nothospecies (11).

10(9). Plants in compact (ultimately hemispheric) mounds, cloaked in straw-colored or silvery spines obscuring the green surfaces of stems as seen from a distance in the field (usually on hot, steep, rocky slopes); average 12 ribs per stem; tetraploid

E. stramineus var. stramineus, p. 157

10. Plants loosely clumped, with gray, dark gray to reddish, straw-colored, ashy-white, or gray-brown spines, widely spaced, either obscuring the stem surfaces or not (usually in deep soils of bajadas and valley floors); average 8–9 ribs per stem; diploid

E. enneacanthus, p. 152

11(9). Central spines 4–6; radial spines 13–16; spine tips bright purplish-red, giving the plant a red-and-white aspect (similar to some E. viridiflorus); flowers produced at variable heights on the stem (as in E. viridiflorus), relatively small (ca. 5 cm long, 2.5–3 cm in diameter); inner tepals red at least distally, or greenish-yellow proximally, or entirely brownish-pink

E. X neomexicanus, p. 143

11. Central spine, radial spine, and plant aspect very diverse, exceeding the limits of the rare F_1 hybrid in the opposing lead; flowers produced on distal parts of the stems (not like E. viridiflorus), relatively large and showy

E. X roetteri, p. 136

Echinocereus coccineus
Claret-Cup Cactus, Texas Claret-Cup Cactus

Herein we treat two morphologic (and geographic) extremes of a widespread and morphologically variable tetraploid species, E. coccineus, as two arbitrarily delimited varieties, vars. rosei and paucispinus, pending further study. Benson (1982) included E. coccineus within E. triglochidiatus Engelm., as did some other authors. Ferguson (1989) separated the presumably diploid E. triglochidiatus, ranging from south-central New Mexico north to Colorado and Utah and west to California, as a distinct species. See Powell and Weedin

(2004) for a review of various taxonomic interpretations concerning *E. coccineus*.

Echinocereus coccineus hybridizes occasionally with several different species, including *E. dasyacanthus*, *E. viridiflorus*, and reportedly *E. fendleri*. Two Trans-Pecos hybrid species, *E.* X *roetteri* and *E.*

X *neomexicanus*, have *E. coccineus* as one of the parents.

The specific epithet is after the Latin *coccineus*, "scarlet," the flower color.

Full Name. *Echinocereus coccineus* Engelm.

Key to the Varieties of *Echinocereus coccineus*

1. Central spines 3–4; radial spines 8–11; ribs 9–12; El Paso and Hudspeth counties

E. coccineus var. *rosei*, p. 131

1. Central spines 0–1(–4); radial spines 4–7; ribs 7–9; Hudspeth Co. E to Terrell Co., and Val Verde Co. E across the Edwards Plateau

E. coccineus var. *paucispinus*, p. 133

Echinocereus coccineus var. rosei
Claret-Cup Cactus

PLATES 115, 116

Distribution. Infrequent to common in arid mountains, on rocky outcrops, and in alluvial deposits. In the Trans-Pecos only in the far western portion, El Paso and Hudspeth counties, from El Paso E to near Sierra Blanca and the grasslands N of Sierra Blanca. 3,700–5,700 ft. Arid lowlands of S-central NM. Replaced by other geographic races of *E. coccineus* N into CO, W to AZ, and into northwestern Mexico. Mexico: northern Chihuahua, from Juarez and Porvenir to SW of Samalayuca. Map 39.

Vegetative Characters. Relatively small plants with relatively few stems mostly resembling larger plants of *E. fendleri* or

Plate 115. *Echinocereus coccineus* var. *rosei* (claret-cup cactus), N-central Hudspeth Co., TX.

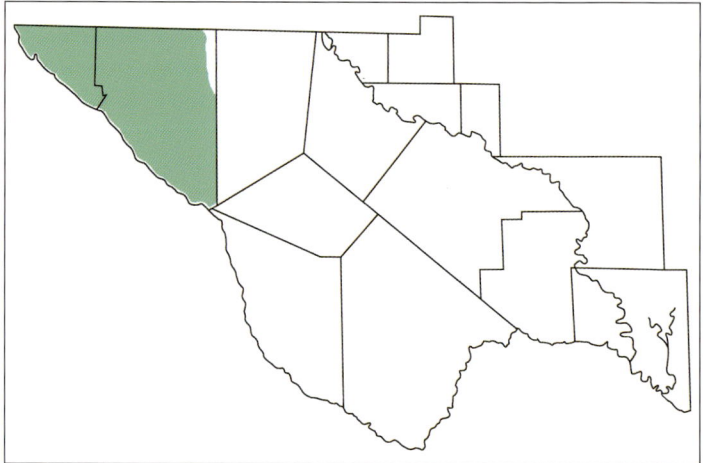

medium-sized plants of *E. enneacanthus.* Typical habit of var. *rosei* loosely cespitose, with 5–20 stems. Oldest plants form hemispheric mounds of potentially 100 or more stems. Stem size 4–22 cm high and 4–10 cm in diameter, typically about 8 cm in diameter, somewhat smaller in cespitose plants with numerous stems. Tubercles usually prominent, coalescent with and projecting from ribs. Spines light (ashy-gray) to dark brown or black in color, darkest at the tips. Radial spine number 7–12 or more in mature areoles, with 1–6, usually 3–4, central spines, 2.5–4.2 cm long.

Flowers. Flowering Mar–Apr. Dimorphic flowers narrowly to broadly funnel-form, 4–7 cm long, 2.5–5 cm in diameter. Crimson tepals typically exhibit "waxy," stiff appearance, with rounded, entire, often mucronulate apexes. Anthers pink or purplish (rarely yellow) and produce distinctive gray-purple pollen in functional anthers of male flowers. Stigma lobes 5–10, green.

Typically in *Echinocereus* flowers developing on sides of stem from areoles already a year old, in some species lower (rarely down to soil level) on the stems, rarely terminal (*E. poselgeri*); buds of most species erumpent (rupturing the stem epidermis above the areole), otherwise superficial from the outset (developing at upper edge of the areole like normal cactus buds).

Fruits. Fruits only on female plants, green, with spine clusters on pericarpel until they ripen and turn pale orange, dull red, brick-red, to pinkish. Fruits fleshy and juicy with white pulp. Seeds black, globular, strongly papillate, 1.5–2 mm long. In *Echinocereus*, spiny areoles either somewhat persistent, deciduous, or easily brushed away; the dried floral remnant persistent at fruit maturity.

Other Distinctions. Variety *rosei* (type locality: Las Cruces, New Mexico) occurs in extreme western Texas, in El Paso County east to about Hudspeth County, where we arbitrarily delimit our concept of var. *paucispinus* (type locality: lower Devils River region). Variety *paucispinus*, as thus defined, is the only variety of *E. coccineus* across most of the Trans-Pecos, from the arbitrary eastern limit of var. *rosei* east across the Edwards Plateau nearly to Austin.

Full Name and Synonyms. *Echinocereus coccineus* var. *rosei* (Wooton & Standl.) A. D. Zimmerman (forthcoming). *E. rosei* Wooton & Standl.; *E. triglochidiatus* var. *rosei* W. T. Marshall; *E. polyacanthus* Engelm. var. *rosei* (Wooton & Standl.) Weniger nom. nud.; *E. polyacanthus* Engelm. var. *neomexicanus* (Standl.) Weniger, in part.

Echinocereus coccineus var. paucispinus
Texas Claret-Cup Cactus
PLATES 117, 118

The populations of *E. coccineus* in southeastern New Mexico, south into the central Trans-Pecos are intermediate between the densely spine-covered western *E. coccineus* var. *rosei* and the green-looking, few-spined eastern plants. The geographic cline between these two

Plate 117. *Echinocereus coccineus* var. *paucispinus* (Texas claret-cup cactus), N Brewster Co., TX.

Plate 118. *Echinocereus coccineus* var. *paucispinus* (Texas claret-cup cactus), N Brewster Co., TX; mature fruits; cultivated.

extremes is sporadically interrupted by populations having (on average) more ribs and shorter spines than any other claret-cup cacti. One of the variable populations in the central region was described as var. *gurneyi*. Because var. *gurneyi* has been interpreted as part of the nothospecies *E.* X *roetteri* and has been brought into synonymy with *E.* X *roetteri*, the name *gurneyi* is no longer available for use in reference to any claret-cup populations, only to products of hybridization. In the current treatment tentatively we place the central Trans-Pecos claret-cups with var. *paucispinus*, and we interpret var. *paucispinus* as the variable entity that extends west to Culberson and Hudspeth counties, where it is replaced by var. *rosei*.

The epithet *paucispinus* refers to the characteristically fewer spines (no central spines in the type specimen) compared to other varieties of *E. coccineus*.

Distribution. Mountains, hills, and mesas, igneous and limestone, oak-juniper-pinyon woodland or juniper woodland on limestone mesas, mostly rocky habitats but also in alluvial basins, grasslands, or among mesquite or other shrubs. In every county of the Trans-Pecos from Culberson Co. SE to Pecos and Val Verde counties. 500–7,500 ft. Southeast NM; E across the Edwards Plateau to San Saba, Burnet, Williamson, Blanco, Hays, Comal, and Uvalde counties. Mexico: NE Coahuila. Map 40.

Vegetative Characters. Mature plants usually multistemmed, forming loosely aggregated or tightly packed mounds 1–10 dm across and 1–4 dm tall. Most mature plants have fewer than 20 stems, but apparently older plants with more than 100 stems or as many as 500 stems are not uncommon. Closely resembles var. *rosei* except usually fewer spines and ribs in var. *paucispinus*; typically (mostly east of the Pecos River), 0–1 central spine and 4–8 radial spines. In claret-cups there is notable variation in spine length throughout the Trans-Pecos. Central spines of var. *paucispinus*, when present, 1.2–5.5 cm long; radial spines 1.5–3.2 cm long. Plants

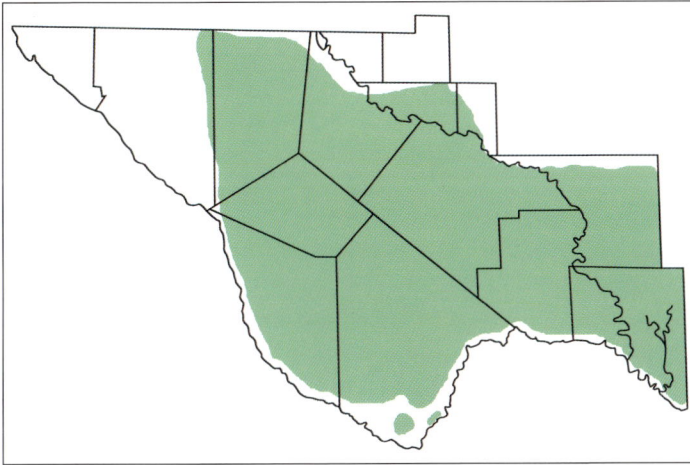

Map 40. *Echinocereus coccineus* var. *paucispinus* (Texas claret-cup cactus)

Plate 119. *Echinocereus coccineus* var. *paucispinus* (Texas claret-cup cactus); male flower.

of more westerly distribution in the Trans-Pecos occasionally with two central spines and 8–10 radials. In eastern Pecos and Terrell counties it is common to find claret-cup plants in close proximity with no central spines and with one central spine.

Flowers. Flowering Mar–Apr. Flowers crimson with a stiff, waxy appearance and rounded tepals. Orange-flowered plants in the central Trans-Pecos, populationally predominant in the Marathon Basin, sporadic elsewhere, possibly are introgressed from *E.* X *roetteri*. Variety *paucispinus* is gynodioecious apparently throughout its range, with dimorphic flowers, those of male plants (Plate 119) being slightly larger than those of female plants (Plate 120).

Fruits. Fruits dull to bright red at maturity. Pulp white or reddish near fruit wall. Spine clusters on ripe fruits easily

Plate 120. *Echinocereus coccineus* var. *paucispinus* (Texas claret-cup cactus); female flowers.

dislodged. Fruits produced only on female plants that have received cross-pollination from male plants. Seeds black, irregularly globose, 1.3–2 mm in diameter, strongly papillose.

Other Distinctions. The claret-cup cacti on granite outcrops and elsewhere in the northern Edwards Plateau region (e.g., in San Saba County) deserve further study; some of these are the basis for the taxon *roemeri*. The distinction is not strictly environmental; the endemic *roemeri* remains distinctive in horticulture and may deserve varietal, subspecific, or other rank.

Full Name and Synonyms. Echinocereus coccineus var. *paucispinus* (Engelm.) D. J. Ferguson. *E. roemeri* Rümpler; *E. coccineus* Engelm. subsp. *aggregatus* (Engelm. ex S. Watson) W. Blum, Mich. Lange, & Rutow (in part, excluding type locality and other western populations); *E. octacanthus* (Muehlenpf.) Britton & Rose.

Other Common Name. Langtry claret-cup cactus.

Echinocereus X roetteri
Roetter's Hybrid Hedgehog Cactus
PLATE 121

The nothospecies (a hybrid "species") *E.* X *roetteri* exists as a series of scattered individual plants, mostly small populations of relatively recent hybrid origin. The total known distribution of *E.* X *roetteri* lies within the sympatric ranges of the two parental species, *E. coccineus* and *E. dasyacanthus.* One of the parents, *E. coccineus*, is gynodioecious, a trait that also occurs in *E.* X *roetteri*. Introgression from fertile hybrids has affected both parental species. The epithet *roetteri* was chosen by Engelmann, who described the taxon for Paulus Roetter, the artist who illustrated cacti for the Mexican Boundary Survey. Two varieties of *E.* X *roetteri* are recognized, western var. *roetteri* and eastern var. *neomexicanus*, each having slightly different parentage.

Full Name. Echinocereus X *roetteri* (Engelm.) Engelm. ex Rümpler.

Key to the Varieties of *Echinocereus* X *roetteri*

1. Western (El Paso Co. and adjacent areas); associated with *E. coccineus* var. *rosei* (all parts smaller and more numerous); stems typically 11–19 cm long, 6–7.5 cm in diameter; ribs usually 12–14; areoles 7–14 mm apart; fruit 2–3 cm long
   E. X *roetteri* var. *roetteri*, p. 137
1. Eastern (see text); associated with *E. coccineus* var. *paucispinus* (all parts larger and fewer); stems typically 19–33 cm long, 8.5–11 cm in diameter; ribs usually 12; areoles typically 13–17 mm apart; fruit 2–5 cm long
   E. X *roetteri* var. *neomexicanus*, p. 139

Echinocereus X roetteri var. *roetteri*
Roetter's Hybrid Hedgehog Cactus
PLATES 122, 123

As a nothotaxon, var. *roetteri* includes all hybrids between *E. coccineus* var. *rosei* and *E. dasyacanthus*.

Distribution. Sporadic in desert hills, grassland, and desertscrub. El Paso Co., near El Paso, E at least to Culberson Co. 3,200–4,500 ft. Adjacent NM, Otero and Doña Ana counties. The best-known population of *E.* X *roetteri* var. *roetteri* is in the Jarilla Mts near Orogrande in Otero Co., NM. Mexico: adjacent Chihuahua. Map 41.

Vegetative Characters. Familiarity with the parental species facilitates recognition of the hybrid taxon var. *roetteri*. Plants are basically intermediate in vegetative and floral characters, or backcross plants may exhibit greater similarity to one parent or the other. Plants single- or multistemmed, with stems ranging 8–27 cm long and 5.5–8.5 cm in diameter. In each areole 4–5 central spines and 10–16 radial spines. Longest central spines 1.1–2 cm; longest

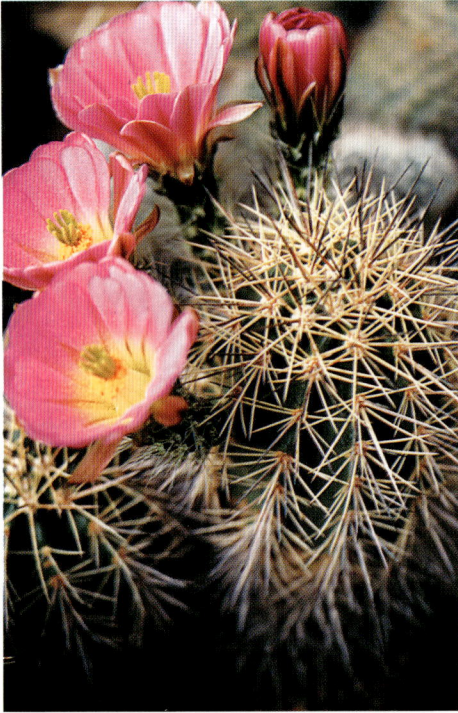

Plate 122. *Echinocereus* X *roetteri* var. *roetteri* (Roetter's hybrid hedgehog cactus), Jarilla Mts, Otero Co., NM.

Plate 123. *Echinocereus* X *roetteri* var. *roetteri* (Roetter's hybrid hedgehog cactus), Jarilla Mts, Otero Co., NM.

radial spines 1.2–1.6 cm. Spine color white to pale pink, or gray to black, brown, or stramineous.

Flowers. Flowering Apr–May. Flowers 5–9 cm long and 4.5–8.5 cm wide. Color from bright yellow to creamy-white or bright magenta to pink or yellow-green, with most of each inner tepal with one of these colors and with or without green, pink, or brown at bases, pink or orange at

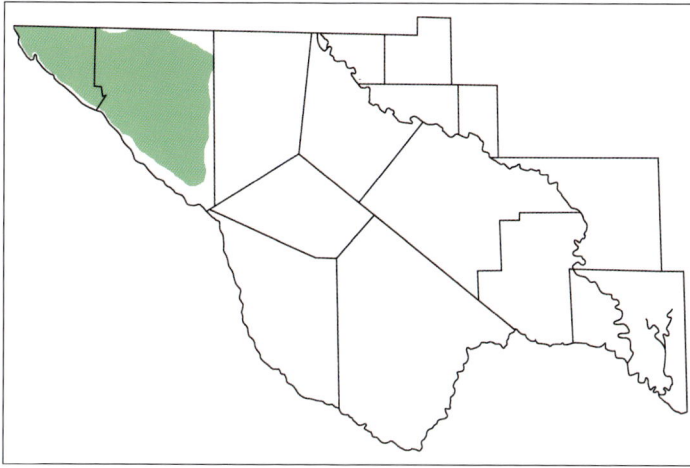

tips or centers, or sometimes with green, pink, or brown midregion.

Fruits. Fruits indehiscent and dull reddish or with brighter pinkish color at base. Juicy pulp white or pinkish near fruit wall. Seeds black, globular, 1.3–2 mm in diameter, papillate.

Other Distinctions. The distinction between var. *roetteri* and var. *neomexicanus* is that each is produced by a different geographic race of *E. coccineus* hybridizing with *E. dasyacanthus* .

Full Name and Synonyms. Echinocereus X *roetteri* var. *roetteri*. Cereus roetteri Engelm.; *E. dasyacanthus* var. *minor* Engelm.

Echinocereus X roetteri var. neomexicanus
Lloyd's Hedgehog Cactus
PLATES 124, 125

In the Trans-Pecos *E.* X *roetteri* var. *neomexicanus* has long been known as *E. lloydii* or *E.* X *lloydii* after its hybrid status was documented (Powell et al., 1991). The nothotaxon *E.* X *roetteri* var. *neomexica-* *nus* includes all hybrids between *E. coccineus* var. *paucispinus* and *E. dasyacanthus*. The epithet *lloydii* is after F. E. Lloyd, who collected the type specimen. Lloyd's hedgehog cactus was accorded federal status in 1979 as an Endangered species and also listed as Endangered by the state of Texas in 1983. Because it was determined to represent an assortment of dynamic hybrid populations rather than an established evolutionary lineage, in June 1999 Lloyd's hedgehog cactus was officially delisted through publication in the Federal Register. See Powell and Weedin (2004) for biosystematic discussion concerning *E.* X *roetteri* var. *neomexicanus*.

Distribution. Rocky hillsides or brushy alluvial habitats, isolated individual plants or localized small populations in Pecos, Brewster, Presidio, Culberson, and Hudspeth counties; a relatively large population in E Pecos Co. 2,500–4,500 ft. Eddy Co., NM. Map 42.

Vegetative Characters. In overall aspect *E.* X *roetteri* var. *neomexicanus* looks more or less intermediate between parental

Plate 124. *Echinocereus* X *roetteri* var. *neomexicanus* (Lloyd's hedgehog cactus), E Pecos Co., TX; orange-flowered; putative F_1 generation.

Plate 125. *Echinocereus* X *roetteri* var. *neomexicanus* (Lloyd's hedgehog cactus), E Pecos Co., TX; mature fruit; cultivated.

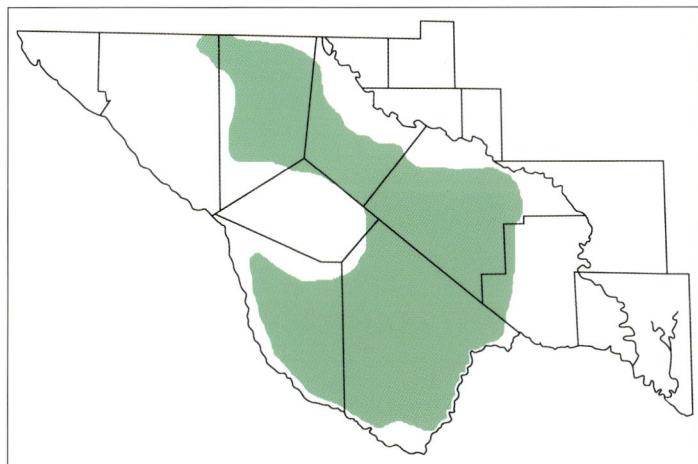

species *E. coccineus* and *E. dasyacanthus*. Details of individual plants depend upon the morphology of the parental species at the particular site where hybridization occurred and upon whether the hybrids are of F_1, F_2 or later, or backcross generations. In habit var. *neomexicanus* is at first single-stemmed, potentially forming clumps of 20 or more stems. Stems about 8–11 cm thick, with reliably 12 ribs, spine clusters of 4–6 centrals, longest 1.2–1.9 cm, and 12–16 radials—unless the plants are products of backcrossing with *E. coccineus* var. *paucispinus*, which has fewer ribs and fewer and thicker spines, or *E. dasyacanthus*, which has more ribs and more spines.

Flowers. Flowering Apr–early May. Flowers of F_1 hybrids orange (Plate 124). Later-generation and backcross hybrids may produce orange, red-orange, red, to pinkish, or yellow flowers (Plates 126–28). Color variations, as well as variation in vegetative morphology, are hallmarks of the extensive population of var. *neomexi-*

canus in eastern Pecos County. Flowers about 4.6–7 cm in diameter, usually smaller in backcrosses to *E. coccineus,* either bisexual or male sterile (with nonfunctional anthers). Anthers usually yellow in bisexual flowers and purplish but abortive in male sterile flowers.

Fruits. Mature fruits purplish-maroon to brick-red, or even greenish-orange to pinkish-green, with juicy pulp white to pink in color. Black seeds strongly papillate and about 1–1.5 mm in largest diameter.

Full Name and Synonyms. Echinocereus X *roetteri* var. *neomexicanus* (J. M. Coult.) A. D. Zimmerman, comb. nov. (forthcoming). *Cereus dasyacanthus* Engelm. var. *neomexicanus* J. M. Coult., non *E.* X *neomexicanus* Standl., pro sp.; *E.* X *roetteri* (Engelm.) Engelm. ex Rümpler; *E. triglochidiatus* var. *gurneyi* L. D. Benson (in part, as to type); *E. coccineus* Engelm. var. *gurneyi* (L. D. Benson) K. D. Heil & S. Brack (in part, as to type); *E. lloydii* Britton & Rose; *E. roetteri* var. *lloydii* Backeb.

Plate 126. *Echinocereus* X *roetteri* var. *neomexicanus* (Lloyd's hedgehog cactus), E Pecos Co., TX; red-orange-flowered; putative F_2 generation.

Plate 127. *Echinocereus* X *roetteri* var. *neomexicanus* (Lloyd's hedgehog cactus), E Pecos Co., TX; pinkish-flowered; putative F_2 generation.

Plate 128. *Echinocereus* X *roetteri* var. *neomexicanus* (Lloyd's hedgehog cactus), Marathon Basin, Brewster Co., TX; yellow-flowered; putative backcross to *E. coccineus*.

Echinocereus X neomexicanus
Triploid Hybrid Hedgehog Cactus
PLATE 129

Echinocereus X *neomexicanus* is a nothospecies of sporadic occurrence, mostly in south-central New Mexico. Sight records of only about 17 plants of *E.* X *neomexicanus*, thought to be sterile first-generation hybrids, have been reported by several cactus experts after many years of fieldwork within the range of the species (Zimmerman, 1993). The specific epithet is a geographic reference to its discovery site in New Mexico.

Distribution. El Paso Co., Franklin Mts along Trans-Mountain Hwy. Culberson Co., Guadalupe Mts, Pine Spring Canyon (sight records by D. Ferguson). 4,000–5,600 ft. New Mexico, Doña Ana, and (sight records) in Otero, Sierra, and Eddy counties. Map 43.

Vegetative Characters. Familiarity with the parental species, *E. coccineus* and *E. viridiflorus*, is a prerequisite for recognizing the very rare and usually isolated individuals of the F_1 hybrid *E.* X *neomexicanus*, which combine the characters of both parents in infertile, sometimes visibly "ill-looking" individuals. Plants single-stemmed or with 2–8 branches, stems 17–25 cm long, 5.5–7 cm in diameter, with 11–12 ribs. Spines partially obscure stem and lend overall red-and-white aspect, much as is characteristic of *E. viridiflorus* var. *cylindricus*. Areoles 10–15 mm apart on ribs, each bearing about 17–22 spines. Central spines 4–6 and radial spines 13–16 in each areole. Lowermost central spine noticeably longer (ca. 3 cm) than the others (shortest ca. 0.6 cm long), angled downward in position, and pale in color. Other central spines usually reddish or pinkish-brown. Radial spines, the longest 1.1–1.5 cm, more slender than centrals.

Flowers. Flowering spring. Flowers are produced, if at all, at different heights along the stem, as in *E. viridiflorus*. Flowers ca. 5 cm long, 2.5–3 cm wide, tubular-funnelform, with inner tepals red distally and greenish-yellow toward the base, or entirely red or entirely brownish-pink.

Plate 129. *Echinocereus* X *neomexicanus* (triploid hybrid hedgehog cactus), Jarilla Mts, Doña Ana Co., NM; tentatively identified (photo by Allan D. Zimmerman).

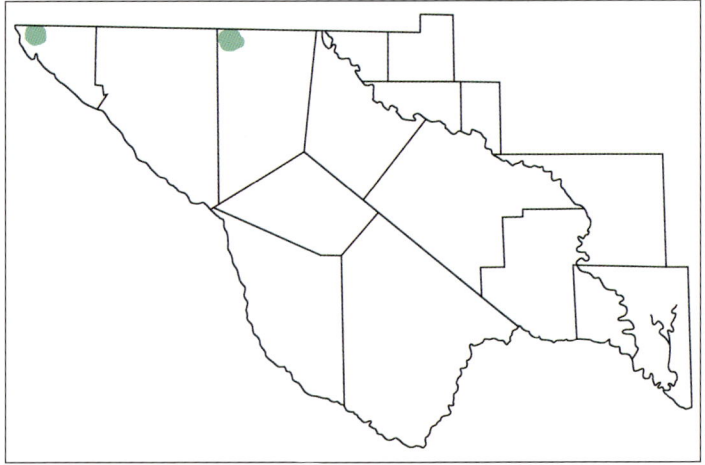

Map 43. *Echinocereus* **X** *neomexicanus* (triploid hybrid hedgehog cactus).

Filaments crowded and support yellow to pinkish anthers, which may be abortive and apparently do not produce pollen.

Fruits. Fruits and seeds have not been reported.

Full Name and Synonyms. Echinocereus X *neomexicanus* Standl., pro sp. *E. neomexicanus* Standl.; *E. triglochidiatus* var. *neomexicanus* (Standl.) W. T. Marshall, in part, as to type.

Echinocereus dasyacanthus
Texas Rainbow Cactus

PLATES 130, 131

Echinocereus dasyacanthus is commonly known in the Trans-Pecos region as the rainbow cactus because of rather subtle rings caused by bands of contrasting spine coloration along the length of the stems. Not all plants or populations of *E. dasyacanthus* exhibit the "rainbow" coloration of the stems. Some previous workers have been confused by flower-color variation associated with *E. dasyacanthus* and related taxa. As interpreted here, *E. dasyacanthus* is a tetraploid, predominantly yellow-flowered species with betacyanic-flowered individuals and populations scattered throughout most of its range, at least in the Trans-Pecos and in southern New Mexico. Recent experimental evidence (Powell et al., 1991; Powell, 1995, 1998c) suggested that the cyanic flower color and possibly some of the spine variation are the result of past or present introgressive hybridization of *E. dasyacanthus* with *E. coccineus.*

The specific epithet is after the Greek *dasys,* "hairy" or "shaggy," and *akantha,* "thorn," or "spine" in this case, which is descriptive in reference to the predominant forms of this species with rather long and divergent, overlapping spines and a "shaggy" aspect.

Distribution. Rocky slopes of arid mountains or desert floor, in desertscrub or desert grasslands, opportunistically in more mesic habitats, occurring in every county of the Trans-Pecos except Val Verde, not extending much E of Sanderson in Terrell Co. 1,700–4,700 ft. In TX extending E to about Mitchell Co., disjunct

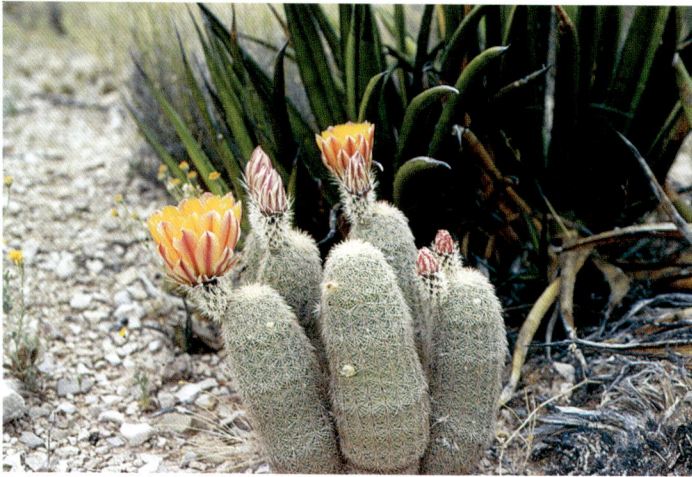

Plate 130. *Echinocereus dasyacanthus* (Texas rainbow cactus), S Brewster Co., TX (photo by Shirley A. Powell).

Plate 131. *Echinocereus dasyacanthus* (Texas rainbow cactus); mature fruit; in cultivation.

in Maverick Co. near Eagle Pass. Southeast and S-central NM; reports from Cochise Co., AZ, are based on specimens of *E. bristolii* W. T. Marshall. Mexico: NE Chihuahua and N Coahuila. Map 44.

Vegetative Characters. Plants typically with solitary stems or 2–3 basal branches. Plants with 3–10 stems not uncommon, however, in populations containing particularly old and vigorous plants. Stems

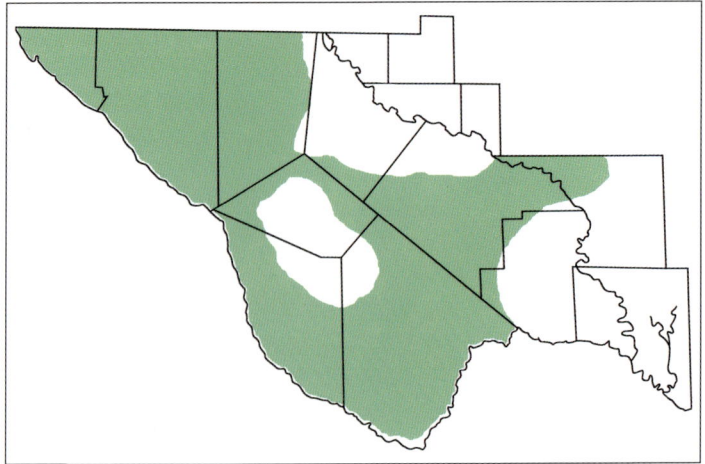

Map 44. *Echinocereus dasyacanthus* (Texas rainbow cactus).

11–24 cm long, 5.5–7 cm in diameter, with 15–18 ribs. Spines in most plants overlapping and obscure the stem surface. Spine colors, numbers, lengths, and position vary considerably, but in general 4–12 diffusely spreading central spines 0.5–1.2 cm long and 14–25 mostly appressed radial spines to 0.7–2 cm long are characteristic of the species in the Trans-Pecos. Basically, spine colors tan or pale yellow to pink, or less often ashy-white to reddish-brown.

Flowers. Flowering Mar–May. Large, showy flowers erupt from above the areoles at the sides of the stem but usually nearer the stem apex, in contrast with the generally lateral flowers of *E. viridiflorus*. Sweet-scented, bee-pollinated flowers are usually 8–12 cm long and 7–11 cm wide. In most Trans-Pecos populations, flowers are yellow with a green throat or, more specifically, the inner tepals have narrow greenish bases and broad blades potentially of different colors: dark to pale lemon-yellow, golden-yellow, or canary-yellow, or rose-pink to deep red or magenta, rarely with an orange tinge, but often salmon-pink, rose-pink, or orange-red with age. Populations containing flowers of various betacyanic colors and orange are in the minority in a mostly yellow-flowered taxon. Usually cyanic-flowered plants are mixed among yellow-flowered plants instead of forming pure populations. Plants with cyanic flowers are sporadic throughout the Trans-Pecos. Extreme flower-size variation seems to be characteristic in some populations that have mostly cyanic-flowered plants, with flowers one to several centimeters smaller or larger than the measurements given above. Tepals relatively thin and soft in comparison with those of *E. coccineus*, but relatively thick and durable compared to *E. reichenbachii* tepals. Spreading stamens have filaments of approximately equal length, resulting in a floral throat filled with a funnel of yellow anthers. Style 1.5–3 mm thick, whitish, with 12–22 deep green stigma lobes.

Fruits. Fruit at first green or greenish-purple but typically maturing dark, dull purplish; juicy pulp may be white to purplish-pink. Globose to ellipsoid fruit relatively large, to 6 cm long and 4.5 cm in diameter. Spiny areoles on the fruits ultimately deciduous. Mature fruits usually indehiscent but may split lengthwise. Black seeds globular, papillose-rugose, 1–1.4 mm in largest diameter.

Other Distinctions. The relatively complex taxonomy concerning *E. dasyacanthus* is discussed in Powell and Weedin (2004). Cyanic-flowered populations of *E. dasyacanthus* are known to occur in several areas of the Trans-Pecos. In different plants the cyanic flower colors include red, orange-red, magenta, pink, and orange. A large, somewhat discontinuous population of these cyanic-flowered plants occurs sympatrically, or in partial sympatry, with *E. coccineus* and in certain sites with *E.* X *roetteri*, from eastern Pecos County (Plate 132) west across southern Reeves County to near Kent in southern Culberson County, where only a few plants with red flowers have been observed. Other cyanic-flowered populations are known from western (Mesa de Anguila) and eastern Big Bend National Park to near Reagan Canyon along the Rio Grande in southeastern Brewster County, and they have been reported from near Ruidosa in Presidio County and south and west of Sanderson in Terrell, Pecos, and Brewster counties.

Full Name and Synonyms. *Echinocereus dasyacanthus* Engelm. *E. pectinatus* (Scheidw.) Engelm. var. *neomexicanus* (J. M. Coult.) L. D. Benson in part, excluding type; *E. hildmannii* Arendt; *E. dasyacanthus* Engelm. var. *hildmannii* (Arendt) Weniger; *E. pectinatus* var. *dasyacanthus* (Engelm.) W. Earle ex N. P. Taylor; *E. pectinatus* var. *ctenoides* (Engelm.) Weniger ex G. Frank.

Other Common Name. Rainbow cactus.

Plate 132. *Echinocereus dasyacanthus* (Texas rainbow cactus), E Pecos Co., TX; introgressed from *E.* X *roetteri* (second- or later-generation backcross = *E.* X *roetteri* var. *neomexicanus* nothospecies).

Echinocereus pectinatus var. wenigeri

Langtry Rainbow Cactus

PLATES 133, 134

Echinocereus pectinatus is a species of north-central Mexico, mostly in the CDR, barely extending into Texas. Two varieties of the species are recognized: the more widespread var. *pectinatus* (Mexico); and the more restricted var. *wenigeri* (near the Rio Grande in Texas and adjacent Mexico). Some previous workers included the similar *E. dasyacanthus* as a variety of *E. pectinatus* (see biosystematic discussion in Powell and Weedin, 2004). The specific epithet is after the Latin *pectinis*, "comb," in reference to the pectinate radial spines. The varietal name honors the late Del Weniger, author of two books on the cacti of Texas and the southwestern United States.

Distribution. Limestone soils, outcrops, slopes, mesas, degraded grasslands, desertscrub. East Pecos Co., Terrell Co. E of Sanderson, Val Verde Co. Reported but needing confirmation in E Brewster Co. 900–3,600 ft. Crockett and Sutton counties. Mexico: adjacent Coahuila. Map 45.

Vegetative Characters. Most likely to be confused with the related *E. dasyacanthus* and the unrelated *E. reichenbachii*. In the Trans-Pecos, *E. reichenbachii* is not known to be commonly sympatric with var. *wenigeri*, but the two may occur together in eastern Pecos and Terrell counties. Ashy-white spines consistent along length of stem and never pink, or in contrasting bands of darker (pink) and lighter (whitish) spines that develop as annual growth increments (from the apex), as characteristic of var. *pectinatus* in Mexico.

Plate 133. *Echinocereus pectinatus* var. *wenigeri* (Langtry rainbow cactus), Terrell Co., TX; cultivated.

Plate 134. *Echinocereus pectinatus* var. *wenigeri* (Langtry rainbow cactus), Terrell Co., TX; mature fruit; cultivated.

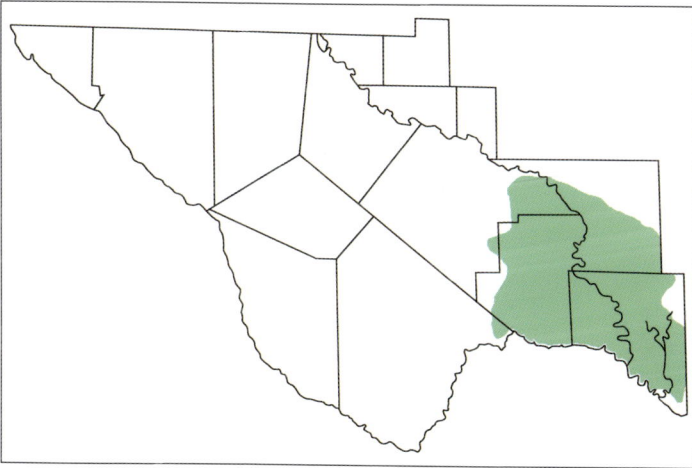

Map 45. *Echinocereus pectinatus* var. *wenigeri* (Langtry rainbow cactus).

Contrasting "rainbow" bands also characterize some individuals and populations of *E. dasyacanthus*.

Plants consist of solitary stems or, in older plants, 1–2 branches at the base. Stems of sexually mature plants cylindrical, 8–21 cm long, 5.5–8 cm in diameter, with 14–18 ribs, and spines that mostly obscure the stem surface. White spines, at least largest ones, usually tipped with pink or purplish-brown. Classic spine configuration is 2–3 short central spines in one vertical row in the center of the areole, with about 14–20 radial spines, pectinate and somewhat recurved, between the ribs.

Flowers. Flowering Mar–May. In the Trans-Pecos, flowers multicolored, giving a banded appearance inside. Inner tepals pink, lavender-pink, to magenta on distal one-third to one-half, then blending or abruptly whitish in the middle and green at the base (Plate 135). Whole flowers 7–9 cm long and 5.5–7.5 cm wide. Cream-yellow anthers on whitish filaments spread across most of the throat. Style whitish with 9–12 dark green stigma lobes.

Hedgehog Cactus | 149

Plate 135. *Echinocereus pectinatus* var. *wenigeri* (Langtry rainbow cactus); the white cross-bands and green bases of the inner tepals are characteristic.

Fruits. Fruits broadly elliptical, 2–3 cm long and nearly as wide, dark dull purplish or reddish, juicy, with whitish pulp. Older fruits may turn bronze or brown, sometimes drying, and pericarpel may split. Pericarpel areoles deciduous after fruit ripens. Seeds black, globular, 1–1.3 mm in diameter, papillate.

Full Name. *Echinocereus pectinatus* (Scheidw.) Engelm. var. *wenigeri* L. D. Benson.

Echinocereus fendleri var. fendleri
Fendler's Hedgehog Cactus
PLATES 136, 137

Several authors (e.g., Benson) have recognized two varieties of *E. fendleri* in Trans-Pecos Texas, var. *fendleri* and var. *rectispinus*. We accept only var. *fendleri* in the Trans-Pecos region (see Powell and Weedin, 2004). *Echinocereus fendleri* var. *kuenzleri* occurs near the northern border of the Trans-Pecos in the Sacramento Mountains of New Mexico (Otero County) and recently has been applied to the disjunct populations long known from north-

ern Chihuahua, Mexico (Taylor, 1985), based on *E. hempelii* Fobe, but it is not expected in the Trans-Pecos (see the section "Other *Echinocereus* Species," below). The specific epithet honors the early collector in New Mexico, Augustus Fendler.

Distribution. Upper desert grasslands and plains grasslands in the alluvial mountain basins, rare near Panther Junction in the Chisos Mts of southern Brewster Co., NW Brewster Co. W across N Presidio Co., Jeff Davis Co. and adjacent Reeves Co., W across SE and central Culberson and Hudspeth counties, and slopes of the Franklin Mts in El Paso Co. 3,500–5,500 ft. Western two-thirds of NM, S CO, NE and SE AZ. Mexico: possibly NE Sonora. Map 46.

Vegetative Characters. Grassland-dwelling *E. fendleri* is characterized by solitary or few stems, but often loose clumps of 5–10 or more erect, rather flaccid stems. Stems dark green, somewhat wrinkled, 7.5–17 cm long, 3.8–7.5 cm in diameter, with 8–10 ribs. Circular areoles about 15–17 mm apart with conspicuous swellings around areoles. Spines promi-

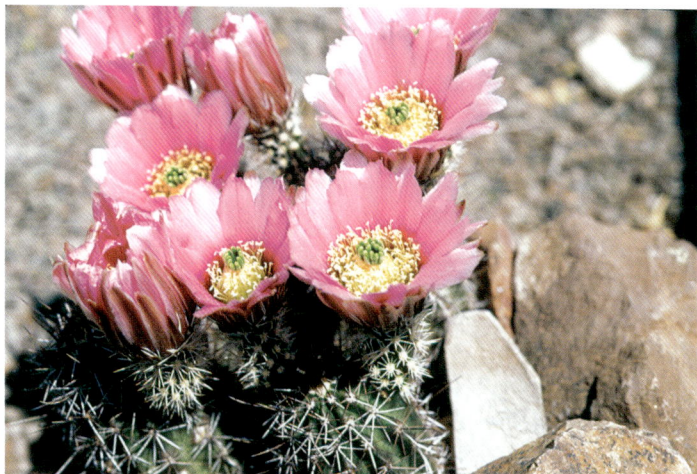

Plate 136. *Echinocereus fendleri* var. *fendleri* (Fendler's hedgehog cactus); in cultivation.

Plate 137. *Echinocereus fendleri* var. *fendleri* (Fendler's hedgehog cactus), Presidio Co., TX; mature fruit (cultivated).

nently bulbous at the base, opaquely white or ashy-gray, often with contrasting black or brown spines in the same areoles, weathering to gray. Alternatively, central spine and largest radial spines may be nearly black proximally and whitish dis-tally except for reddish-brown tips. Whit-ish spines often have a longitudinal dark stripe on the underside, or the dark stripe is brought to the upper surface in twisted spines, or spines sometimes variegated. Best identifying feature of var. *fendleri*,

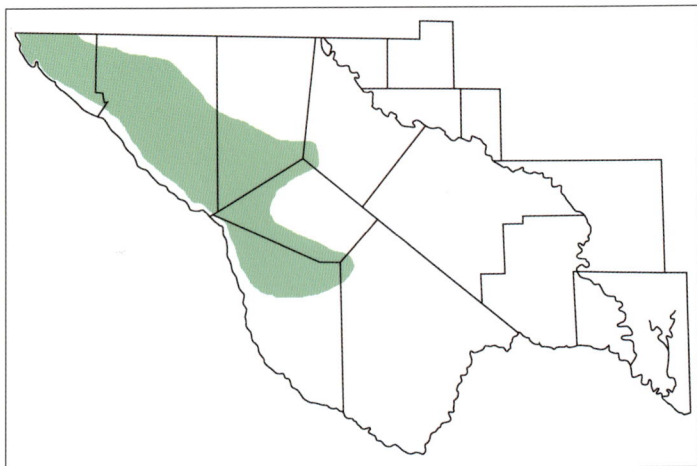

for example, when superficially similar but larger-stemmed *E. coccineus* might be in the area, is the single central spine to 5 cm long and curving upward, at least in some areoles on some stems. Usually 5–9 radial spines in the areoles of var. *fendleri*.

Flowers. Flowering Apr–May. Flowers relatively large and showy, 5–11 cm long, 5–11 cm wide. Inner tepals magenta, often darker purplish-red at base. Pink- and white-flowered forms less common or rare. Flower buds conspicuously erumpent above areoles on sides of the stems near the apex. Stamens with greenish filaments, light yellow anthers. Style whitish, only slightly longer than anthers, supporting 9–16 dark green stigma lobes.

Fruits. Bright red to brick-red fruits ellipsoidal, sometimes almost spherical, 2–3 cm long. Fruits juicy with magenta pulp. At maturity, fruits may rupture. Areoles with whitish spines and white wool, readily dehiscent on ripe fruits. Seeds black, irregularly globular or obovoid, reticulate-punctate, 1–1.5 mm in largest diameter.

Full Name. *Echinocereus fendleri* (Engelm.) Rümpler var. *fendleri*.

Other Common Name. Fendler's hedgehog.

Echinocereus enneacanthus
Strawberry Cactus

This treatment follows Taylor (1985) in recognizing *E. enneacanthus* as the correct name for a taxon that includes as synonyms *E. dubius* and *E. enneacanthus* var. *dubius*, familiar names in the Trans-Pecos, and *E. enneacanthus* var. *brevispinus*, a familiar name in southern Texas. *Echinocereus enneacanthus* is one of the two (or more) species commonly known as strawberry cactus. In the Trans-Pecos, the best-known strawberry cactus is *E. stramineus*. The *E. enneacanthus* complex, traditionally involving four or more taxa, is one of the most thoroughly investigated cactus groups in the Trans-Pecos (see discussions in Powell and Weedin, 2004). The specific epithet is after the Greek *ennea*, "nine," and *akantha*, "thorn," presumably in reference to the ca. nine radial spines per areole.

Full Name. *Echinocereus enneacanthus* Engelm.

Key to the Varieties of *Echinocereus enneacanthus*

1. Stems 5–10 cm in diameter; areoles 2.5–4.5 cm apart on ribs; central spines 1–4, to 9.2 cm long, divergent, often curved, stout, 1.5–2 mm at the base; radial spines to 4.2 cm long; distribution Big Bend region of the Trans-Pecos, and W to El Paso Co.
E. enneacanthus var. *enneacanthus*, p. 153

1. Stems to 5 cm in diameter; areoles 1–1.5 cm apart on ribs; central spines 1–3, to 4.5 cm long, porrect, straight, relatively slender, 1.0–1.5 mm at the base; radial spines 1–2 cm long; distribution mostly in S TX and Edwards Plateau, extending W along the Rio Grande to the Big Bend region
E. enneacanthus var. *brevispinus*, p. 155

Echinocereus enneacanthus var. enneacanthus
Strawberry Cactus

PLATES 138, 139

This is the entity that in Trans-Pecos Texas has long been known as *E. enneacanthus* var. *dubius* or *E. dubius* (see Powell and Weedin, 2004).

Distribution. Gravel hills and benches, alluvial outwash fans and bolsones, also desert mountains, exposed to the sun or under shrubs, El Paso Co. along and near the Rio Grande, SE to Brewster Co. where it extends to ca. 50 mi N of the Rio Grande (also reported in the Guadalupe Mts where disjunct). 1,800–4,000 ft. Mexico: E Chihuahua, central and S Coahuila, NE Durango, N Zacatecas, N San Luis Potosí, W Nuevo León, and SW Tamaulipas. Map 47.

Vegetative Characters. Sexually mature plants with several to numerous stems in loose or rather tight clumps. Larger plants may have 50–100 or even several hundred stems. Stems about 15 cm long, longer in shade forms. Stems medium to light green

Plate 138. *Echinocereus enneacanthus* var. *enneacanthus* (strawberry cactus), S Brewster Co., TX.

Plate 139. *Echinocereus enneacanthus* var. *enneacanthus* (strawberry cactus); mature fruit; in cultivation.

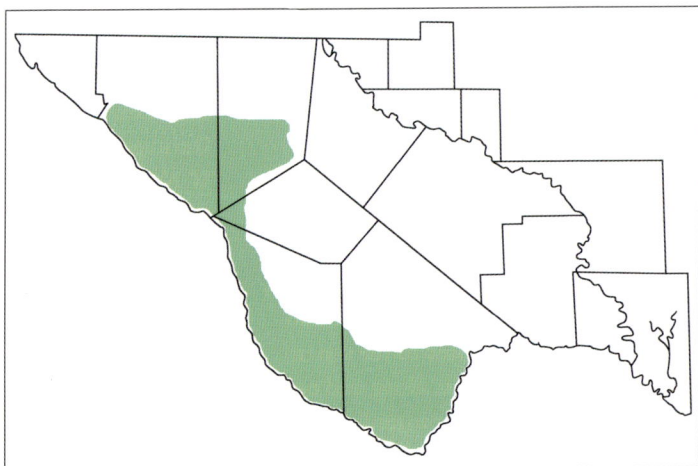

Map 47. *Echinocereus enneacanthus* var. *enneacanthus* (strawberry cactus).

or somewhat yellow-green. Ribs 7–10, but on average 8–9. Central spines flattened or angular. Radial spines 6–9.

Flowers. Flowering Apr–May. Magenta flowers funnelform, to 8 cm long, 7–9 cm in diameter. Tepals in 2–3 series. Throat of magenta flowers deepens in color to reddish. Filaments greenish to pink, anthers yellow. Style about 3 cm long, 1.5–2 mm in diameter, supporting 6–10 green stigma lobes.

Fruits. Fruit globular to ovoid, to 3.8 cm long and 2.5 cm in diameter, green at first but maturing red or brick red, pink inside, with strawberry smell and flavor. Seeds black, ovoid, tuberculate, 1–1.4 mm long.

Other Distinctions. In the southern Big Bend region of the Trans-Pecos, *E. enneacanthus* commonly occurs in a variety of substrates, including alluvial gravels and soils in exposed areas and in the shade of shrubs and small trees. *Echinocereus enneacanthus* var. *enneacanthus* (= *dubius*) occurs in habitats along and near the Rio Grande and in desert areas to about 50 miles north of the river. *Echinocereus enneacanthus* var. *brevispinus* occurs in southeastern Brewster County near the Rio Grande, usually in the shade of riparian vegetation. Farther east var. *brevispinus* occurs as far north as Upton County. Variety *brevispinus* also grows to the east on the wooded Edwards Plateau and south to near Brownsville.

Echinocereus enneacanthus var. *enneacanthus* and *E. stramineus* are regionally sympatric in the southern Trans-Pecos, where they occur in close proximity in many locations. The two taxa usually are ecologically separated, with var. *enneacanthus* characteristically occupying alluvial habitats and *E. stramineus* usually found in various exposed rocky habitats.

Full Name and Synonyms. *Echinocereus enneacanthus* var. *enneacanthus*. *E. dubius* (Engelm.) Rümpler; *E. enneacanthus* var. *dubius* (Engelm.) L. D. Benson.

Other Common Names. Strawberry hedgehog; pitaya; pitahaya.

Echinocereus enneacanthus var. brevispinus
Strawberry Cactus
PLATES 140, 141

Echinocereus enneacanthus var. *carnosus*, recognized by Weniger (1984), found near Laredo and Eagle Pass, is treated here as an unusually large and flabby-stemmed form of var. *brevispinus* (Powell and Weedin, 2004). The name *brevispinus* refers to the short spines, as opposed to the long-spined var. *enneacanthus*. The root word *brevi* is after the Latin *brevis*, which means "short."

Distribution. Mostly alluvial substrates among riparian vegetation along and near the Rio Grande in SE Brewster Co., occasionally in limestone rocks in various substrates, including open rocky habitats. 1,000–3,000 ft. Edwards Plateau, Tamaulipan scrublands, and on the lowland plains along and near the Rio Grande to Cameron Co., at near sea level. Mexico: E Coahuila, N Nuevo León, Tamaulipas. Map 48.

Vegetative Characters. Distinguished

Plate 140. *Echinocereus enneacanthus* var. *brevispinus* (strawberry cactus), W Val Verde Co., TX; cultivated.

Plate 141. *Echinocereus enneacanthus* var. *brevispinus* (strawberry cactus), W Val Verde Co., TX; mature fruits; cultivated.

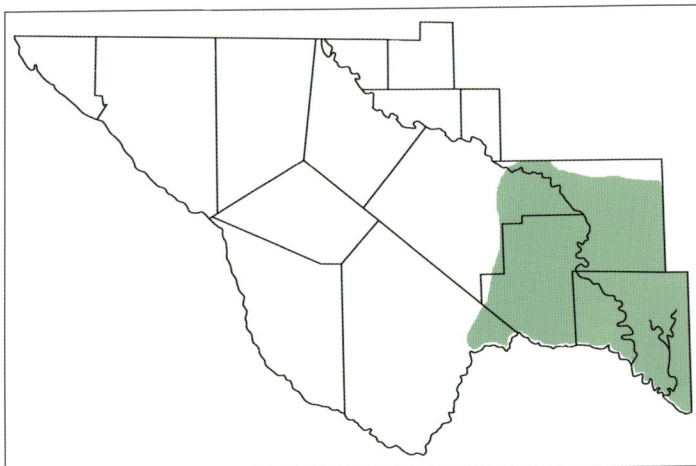

Map 48. *Echinocereus enneacanthus* var. *brevispinus* (strawberry cactus).

from *E. stramineus*, with which it is sympatric in part of its range in the eastern Trans-Pecos, by less mounded habit, fewer ribs, and shorter spines. Mature plants form flat-topped or rounded clumps with 15–100 or more stems. In most populations, stems slender and upright, to ca. 5 cm in diameter, but in some populations, particularly in the form recognized as var. *carnosus*, prostrate, decumbent stems 8–10 cm in diameter. Stems typically produce 8–9 ribs, with circular areoles 1–2 cm apart. Typically in each areole 1–2 yellowish, brownish, or gray-blue central spines, to 4 cm long, and 8–9 whitish to tan or brownish radial spines. Central spines porrect and straight compared to the typically curving central spines of var. *enneacanthus*.

Flowers. Flowering Apr–May. Magenta flowers similar in size to those of var. *enneacanthus*. Tepals in flowers of var. *brevispinus* in 2–3 series. The throat a darker reddish color than in var. *enneacan-*

thus. Filaments in var. *brevispinus* greenish to pink, ca. 1 cm long, and yellow anthers are spread across the throat. Style ca. 3 cm long, 1.5–2 cm thick, supporting 6–12 green stigma lobes 5–6 mm long.

Fruits. Fruit globular to ovoid, 2.5–3.8 cm long, ca. 2.5 cm in diameter. Mature fruits greenish to reddish-brown or ultimately red. Areoles with spines easily detached from mature fruits. Pulp pink, with strawberry smell and flavor. Seeds black, ovoid, 1–1.4 mm long, prominently tuberculate.

Full Name and Synonyms. Echinocereus enneacanthus var. *brevispinus* (W. O. Moore) L. D. Benson. *E. enneacanthus* Engelm. f. *brevispinus; E. enneacanthus* var. *carnosus* (Rümpler) Quehl; *E. enneacanthus* var. *enneacanthus,* sensu L. D. Benson.

Echinocereus stramineus var. stramineus
Strawberry Cactus
PLATES 142, 143

In Trans-Pecos Texas *E. stramineus* is the "real" strawberry cactus, as opposed to *E. enneacanthus,* because of its larger, juicier fruits that have the color, smell, and flavor of strawberries. Both *E. stramineus* and *E. enneacanthus* var. *enneacanthus* (= *dubius*) are of about equal abundance in deserts of the lower and eastern Big Bend area. The specific epithet is after the Latin *stramineus,* "made of straw," a reference to the dense covering of divergent, straw-colored spines that characterize the plants. In age the spines may bleach to glassy white.

Distribution. Mostly on exposed limestone rock outcrops, in the desert

Plate 142. *Echinocereus stramineus* var. *stramineus* (strawberry cactus), S Brewster Co., TX.

Plate 143. *Echinocereus stramineus* var. *stramineus* (strawberry cactus), S Brewster Co., TX; nearly mature and mature fruits.

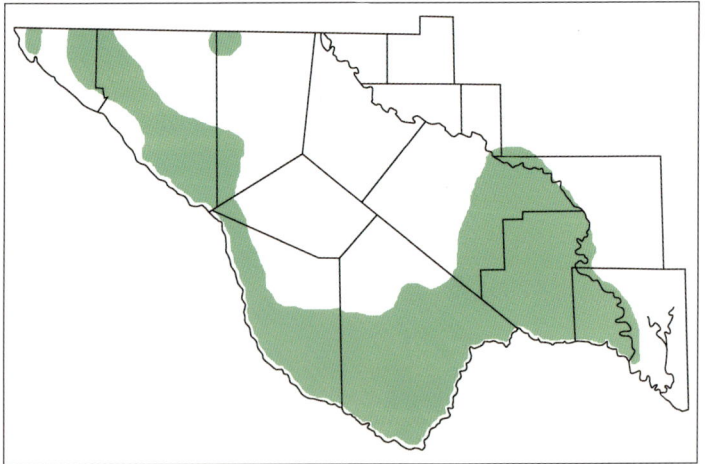

Map 49. *Echinocereus stramineus* var. *stramineus* (strawberry cactus).

mountains and arid S slopes of higher mountains, El Paso Co. SE through the southern Trans-Pecos and Big Bend region, SE to Terrell Co. and the lower Pecos River, possibly in Val Verde Co. 2,500–5,000 ft. East to Upton Co. across the Pecos River; reported from Tom Green Co. South-central NM, Doña Ana, Otero, and Eddy counties. Mexico: E Chihuahua, Coahuila, W edge of Nuevo León, NE Durango, N Zacatecas, and N San Luis Potosí. Map 49.

Vegetative Characters. Plants cespitose with numerous stems forming hemispheric mounds. In sexually mature plants, number of stems may be 10–50, or in largest plants, 100–350. Plants ca. 1 m in diameter with up to 500 stems have been reported. Number of ribs per stem is one of the best distinguishing features of *E. stramineus*, especially when comparing this species with *E. enneacanthus*. *Echinocereus stramineus* has 11–17 ribs per stem, but a populational average of ca. 12 ribs per stem, compared to 8–9 ribs per stem for *E. enneacanthus*. In *E. stramineus*, the circular areoles are 7–15 mm apart on the ribs, with stem surfaces mostly obscured by spines. Areoles with 2–4 central spines, 5–9 cm long, terete to somewhat flattened, straight or slightly curved, strongly projecting. Radial spines 7–10 in number, 2–3 cm long, mostly acicular. Radial spines to 4.6 cm long observed in plants from lower Maravillas Canyon.

Flowers. Flowering late Mar–May. Magenta flowers noticeably larger than those of *E. enneacanthus*, at least in the Trans-Pecos. Flowers 8.5–12.5 cm in length and diameter. Inner tepals bright magenta to base of throat, or in some plants throat base may be darker, approaching deep red, or sometimes even lighter in color. Filaments 0.8–1 cm long, reddish; anthers yellow. Style ca. 2.7 cm long, ca. 2.5 mm thick, reddish, supporting 10–13 green stigma lobes, ca. 8 mm long. Sporadic individual plants with pure white flowers reported from Big Bend National Park. Cespitose plants of *E. stramineus* are spectacular when in full sun and covered with completely open large flowers.

Fruits. Fruits globular to broadly oval, 5–6 cm long and nearly as broad, reddish-brown when ripe. Bristly areoles on fruit surface easily dislodged. Copious pink to white flesh is a refreshingly cool treat on hot summer days of the fruiting season. Numerous seeds black, ovoid, 1.2–1.5 mm long, tuberculate.

Other Distinctions. *Echinocereus stramineus* is most frequently confused with densely clumped and mounded specimens of *E. enneacanthus* var. *enneacanthus*. Where *E. stramineus* overlaps with *E. enneacanthus* var. *brevispinus* in the southeastern and eastern Trans-Pecos, identification of taxa in the field usually is easy because of distinct spine patterns that obscure the stem surface in *E. stramineus* and reveal the stem surface in var. *brevispinus*. *Echinocereus stramineus* is only superficially similar to *E. coccineus*, which has distinct spination but forms equally large mounds of stems; these two species occasionally may be found together, but the differences are obvious in the field.

Full Name and Synonyms. *Echinocereus stramineus* (Engelm.) Rümpler var. *stramineus*. *E. enneacanthus* Engelm. var. *stramineus* (Engelm.) L. D. Benson.

Other Common Names. Pitaya; pitahaya.

Echinocereus chisoensis var. chisoensis
Chisos Hedgehog Cactus
PLATES 144, 145

Two varieties of *E. chisoensis* are recognized: ours is var. *chisoensis;* the other is var. *fobeanus* (Oehme) N. P. Taylor, endemic to Mexico, documented only in the vicinity of San Pedro de las Colonias in southwestern Coahuila (Powell and Weedin, 2004). The specific epithet is a reference to the Chisos Mountains, whose impressive main peaks loom approximately 10 miles to the northwest of the only known range of var. *chisoensis*. *Echinocereus chisoensis* is listed as Threatened by both federal and state agencies.

Distribution. Typically associated with desertscrub, in gravel to sandy alluvium, flats, benches, small and larger drainages, near the margins or in the drainages. Known only from southern Brewster Co., Big Bend National Park, SE of the Chisos Mts. 1,900–2,600 ft. Although seemingly

Plate 144. *Echinocereus chisoensis* var. *chisoensis* (Chisos hedgehog cactus), Big Bend National Park, TX.

Plate 145. *Echinocereus chisoensis* var. *chisoensis* (Chisos hedgehog cactus), Big Bend National Park, TX; immature fruits.

suitable localities abound in adjacent Coahuila, Mexico, *E. chisoensis* var. *chisoensis* has yet to be reported there. Map 50.

Vegetative Characters. Plants almost always associated with nurse plants, either under shrubs, among grasses, or in mats of dog cholla, often among decaying stems toward the center of the mats. Stems of older plants typically branched, while smaller, presumably younger plants may have solitary stems. Stems 5–20 cm long, 3–5 cm in diameter, cylindroid or slightly conical. Zones of darker and lighter spines lend banded aspect to stems of some

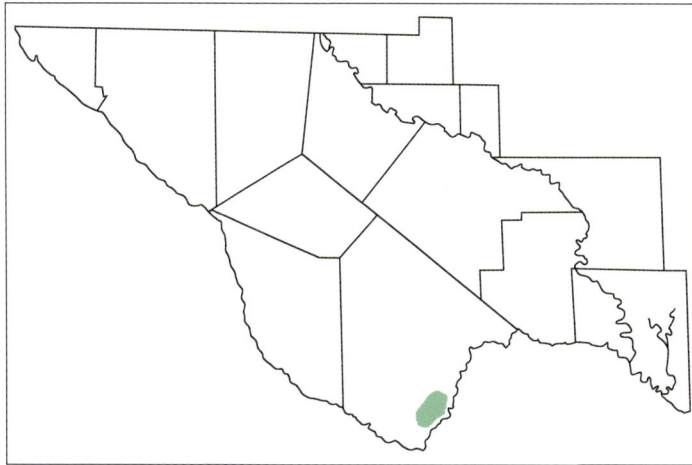

plants. Stems have 11–16 ribs with slender tubercles 6–8 mm apart, supporting small areoles with mostly divergent spines that only partially obscure the stem surface. Areoles at stem apex usually densely white-woolly, a good field character, but most areoles down the stem are naked, or with some persistent wool. Scattered areoles downstem may be densely woolly, especially in enlarged flower-bearing areoles. Central spines 2–4 in number, with main 1–2 centrals dark reddish-brown, 6–12 mm long, especially on distal portions. Radial spines usually 10–16 in number, 6–12 mm long, tan, ashy-white, or pinkish-gray, some with red-brown tips.

Flowers. Flowering mid-Mar to mid-Apr. Magenta, funnelform flowers ca. 6 cm long, ca. 5 cm in diameter, and in the field sometimes do not open widely. Receptacular tube has evenly spaced areoles with much white wool and 8–14 bristle-like to hairlike spines, white or red-brown at the tips, resembling the closely related *E. reichenbachii*. Delicate inner tepals usually in two whorls. Tepal color may be deep magenta or a lighter rose or pinkish,

particularly toward the apex, often with a darker midstripe. Throat of flower usually with a broad or narrow, whitish (or at least pale) band above the dark red base. Stamens with white to pinkish filaments ca. 1.3 cm long and pale yellow anthers. Style ca. 2.5 cm long, whitish, exserted 4–5 mm above stamens, with ca. 10 green stigma lobes, 3–6 mm long.

Fruits. Fruit oblong to narrowly obovoid, 2.5–3.7 cm long, ca. 1.4 cm in diameter, more or less covered with wool and bristlelike spines. Maturing fruits remain covered by woolly areoles, but ultimately areoles are deciduous. Fruit green to dull red and fleshy when ripe, but drying dull red or brownish-green and sometimes eventually splitting open on one side. Fruit pulp whitish, somewhat viscid or nearly dry. Seeds ovoid, ca. 1.2 cm long, black, strongly tuberculate.

Full Name and Synonym. *Echinocereus chisoensis* W. T. Marshall var. *chisoensis*; *E. reichenbachii* var. *chisosensis* (W. T. Marshall) L. D. Benson.

Other Common Names. Chisos pitaya; Chisos Mountain hedgehog cactus.

Echinocereus reichenbachii

Lace Cactus

PLATES 146, 147

Echinocereus reichenbachii was not reported by Benson (1982) and other workers to occur in the Trans-Pecos, although Weniger (1984) stated that var. *perbellus* (Britton & Rose) L. D. Benson occurred in Texas west to the Pecos River. The *E. reichenbachii* complex, ranging from Colorado south through Texas and into Mexico from sea level to elevations above 4,000 feet, needs additional study (Zimmerman and Parfitt, 2003; Powell and Weedin, 2004; Blum et al., 2004). Recent workers list up to 13 taxa for the complex. The photographs of *E. reichenbachii* (Plates 146–49) represent flower, fruit, and vegetative morphology typical of the complex, as recognized by some authors: Plate 146 resembles var. *reichenbachii* (south-central Texas north to the Panhandle, adjacent Oklahoma, northeastern Mexico); Plate 147 is similar to var. *perbellus* (central and northern Texas, eastern New Mexico, and southern Colorado); Plate 148 is var. *fitchii* (South Texas); and Plate 149 is var. *baileyi* (eastern Texas Panhandle and adjacent southern Oklahoma). The specific epithet honors the German naturalist H. G. L. Reichenbach.

Distribution. Eastern Trans-Pecos in rocky limestone outcrops, or in shallow, rocky to sandy soils among shrubs, junipers, or grasses. Eastern Reeves Co. Pecos Co., at extreme S point of the county and from Fort Stockton E to the Pecos River. Terrell Co., S of Sheffield, limestone mesas. Val Verde Co. near Pandale. 1,800–4,000 ft. In TX from Val Verde, Crockett, and Sutton counties NE to Taylor Co., N across the plains country and the Panhandle, and other varieties to sea level. Southeastern NM, SE CO, SW OK in the Great Plains grassland. Adjacent Mexico. Maps 51a and b.

Vegetative Characters. Plants are relatively small compared to most species in *Echinocereus*, with solitary or branching stems usually to 7–10 cm long, occasionally

Plate 146. *Echinocereus reichenbachii* var. *reichenbachii* (lace cactus); cultivated.

Plate 147. *Echinocereus reichenbachii* var. *perbellus* (lace cactus); mature, dehiscent fruit; cultivated.

Plate 148. *Echinocereus reichenbachii* var. *fitchii*; cultivated.

Plate 149. *Echinocereus reichenbachii* var. *baileyi*; cultivated.

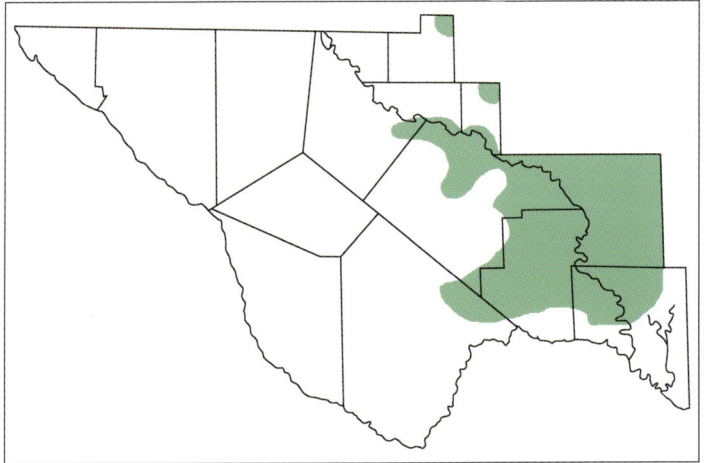

Map 51a. *Echinocereus reichenbachii* (lace cactus) in the Trans-Pecos.

longer, and 3–6.5 cm in diameter. Mature plants often with 3–12 or more stems. Stems subglobose to cylindroid and sometimes tapered toward the apex. Usually the dark green surfaces are partially obscured by spines, especially in dehydrated plants, in which radial spine tips are drawn into interlocking position. In *E. reichenbachii* as a whole, 13–15 narrow ribs with low tubercles. Areoles elliptic to oval, 2–3 mm long, 3–5 mm apart. Areoles at and near the stem apex are conspicuously white-woolly.

Areoles downstem are naked, or mostly so. A distinctive vegetative feature is the pectinately arranged radial spines in a single row on each side of the elongated areole and tightly appressed against the stem. Radial spines number 15–21; central spines 0–2. In most Trans-Pecos specimens examined, central spine absent, but in extreme south Pecos County, 1–2 centrals present in some plants; when two central spines are present, lower one is porrect and an upper one is ascending and often slightly longer at 1–2.2

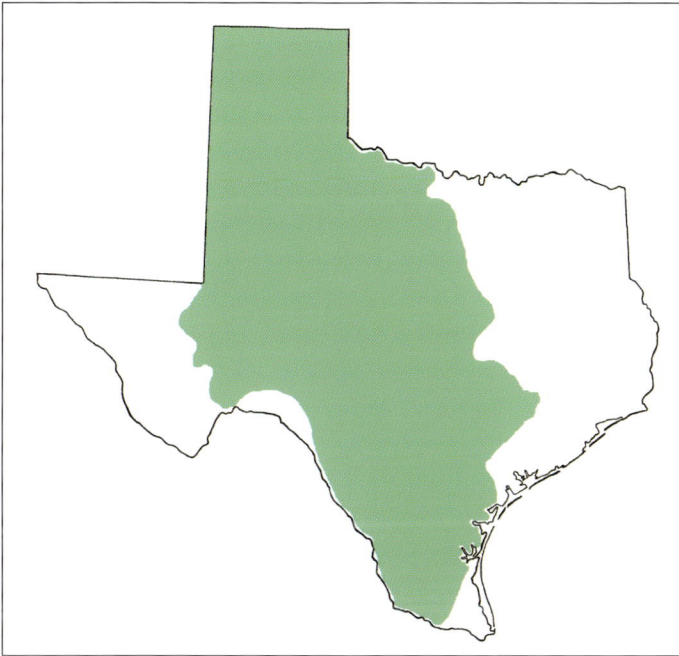

mm long. Longest radial spines 6–8 mm. Typically, spines ashy-white with pinkish, reddish, or black tips.

Flowers. Flowering Apr–May. Funnelform flowers 5–7.5 cm long, 5–6 cm wide. Receptacular tube obscured by closely spaced areoles bearing much diffuse wool and dark brown, bristlelike spines. White wool also present on flower buds. Inner tepals, in 2–4 whorls, magenta or deep pinkish with dark red bases. Anthers cream-yellow, on filaments to 2 cm long. Style pinkish, to 3.5 cm long and ca. 3 mm in diameter, with 8–20 stigma lobes, 5–6 mm long.

Fruits. Fruits globose to ovoid, 1.5–2.5 cm long, covered with wool and slender spines. Ripened pericarpel remains permanently green, ultimately splitting longitudinally to reveal fleshy white pulp that quickly dries. Areoles on fruit wall ultimately deciduous. Seeds ovoid, ca. 1.5 mm long, strongly tuberculate.

Other Distinctions. In the southeastern Trans-Pecos *E. reichenbachii* might be confused with *E. pectinatus* var. *wenigeri*, which also has pectinate white spines and woolly areoles at the stem apex. Without its woolly, finely bristly, thin-walled, ephemeral flowers, our geographic race of *E. reichenbachii* usually can be distinguished from *E. pectinatus* by its complete lack of central spines (or 1–2 very short ones present), thinner and shorter radial spines, and narrower, elliptic-oblong areoles. In our region the radial spines of *E. reichenbachii* are appressed, but elsewhere they sometimes project strongly, as in the Oklahoma populations. The largest plants we have seen in the Trans-Pecos were from extreme south Pecos County, where one stem measured 15 cm long and 7.5 cm in diameter. According to Blum et al. (2004), two subspecies of *E. reichenbachii* subsp. *caespitosus* (Engelm.) Blum & Lange and subsp. *perbellus* (Britton & Rose) N. P. Taylor, occur in the Trans-Pecos.

Full Name and Synonyms. *Echinocereus reichenbachii* (Terscheck) F. Haage. *E. reichenbachii* Terscheck; *E. caespitosus* Engelm.; *E. perbellus* Britton & Rose; *E. reichenbachii* (Terscheck) Britton & Rose var. *perbellus* (Britton & Rose) L. D. Benson; *E. caespitosus* (Engelm.) Engelm. var. *perbellus* (Britton & Rose) Weniger; *E. reichenbachii* subsp. *perbellus* (Britton & Rose) N. P. Taylor; see Blum et al. (1998) for additional synonymy.

Other Common Name. Purple candle.

Echinocereus davisii
Dwarf Hedgehog Cactus
PLATES 150–52

The substrate in which *E. davisii* grows is derived from (and filled with chips of) novaculite, a pale, very dense, fine-grained rock material that is almost pure silica and is essentially chert. Novaculite is late Devonian (about 370 million years) in age and is thus much older than the limestone and igneous formations that overlie it. In geologic history the novaculite stratum was tilted upward during the formation of a large dome and subsequently exposed as the dome eroded. The Marathon Basin is the name of the area approximately 30–40 miles in diameter inside the eroded dome. Caballos Novaculite is the geologic name for the whitish, tan, or rusty-colored formation that is exposed at various sites in the Marathon Basin, often as ridges, steep "hogback" ridgetops, or strata in hillsides.

The specific epithet chosen by A. D. Houghton honors A. R. Davis, who discovered the taxon. *Echinocereus davisii* is listed as Endangered by the state of Texas and the federal government.

Distribution. Trans-Pecos endemic, restricted to the Caballos Novaculite (and adjacent exposures) in the Marathon Basin in northern Brewster Co., TX; the small plants are often found growing in and partially hidden beneath mats of prostrate *Selaginella* P. Beauv. (inaccurately called "moss" in some of the literature). 3,900–4,400 ft. Map 52.

Plate 150. *Echinocereus davisii* (dwarf hedgehog cactus), Marathon Basin, Brewster Co., TX; three plants in flower in a *Selaginella* mat, along with an immature plant of *Thelocactus bicolor* var. *flavidispinus*.

Plate 151. *Echinocereus davisii* (dwarf hedgehog cactus), Marathon Basin, Brewster Co., TX; one multistemmed plant in flower.

Plate 152. *Echinocereus davisii* (dwarf hedgehog cactus), Marathon Basin, Brewster Co., TX; nearly mature fruit.

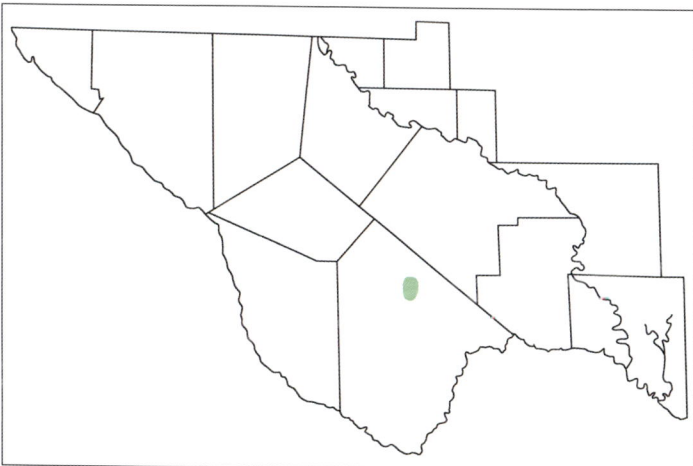

Map 52. *Echinocereus davisii* (dwarf hedgehog cactus).

Vegetative Characters. One of the smallest cacti in the world; sterile specimens are difficult to locate in the field because globose to ovoid, usually solitary stems aboveground are mostly only 1–2 cm tall. During dry periods, stems shrink, pulling the plants even farther into the ground, often under mats of *Selaginella*. Spines partially or mostly obscure the dark green stem surface. Usually all spines (8–15, the longest 0.8–2.5 cm) radial, although sporadic solitary central spines (to 1–1.2 cm long) are present in areoles of some older plants. Spines stout for such small stems, ca. 0.5 mm in basal diameter, terete or slightly flattened, in older plants often curving in any direction, gray to ashy-white with brown or reddish tips or whitish throughout. Some spines in apical areoles reddish throughout or nearly black.

Flowers. Flowering Mar–Apr. Greenish-yellow, lemon-scented flowers 1.8–2.5 cm long, 1.5–2 cm wide, frequently larger than the stems. Flowers may not open wider than 1.5–2 cm, or glossy tepals may recurve and become revolute, apparently only in full sun. Filaments pale green, 5–9 mm long; anthers yellow, 0.5–1.2 mm long. Style pale green, ca. 1.4 cm long, supporting 5–7 short (1.5–2 mm long), green stigma lobes.

Fruits. Mature fruits green or reddish-brown, 6–9 cm long, 4–5.5 mm in diameter. At maturity, fruits split longitudinally and immediately dry out; areoles are deciduous. Black seeds 0.9–1 mm long, with finely tuberculate surfaces, 40 or fewer in each small fruit.

Other Distinctions. Only the healthy adults bearing traces of flowers or fruits can be recognized as the truly dwarf plants that they are. Otherwise, they are just plain small cacti, like seedlings of some ordinary larger species. In the novaculite substrates where it grows there are two other diminutive endemic species, *Coryphantha minima* and *C. hesteri*, easily distinguished by spine characters. Otherwise, identification requires comparison with the seedlings and immatures of all the other, larger species.

Full Name and Synonym. Echinocereus *davisii* Houghton. *E. viridiflorus* Engelm. var. *davisii* (Houghton) W. T. Marshall.

Other Common Names. Davis' dwarf hedgehog cactus; Davis' green pitaya.

Echinocereus viridiflorus
Green-Flowered Hedgehog Cactus

About 12 varieties of *E. viridiflorus* are known, seven of them in the Trans-Pecos. Taxa of the *E. viridiflorus* complex not known to occur in the Trans-Pecos are listed in the section "Other *Echinocereus* Species," below, including var. *viridiflorus*, which is found in the Texas Panhandle. See Powell and Weedin (2004) for a taxonomic review of the complex. The specific epithet is after the Latin *viridis*, "green," and *floris*, "flower," in reference to the green flowers of the typical variety.

Full Name. Echinocereus viridiflorus Engelm.

Key to the Varieties of *Echinocereus viridiflorus*

1. Central spines 0–3, stout or relatively stout, when present, with prominent bulbous bases; radial spines 14–23, pectinately arranged, relatively stout, usually laterally compressed at their bases (2).
1. Central spines 6–12, relatively thin and flexible, with smallish-bulbous bases; radial spines 30–45, appressed or spreading, slender, flexible, usually terete at their bases (4).

2(1). Central spines three or more per areole [will also key under second lead 1 above]; lower central spine usually white, mostly white or white on the upper surface, 2.5–4.3 cm long, directed downward but usually curving upward; Culberson Co. W to El Paso Co. and S NM; (in part)
  E. viridiflorus var. *chloranthus*, p. 175
2. Central spines 0–3; lower central, when present, colored like the other central spines (yellow, reddish, or reddish-brown, rarely whitish), 0.7–1.4 cm long, directed downward or porrect, straight or slightly curved; mostly Jeff Davis and northern Brewster counties (3).

3(2). Plants all or mostly with a red or red-brown and white aspect; flowers usually unscented, typically opening only about 45°, of any color; radial spines 14–23; a familiar cactus of the Davis Mts and elsewhere
  E. viridiflorus var. *cylindricus*, p. 170
3. Plants all with a greenish-yellow and tan or ashy-white aspect; flowers lemon-scented, widely opening, all yellow-green; radial spines 19–28; mostly restricted in the northern Marathon Basin
  E. viridiflorus var. *correllii*, p. 172

4(1). Stems clothed in white spines; colored spine tips, when present, all red-purple (no yellow color-form present); seedlings with long, white, hairlike spines and not stiff spines in the areoles; flowers bright, light green; distribution Solitario Dome in SE Presidio Co.
  E. viridiflorus var. *canus*, p. 186
4. Not as above; either spines are fewer, or yellow color-forms present, or seedlings normal, or flowers varying to red/brown; distribution not in the Solitario (5).

5(4). Central spines 5–12 or fewer [will also key under first lead 1 above], relatively thin and flexible, acicular, with smallish-bulbous bases; radial spines 30–45, appressed or spreading, slender, flexible, usually terete at the base; (in part when spines are numerous)
  E. viridiflorus var. *chloranthus*, p. 175
5. Not as above (6).

6(5). Flowers greenish-brown; seedlings with white, hairlike spines and not stiff spines in areoles; stems usually with sharply defined horizontal bands of darker/lighter spines; yellow spine-color phase present in every population; in Marathon Basin

E. viridiflorus var. *neocapillus*, p. 180

6. Not as above; horizontal bands of contrasting spine colors vague or none (7).

7(6). Flowers dull green or greenish-brown; stems with yellow, white, or red-tipped spines; higher elevations in Davis Mts and at Cattail Falls, Chisos Mts

E. viridiflorus var. *weedinii*, p. 183

7. Flowers reddish to reddish-brown; stems with reddish, brownish, white, or black spines, yellow spine-color phase absent except on W slope of Chisos Mts (where in contact with var. *weedinii*); widespread in southern Brewster Co.

E. viridiflorus var. *russanthus*, p. 177

Echinocereus viridiflorus var. cylindricus
Small-Flowered Hedgehog Cactus
PLATES 153, 154

No modern authors have given full specific rank to this taxon. The taxonomic "splitters" who recognize *E. chloranthus* at the level of species have either placed var. *cylindricus* there (see Powell and Weedin, 2004) or have left it as a variety of *E. viridiflorus* instead. The varietal epithet *cylindricus* is from the Greek *kylindros*, "cylinder," alluding to the characteristic stem shape in the taxon.

Distribution. Widespread in both rocky and alluvial substrates of the Davis Mts (Jeff Davis, Brewster, and Presidio counties), SE to the Glass Mts, S through Alpine to the northern Del Norte Mts, W to the Chinati Mts and Sierra Vieja, with outliers at Reagan Canyon (2,100 ft) in Brewster Co. and the Guadalupe Mts, Culberson Co. 4,000–5,800 ft. Also Eddy, Chavez, Lincoln, Otero, and Doña Ana counties, S NM. Mexico: NE Chihuahua. Map 53.

Vegetative Characters. Plants superficially resemble those of *E. dasyacanthus*, and these taxa may occur together at intermediate elevations in and near the Davis Mountains. The more brightly red-and-white–banded aspect of typical var. *cylindricus* usually allows easy distinction from a distance. Central spines are much more diagnostic: 0–3 (individually conspicuous) in a vertical row in var. *cylindricus* and 2–5 or more (individually inconspicuous), either very short or spreading in all directions, in *E. dasyacanthus*. Radial spines about the same number in var. *cylindricus* and *E. dasyacanthus* (14–25), but in var. *cylindricus* radials more distinctly pectinate, 8–11 mm long, very variable in color but usually having relatively brighter, more distinctly purplish-red tips.

Stems of var. *cylindricus* usually unbranched except in relatively old age (then up to 10 or more branches), to 20 cm long and 5–8 cm in diameter. Ribs number 12–17. Largest, "main" central spine in each areole 0.6–2 cm long. Central spine

Plate 153. *Echinocereus viridiflorus* var. *cylindricus* (small-flowered hedgehog cactus), Davis Mts, Jeff Davis Co., TX; cultivated.

Plate 154. *Echinocereus viridiflorus* var. *cylindricus* (small-flowered hedgehog cactus), 9 mi SE of Alpine, Brewster Co., TX; mature fruits.

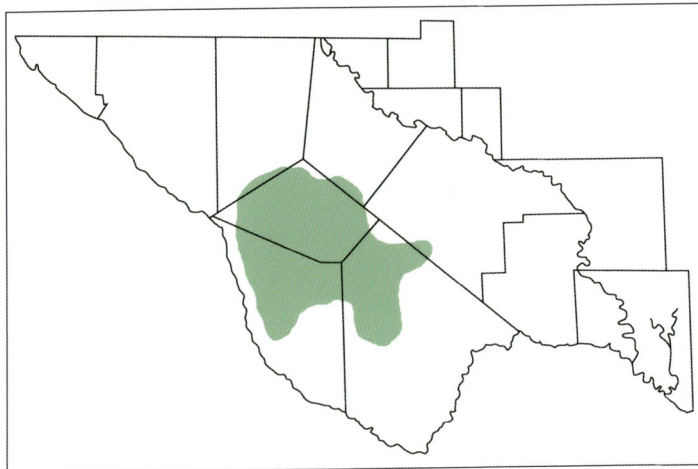

colors gray or yellowish at base with reddish distal halves, or main central may be entirely whitish to yellow. Radial spines 14–23 per areole, longest 8–11 mm, tan, yellowish, or reddish, or yellowish proximally and red on the distal half, or whitish with an ashy-white coating and red tips. Individual plants in the Davis and Guadalupe mountains have all-yellow radial spines and no central spines.

Flowers. Flowering Apr–May. As in all other varieties of *E. viridiflorus*, often numerous flowers are produced at midstem, or between midstem and apex. Flowers funnelform, usually not opening fully, varying in color from amber or sulfur-yellow to greenish-brown, reddish-brown, or carmine. Flowers 2–3.3 cm long, filaments 7–8 mm long, yellow anthers 0.8–1.5 mm long. Style green, 7–16 mm long, supporting ca. nine green stigma lobes 1.5–3 mm long. Flowers unscented, (mostly if not always) in contrast with the lemony scent of *E. davisii*, *E. viridiflorus* var. *viridiflorus*, var. *correllii*, and other taxa in this group.

Fruits. Fruits mostly remain dark green, but those of some plants may turn dull red; somewhat fleshy with sugary pulp, oval to nearly round, 0.9–1.7 cm long, 0.8–1.2 cm in diameter, indehiscent. Seeds about 1–1.2 mm long.

Full Name and Synonyms. *Echinocereus viridiflorus* var. *cylindricus* (Engelm.) Engelm. ex Rümpler. *E. standleyi* Britton & Rose; *E. viridiflorus* Engelm. var. *standleyi* (Britton & Rose) Orcutt.; *E. chloranthus* (Engelm.) F. A. Haage var. *cylindricus* (Engelm.) N. P. Taylor; *E. chloranthus* (Engelm.) Hort. F. A. Haage subsp. *cylindricus* (Engelm.) W. Blum and Mich. Lange.

Echinocereus viridiflorus var. correllii
Correll's Green-Flowered Hedgehog Cactus
PLATES 155, 156

The type locality of *E. viridiflorus* var. *correllii* is on private land near Fort Peña Colorado, ca. four miles southwest of Marathon. The varietal epithet honors D.

Plate 155. *Echinocereus viridiflorus* var. *correllii* (Correll's green-flowered hedgehog cactus), Marathon Basin, Brewster Co., TX.

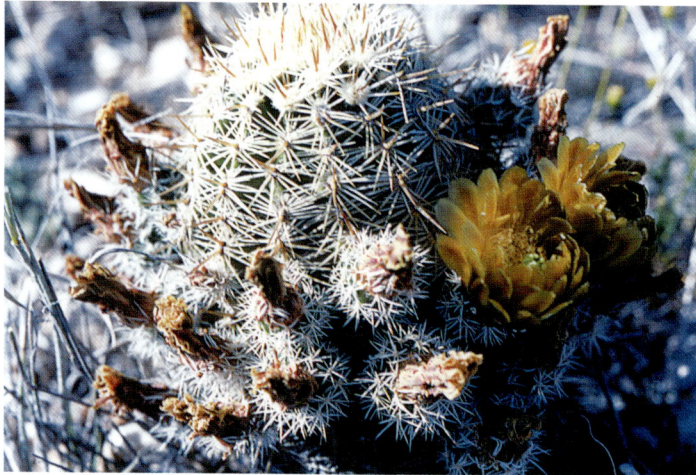

Plate 156. *Echinocereus viridiflorus* var. *correllii* (Correll's green-flowered hedgehog cactus), Marathon Basin, Brewster Co., TX; mature fruits.

S. Correll, a famous Texas botanist who accompanied L. D. Benson in collecting the holotype.

Distribution. Endemic to the Trans-Pecos. Caballos Novaculite, rocky hills and slopes, and in alluvial grassland with novaculite rubble, near Marathon, NE Brewster Co., TX; also in limestone at a site N of Fort Stockton. 3,000–4,000 ft. Map 54.

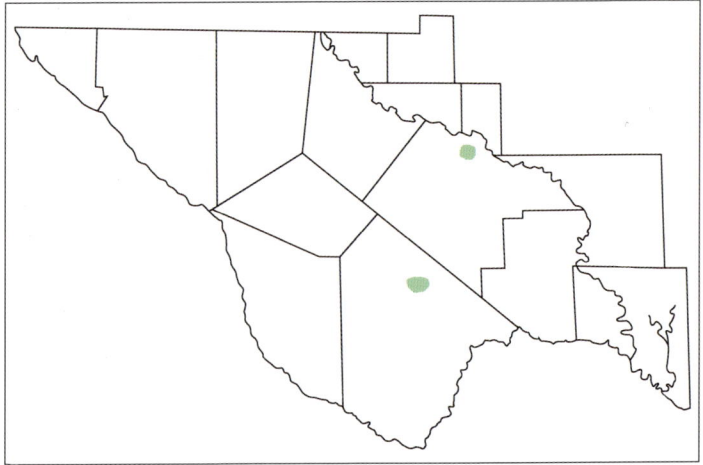

Map 54. *Echinocereus viridiflorus* var. *correllii* (Correll's green-flowered hedgehog cactus).

Vegetative Characters. In habit, very similar to var. *cylindricus*. In vegetative characters, var. *correllii* is best distinguished from var. *cylindricus* by greenish-yellow and ashy-white horizontal bands on the stems, or a yellowish aspect of the stems, and by characteristically higher number, 20–27, of radial spines. In var. *correllii*, 1–2 central spines, and if more than one central present, one much longer (0.7–1.3 cm) than the other(s), as in var. *cylindricus*, and yellowish or white, often with a reddish tip.

Stems of var. *correllii* with 14–17 ribs. Radial spines pectinate and appressed. Longest radial spines 8–9.5 mm long, in lateral position in the areole, and the upper ones are shortest. Radial spines yellowish or ashy-white, seemingly with a thin whitish surface layer, with darker yellow or reddish tips.

Flowers. Flowering Mar–May. Lemon-scented flowers of var. *correllii* are typically greenish-yellow but may vary to (1) more greenish than yellow, (2) golden-bronze, (3) yellowish-brown, or (4) pale reddish-brown in a greenish-yellow background. Campanulate corollas open widely compared to those of var. *cylindricus;* tepals of var. *correllii* often reflexed at maximum expansion, as in *E. davisii.* First-day flowers 1.5–2 cm wide, 3.5–3.8 cm long, and 2.5–3 cm across when fully open. Anthers light yellow. Style greatly exserted from the stamens, with 6–10 green stigma lobes.

Fruits. Fruits at maturity remain green or may turn dull red, split open along longitudinal lines, and quickly dry. Fruits ovate, 1–1.3 cm long, 0.7–1 cm in diameter. Black seeds ca. 1 mm long.

Full Name and Synonym. *Echinocereus viridiflorus* var. *correllii* L. D. Benson. *E. viridiflorus* Engelm. subsp. *correllii* (L. D. Benson) W. Blum & Mich. Lange.

Echinocereus viridiflorus var. chloranthus
Western Green-Flowered Hedgehog Cactus
PLATES 157, 158

Typical var. *chloranthus* usually is distinguished by a white, curved, lower central spine; otherwise, it is essentially a

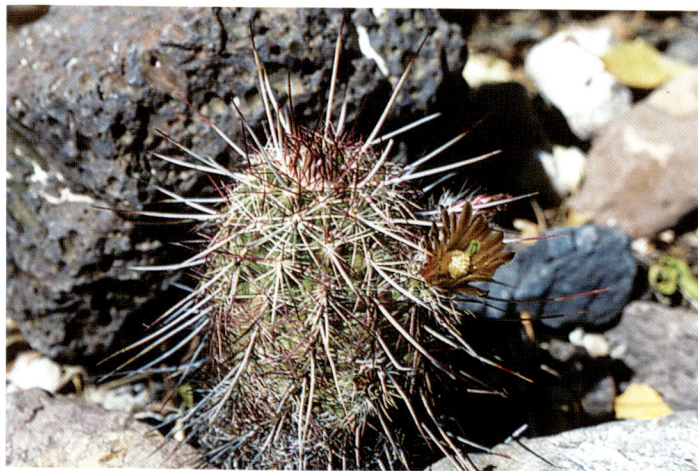

Plate 157. *Echinocereus viridiflorus* var. *chloranthus* (western green-flowered hedgehog cactus), **Jeff Davis Co., TX**; putative hybrid with var. *cylindricus*.

Plate 158. *Echinocereus viridiflorus* var. *chloranthus* (western green-flowered hedgehog cactus), **Hudspeth Co., TX**; mature fruit.

repository for plants intermediate between var. *cylindricus* (fewer central spines) and var. *russanthus* (more central spines). Such plants are abundant from the Franklin and Hueco mountains and adjacent New Mexico east around the Guadalupe Mountains, south through the Delaware Mountains and Rustler Hills to the Apache Mountains (and at least a few miles east of the Apache Mountains), where populations include individuals resembling var. *cylindricus*, and back west through the Baylor Mountains and the Sierra Diablo and to near Sierra Blanca. See Powell and Weedin (2004) for biosystematic discussion concerning var. *chloranthus*. The epithet *chloranthus* is meant to be descriptive of the flowers, after the Greek *chloros*, "green" or "greenish-yellow," and *anthos*, "flower," although the flower colors in this taxon vary from yellowish-green to brown or reddish-brown.

Distribution. On both igneous and sedimentary substrates; in habitats that are more arid than is typical for the related var. *cylindricus*; El Paso Co. E through

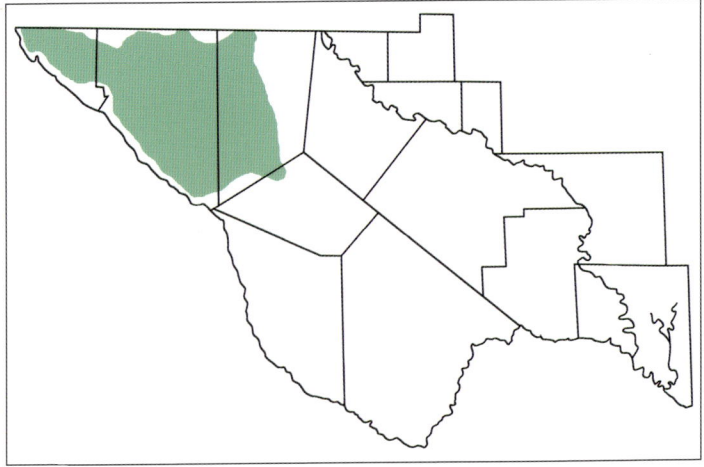

Map 55. *Echinocereus viridiflorus* var. *chloranthus* (western green-flowered hedgehog cactus).

Hudspeth Co. and Culberson Co. 3,900–5,300 ft. Also southern NM, Doña Ana, Otero, Lincoln, and Eddy counties, but not in higher elevations of the Sacramento Mts. Mexico: barely documented from adjacent Chihuahua, near the Rio Grande. Map 55.

Vegetative Characters. Has the general habit of var. *cylindricus* except that stems of var. *chloranthus* are often relatively slender and the plants have a more bristly aspect, the effect of longer and/or more numerous central spines. At least one central spine, usually the one lowest in areole, is white, although often with a reddish tip. Other centrals usually reddish. Lower central usually longest, 2–4.3 cm, typically longer than the centrals of var. *cylindricus*. In all, 2–6 central spines in var. *chloranthus;* when 2 or 3, they are positioned in a single vertical row as in var. *cylindricus;* otherwise, they spread in all directions. Radial spines in var. *chloranthus* 15–23, about the same number as in var. *cylindricus* but usually longer, to 1.2–1.4 cm. Stems may show horizontal bands of more white and more reddish spines, as in var. *cylindricus*, except that this character in

var. *chloranthus* is more obscured by the central spines. Radial spine color may be different in separate areoles, with all or almost all ashy-white, almost all reddish or reddish-black, or some in each areole reddish and some whitish, or spines multicolored, cream, whitish, or gray at the base with reddish distal portions or tips, as may be the case in var. *cylindricus*. Variety *chloranthus* is easily distinguished from var. *russanthus* by consistently fewer central spines and radial spines, by the larger caliber of spines, and by the appressed angle of radial spines.

Flowers. Flowering Mar–May. Flowers, in overall aspect, dark green to yellowish-green or greenish-brown. More specifically, inner tepals described as brownish-green or rusty distally and green proximally, or as having dark green midlines with margins lighter green and suffused with brown or olive-yellow. Almost chocolate-brown and reddish-brown flower colors have also been reported. Flower morphology in var. *chloranthus*, var. *cylindricus*, and var. *russanthus* is virtually identical: funnelform, 2–3.4 cm long, with styles 1.2–2 cm long, usually eight green stigma lobes, pale yellow anthers on filaments 5–8 mm long.

Fruits. Mature fruits of var. *chloranthus* green (or dull red), almost round, with areoles that fall away, leaving a glistening fruit wall, at least in plants from the Franklin Mountains. Fruits indehiscent. Black seeds 1–1.3 mm long and broad, papillate.

Full Name and Synonyms. Echinocereus viridiflorus var. *chloranthus* (Engelm.) Backeb. *E. chloranthus* (Engelm.) F. A. Haage; *E. viridiflorus* subsp. *chloranthus* (Engelm.) N. P. Taylor.

Echinocereus viridiflorus var. russanthus
Rusty Hedgehog Cactus
PLATES 159–62

This distinctive taxon remained submerged under *E. chloranthus* until it was described by Weniger (1969) as *E. russanthus*. It consistently differs from everything else formerly included in var. *chloranthus* and is geographically isolated from those, in the southern Big Bend region. The epithet *russanthus* is after the Latin *russus*, "reddish," and Greek *anthos*, "flower," alluding to the characteristic flower color of var. *russanthus*.

Distribution. Endemic to the Trans-Pecos, arid mountains, desert grasslands, among desertscrub, mostly igneous but also in limestone substrates, and on the southernmost novaculite hills of the Marathon Basin, extending upslope to mesic habitats in the Chisos Mts. 2,300–6,400 ft. Map 56.

Vegetative Characters. Stems cylindrical, often branched at the base, 5–25 cm long, 4–9 cm in diameter, with 12–18 ribs. Bristlelike central spines flexible, almost as slender as the radial spines. Typically 9–12 central spines, the longest 1–3.7 cm, and 30–45 radial spines, the lateral ones 0.8–1.6 cm long. Diffusely spreading spines typically reddish-brown to reddish-black or white with grayish-purple (or reddish-brown) tips; rarely, spines whitish-yellow. Different populations of var. *russanthus* vary in spine length and color (e.g., see Plate 162; Powell and Weedin, 2004).

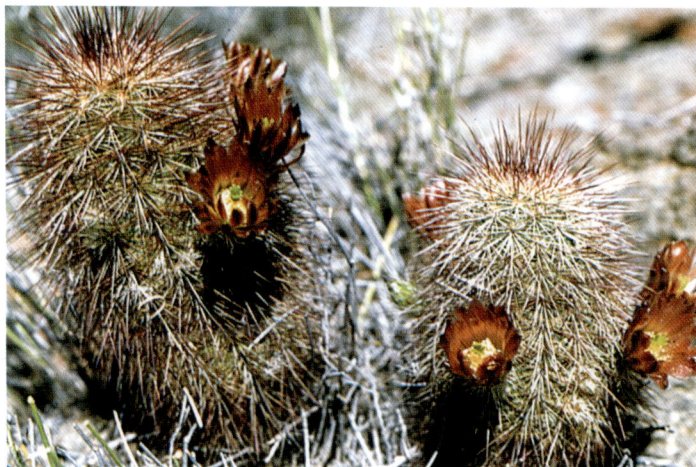

Plate 159. *Echinocereus viridiflorus* var. *russanthus* (rusty hedgehog cactus), Christmas Mts, Brewster Co., TX.

Plate 160. *Echinocereus viridiflorus* var. *russanthus* (rusty hedgehog cactus), W Chisos Mts, Brewster Co., TX.

Plate 161. *Echinocereus viridiflorus* var. *russanthus* (rusty hedgehog cactus), S Brewster Co., TX; mature fruits; cultivated.

Plate 162. *Echinocereus viridiflorus* cf. var. *russanthus* (rusty hedgehog cactus); Rosillos Mts, Brewster Co., TX.

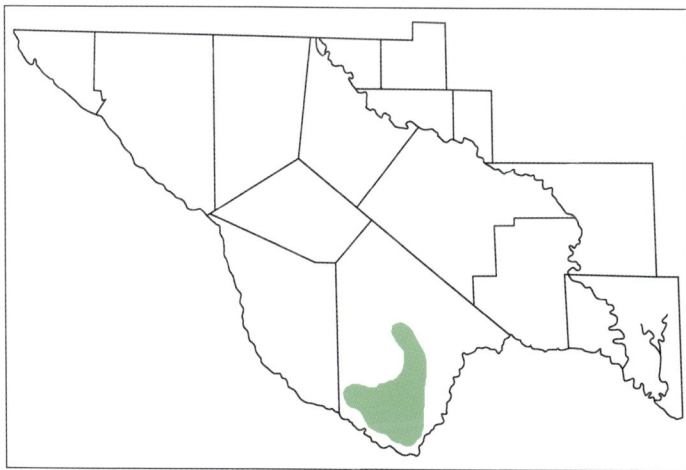

Map 56. *Echinocereus viridiflorus* var. *russanthus* (rusty hedgehog cactus).

Flowers. Flowering Feb–Apr. Flowers rather narrowly funnelform, 2–3.5 cm long, most like those of var. *chloranthus* or var. *cylindricus*. Predominant flower color appears to be rusty-red or russet-red. Flower centers usually greenish, which may account for one report of greenish-brown flowers in the variety. Filaments 5–6 mm long, with pale yellow anthers ca. 1 mm long. Yellowish style 1.3–1.4 cm long, supporting 8–10 green stigma lobes 2.5–3.5 mm long.

Fruits. Fruits oval to almost spherical, 1–1.3 cm long, green or dull red or reddish-brown, succulent, not sugary, indehiscent. At least in some plants fruits remain green at maturity and may not turn reddish, except that they develop a tinge of red or brown. Fruits partially obscured by clusters of slender white spines, 10–20 in each areole. Seeds ca. 1 mm long and broad.

Full Name and Synonyms. Echinocereus viridiflorus var. *russanthus* (Weniger) A. D. Zimmerman, comb. nov. (forthcoming). E. *russanthus* Weniger; E. *chloranthus* F. A. Haage var. *russanthus* (Weniger) Lamb ex G. Rowley.

Echinocereus viridiflorus var. neocapillus
Weniger's Small-Flowered Hedgehog Cactus
PLATES 163, 164

Variety *neocapillus* is one of five remarkable Marathon Basin endemic taxa (counting *Coryphantha hesteri*). All of the Marathon Basin endemics are either restricted to or most common upon Caballos Novaculite substrates. The most remarkable feature of var. *neocapillus* is that the seedlings produce only long, soft, hairlike white spines in the areoles (Plate 165). The seedling "hairs" are the basis for the varietal epithet, after the Greek *neos*, "new" or "young," and the Latin *capillas*, "hair."

Plate 163. *Echinocereus viridiflorus* var. *neocapillus* (**Weniger's small-flowered hedgehog cactus**), Marathon Basin, Brewster Co., TX; cultivated.

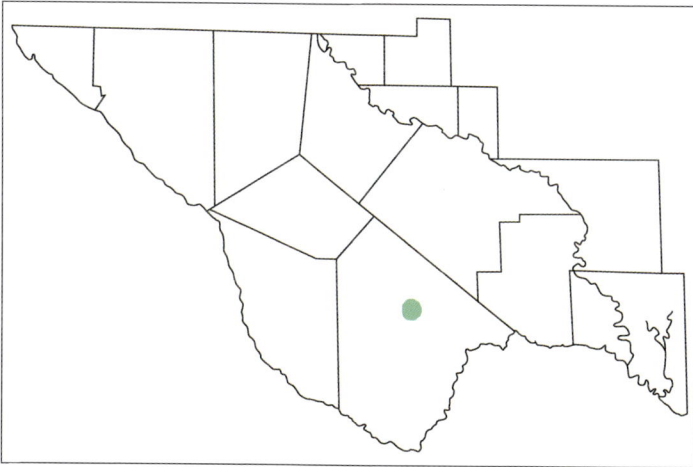

Distribution. Endemic to the Trans-Pecos, on certain novaculite hills of the Marathon Basin, from ca. 10 to 17 mi S of Marathon. 3,700–4,500 ft. Map 57.

Vegetative Characters. Cylindrical stems of var. *neocapillus* are usually single but occasionally branched, 8–25 cm long and 3–7 cm in diameter, with 12–16 ribs. In aspect, stems are pale yellowish-green or with horizontal bands of white spines alternating with yellowish, tan, or light brown spines. In larger stems, 5–7 or more annual bands are visible aboveground, and the previous years' growth is compressed and weathered at the base of the plant. White spines are mixed with some reddish or reddish and white. The horizontal banding characteristic in populations of var. *neocapillus* is much more prominent than the annual growth markings of var. *russanthus*.

In var. *neocapillus*, 5–11 slender,

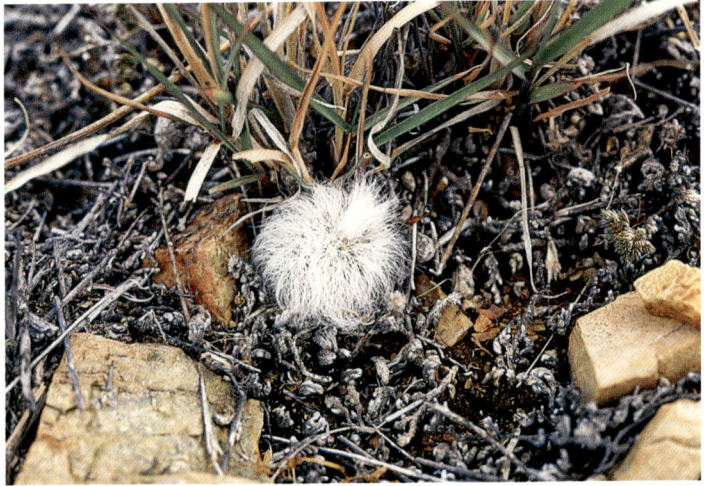

Plate 165. *Echinocereus viridiflorus* var. *neocapillus* (Weniger's small-flowered hedgehog cactus), Marathon Basin, Brewster Co., TX; "hairy" seedling.

Plate 166. *Echinocereus viridiflorus* var. *neocapillus* (Weniger's small-flowered hedgehog cactus), Marathon Basin, Brewster Co., TX; "hairy" juvenile.

spreading central spines, 1.2–2.0 cm long, yellow, often reddish distally or all reddish. Slender radial spines number 30–38, usually crowded and interlocking with spines from adjacent areoles. Radial spines 6–12 mm long, yellow or white, usually concolorous in the same areole.

Flowers. Flowering Mar–May. Flowers 2.8–3 cm long, 1.5–2 cm in diameter, cylindrical-funnelform, campanulate, or rarely subrotate with usually greenish-yellow, reflexed tepals. Flower color variation described as "yellowy," "greenish," "brown," or greenish-brown to bronze, and with greenish bases inside the flower with brownish distal portions of the inner tepals. Filaments 7–12 mm long; light yellow anthers ca. 1.2 mm long. Greenish-yellow style ca. 1.6 cm long, topped by 8–10 green stigma lobes, 2.5–4 mm long.

Plate 167. *Echinocereus viridiflorus* var. *neocapillus* (Weniger's small-flowered hedgehog cactus), Marathon Basin, Brewster Co., TX; young adult.

Flowers resemble those of var. *correllii*, but lack the lemon scent of var. *correllii* and only rarely open as widely.

Fruits. Fruits at maturity oval to subglobose, 1.2–1.4 cm long, 0.7–0.9 cm in diameter, covered with clusters of white spines, 20–25 in each areole. Mature fruits remain green or turn dull purplish-red, splitting longitudinally and quickly drying out. Black seeds are 1–1.2 mm long, finely papillate.

Other Distinctions. In each areole of seedlings and juvenile plants there are about 40 flexuous "hairs," to 3 cm or more long, in place of normal spines. In the field, at first glance, seedlings resemble a clump of seeds of milkweed (*Asclepias* L.) among novaculite rocks or on mats of *Selaginella*.

The tuft of flexible, white spines may be the only visible part of smaller seedlings, which are small enough to be completely obscured by the hairs. Older juveniles with stems 2–5 cm high (Plate 166) typically are still covered with white hairs, but at about this size true spines form at the stem apex. The juvenile hairs may persist at the base of semimature plants (Plate 167). See Powell and Weedin (2004) for biosystematic discussions concerning var. *neocapillus*.

Full Name and Synonyms. Echinocereus viridiflorus var. *neocapillus* (Weniger) Leuck ex A. D. Zimmerman, comb. nov. (forthcoming). *E. chloranthus* (Engelm.) F. A. Haage var. *neocapillus* Weniger; *E. neocapillus* (Weniger) W. Blum & Mich. Lange.

Echinocereus viridiflorus var. weedinii
Weedin's Small-Flowered Hedgehog Cactus

PLATES 168, 169

Echinocereus viridiflorus var. *weedinii* was recognized as distinct by Kolle (1978), who discussed the taxon under a provisional name referring to Mount Livermore, the primary locality, and was reported but never formally published as var. *weedinii* by Leuck (1980). Both authors indicated that var. *weedinii* was geographically widespread, disjunct between higher elevations of the Davis, Chinati, and Chisos mountains. The epithet *weedinii* is after James F. Weedin, second author of this guide, who also recognized the taxon as distinct at the time Kolle was studying the *E. viridiflorus* complex (Powell and Weedin, 2004).

Distribution. Trans-Pecos endemic. Exposed rocky slopes and ridge crests, higher elevations in the Davis Mts (Jeff Davis Co.), Chinati Peak (Presidio Co.) and western Chisos Mts (Brewster Co.). 5,300–8,382 ft. (summit, Mt Livermore).

Similar plants from a few high peaks and grasslands in SW NM (Catron, Luna, Sierra, Doña Ana, and Socorro counties), treated either as var. *weedinii* (Zimmerman et al., forthcoming) or "*Echinocereus chloranthus* subsp. *rhyolithensis*" (Blum et al., 1998). Map 58.

Vegetative Characters. Stems solitary or branched, cylindroid, tapering slightly at the apex, 7–20 cm long, 3.5–6 cm in diameter, with 15–17 ribs. Characteristically, covered with golden-yellow spines (at Timber Mountain, the type locality), both centrals and radials, but individuals with red-tipped white spines are common within the population on Mount Livermore and elsewhere in the range. Compared with var. *russanthus*, var. *weedinii* has fewer central spines (5–12; 1–2.7 to ca. 4 cm long), and the fewer radial spines (20–35; 0.7–1.2 cm long) are more appressed; sometimes the spines are slightly larger in caliber than is typical of var. *russanthus*. In the Davis and Chinati mountains var. *weedinii* is mostly restricted to the highest peaks, where it tends to replace the much more common

Plate 168. *Echinocereus viridiflorus* var. *weedinii* (Weedin's small-flowered hedgehog cactus), Davis Mts, Jeff Davis Co., TX; cultivated.

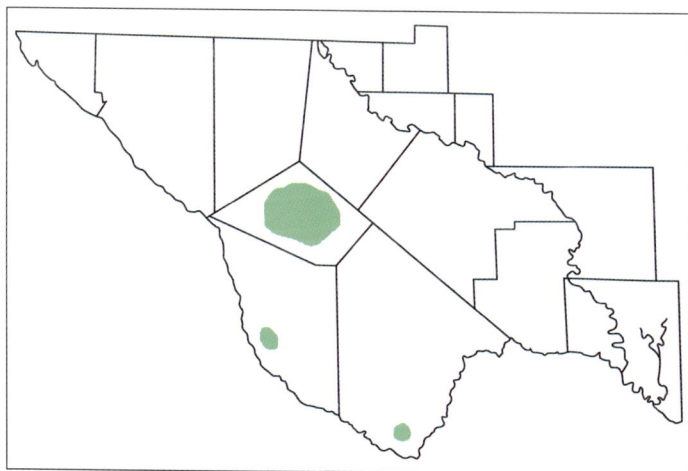

Map 58. *Echinocereus viridiflorus* var. *weedinii* (Weedin's small-flowered hedgehog cactus).

and widespread *E. viridiflorus* var. *cylindricus*.

Flowers. Flowering Apr–May. Typically distinguished from var. *russanthus* by yellow-green inner tepals and presence of floral scent, whereas in var. *russanthus*

the flowers are reddish. Flower color of var. *weedinii* varies widely from greenish to olive-yellow, yellow-orange, brownish-green, and even reddish-brown, as described by various researchers. A potted specimen from the Davis Mountains produced yellow-green flowers, while a separate plant from the same locality formed greenish-brown to rusty-brown flowers. Flowers also have been observed to change color over a period of several days, from greenish-brown or rusty-brown, when first open, to yellow-green. Flowers cylindroid to campanulate, 1.4–3 cm long, 1.5–3 cm in diameter. Style 1–2 cm long, with 6–8 green stigma lobes. Filaments 5–10 mm long, white-greenish, supporting yellow anthers ca. 1 mm long.

Fruits. Fruits at maturity reddish (dark purple) or rarely green, globose to ovoid, 5–8 mm long, 5–7 mm in diameter, fleshy with nonsugary pulp, indehiscent. Black seeds 1–1.2 mm long, finely papillate.

Full Name and Synonym. Echinocereus viridiflorus var. *weedinii* (Leuck ex W. Blum & Mich. Lange) A. D. Zimmerman, comb. nov. (forthcoming). *E. russanthus* Weniger subsp. *weedinii* W. Blum & Mich. Lange.

Echinocereus viridiflorus var. canus
Graybeard Cactus
PLATES 170, 171

This remarkable white-spined taxon was discovered in 1984 by James Jeff Clark (see Powell and Weedin, 2004). The varietal epithet, *canus*, is a reference to the whitish aspect of the plants, after the Latin *canus*, "grayish-white."

Distribution. Endemic to the Trans-Pecos, exposed Caballos Novaculite ridge crests and mostly S-facing slopes with *Selaginella;* also on chert and igneous rock; W-facing aphyric rhyolite slope; and on "sandstone"; inside the Solitario Dome, a rare form of laccolith and caldera (Hardy, 1997), in SE Presidio Co., TX. 4,400–4,800 ft. Map 59.

Vegetative Characters. Sexually mature stems ovoid-cylindroid, 6–15 cm long, 3.5–6 cm in diameter, with ca. 14–16 ribs. Central spines 8–15, usually 9–13, in each

Plate 170. *Echinocereus viridiflorus* var. *canus* (graybeard cactus), Solitario, Presidio Co., TX; cultivated.

Plate 171. *Echinocereus viridiflorus* var. *canus* (graybeard cactus), Solitario Uplift, Presidio Co., TX; mature fruits.

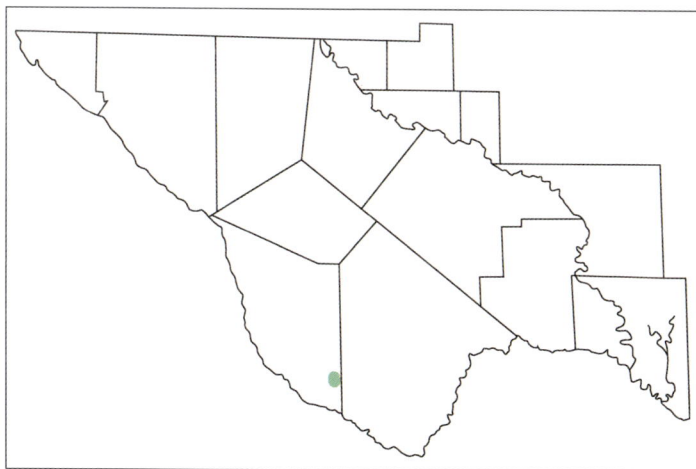

Map 59. *Echinocereus viridiflorus* var. *canus* (graybeard cactus).

areole, slender, flexible, very fragile, the longest per areole 1.7–2.5 cm, spreading in all directions. Very slender radial spines 30–48, maximum of 5–8 mm long, mostly appressed to the stem. Both central and radial spines white, or one or more central spines in each areole reddish (usually on distal half or only at tip), particularly in upper and apical areoles.

Flowers. Flowering Mar–Apr. Flow-

Plate 172. *Echinocereus viridiflorus* var. *canus* (graybeard cactus), Solitario Uplift, Presidio Co., TX; seedlings; cultivated.

ers vary in saturation from light green to bright golden-green, perhaps more nearly pure green than those of var. *viridiflorus* itself. Lemon-scented flowers 2–3.4 cm or more long, at first funnelform, then in a hot greenhouse usually opening widely. Filaments 5–8 mm long, pale yellow anthers ca. 1 mm long. Style slender, 1.5–1.7 cm long, with usually 7–8 dark green stigma lobes, 2.5–3 mm long.

Fruits. Fruits not yet well characterized, but all of ours have remained green. Fruits ovoid to oblong, sometimes splitting open longitudinally on at least one side and quickly drying. Fruits partially obscured by slender white spines, 18–20 or more in each areole. Black seeds subpyriform, 1–1.3 mm long, finely papillate except at the truncate base.

Other Distinctions. Seedlings of var. *canus* less than 1–3 cm tall produce only flexible, hairlike white spines, 0.5–1 cm or more long (Plate 172). The hairs are straight, wavy, or curved and number ca. 40 per areole. They persist for several years as a dense "skirt" (1–2 cm wide) of white hairs on the lower portion of the

stem. They become gray with age until the plants are 3–8 cm high, but usually they are lost on older plants. White and red normal spines are first produced in the apical areoles when juvenile stems are 1–3 cm long.

Full Name. Echinocereus viridiflorus var. *canus* A. M. Powell and J. F. Weedin.

Other Echinocereus Species

Extralimital taxa of *Echinocereus* relevant to the Trans-Pecos flora and not discussed elsewhere in the Trans-Pecos treatment of the genus are summarized below.

Echinocereus fendleri var. kuenzleri
Kuenzler's Hedgehog Cactus

The status of var. *kuenzleri* requires further study, especially with respect to its populational integrity. It is possibly an occasional spine form that is not distinct from var. *fendleri*. Variety *kuenzleri* superficially resembles *E. triglochidiatus* var. *triglochidiatus*.

Distribution. Variety *kuenzleri* is not

known to occur in the Trans-Pecos, but the taxonomic implications of its possible existence there were discussed by Taylor (1985). Its restricted distribution just north of the Trans-Pecos in the Sacramento and Guadalupe mountains (Otero and Eddy counties) of southern NM in woodlands at ca. 5,000 ft suggests that it eventually might be found in the Trans-Pecos.

Vegetative Characters. This variety and its Chihuahuan counterpart, *E. hempelii,* are distinguished by unusually small spine clusters, with 2–6 radials up to 2.5 cm long, thick, angular, flattened, chalky-white, and often twisted, and no central spine (or rarely one central), thus revealing more of the dark green, soft, flabby stem surface and the very tuberculate ribs as compared with those of var. *fendleri.*

Flowers. According to Taylor, var. *kuenzleri* produces flowers to 11 cm in length and diameter (based on Chihuahuan material?), the maximum flower size for the species, and is therefore one of the most desirable ornamental types in the genus.

Full Name and Synonym. Echinocer-eus fendleri var. *kuenzleri* (Castetter, P. Pierce, & K. H. Schwer.) L. D. Benson. *E. kuenzleri* Castetter et al.

Echinocereus viridiflorus var. viridiflorus
Green-Flowered Hedgehog Cactus
PLATE 173

Distribution. South-central NM north to SD and E to the TX Panhandle in Oldham and Potter counties.

Vegetative Characters. Distinguishing features of var. *viridiflorus* include relatively short, solitary or branching stems 3–9 cm long; 10–15 ribs; spines white or reddish, frequently white at the base and reddish distally, with predominant spine colors often in horizontal bands encircling the stem, central spines 0–1, radial spines 12–18, pectinate and appressed.

Flowers. Greenish-yellow flowers, bell-shaped to rotate, having a strong lemony scent; and fruits green, dry, and dehiscent along longitudinal lines.

Full Name. Echinocereus viridiflorus var. *viridiflorus.*

Plate 173. *Echinocereus viridiflorus* var. *viridiflorus* (green-flowered hedgehog cactus), N NM; cultivated.

"Echinocereus carmenensis"

Distribution. Foothills of the Sierra del Carmen, Coahuila, Mexico.

Full Name. "*Echinocereus carmenensis*" W. Blum, Mich. Lange, & Scherer.

Echinocereus "russanthus subsp. fiehnii"

Distribution. Endemic to the Sierra del Nido, central Chihuahua, Mexico.

Full Name. *Echinocereus* "*russanthus* subsp. *fiehnii*" (Trocha) W. Blum & Mich. Lange.

Echinocereus "chloranthus subsp. rhyolithensis"

Distribution. The counterpart of var. *weedinii* in southwestern NM, but lacks the yellow-spined color-form.

Full Name. *Echinocereus* "*chloranthus* subsp. *rhyolithensis*" W. Blum & Mich. Lange.

Echinocereus milleri

Distribution. A population in Coke Co., TX, N of Robert Lee, with features of both var. *correllii* (which it obviously is not) and var. *neocapillus* (which it obviously is not, but it has the "hairy" seedlings). Map 60.

Full Name. *Echinocereus milleri* W. Blum, H. Kuenzler, & T. Oldach.

Echinocereus pentalophus
Alicoche, Lady-Finger Cactus

PLATES 174, 175

Echinocereus pentalophus from South Texas is not cold-hardy enough to be

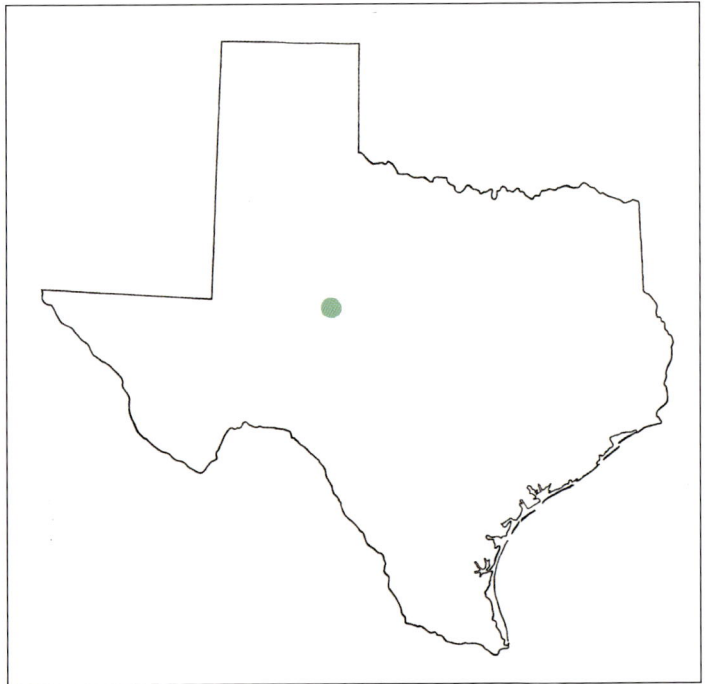

Map 60. *Echinocereus milleri* (small-flowered hedgehog cactus).

grown outside in most of the Trans-Pecos. The epithet *pentalophus* refers to the five-ribbed stems.

Distribution. The TX entity of this species is *E. pentalophus* var. *procumbens* (Engelm.) P. Fourn., found in S TX brush-lands (Map 61). Mexico: more widespread in the E to NE, reaching elevations of ca. 4,300 ft; an endemic variety, *E. penta-lophus* var. *leonensis* (Mathsson) N. P. Taylor occurs in SE Coahuila and Nuevo León, Mexico; and typical *E. pentalophus,* with larger stems, grows in tropical eastern Mexico.

Vegetative Characters. Plants low clumps to ca. 1 m in diameter. Stems branched, 20–65 cm long, 1–6 cm in diameter, erect or prostrate, reddish- to yellowish-green. Ribs 4–5. Central spine 0–1, less than 1–2 cm long, porrect to

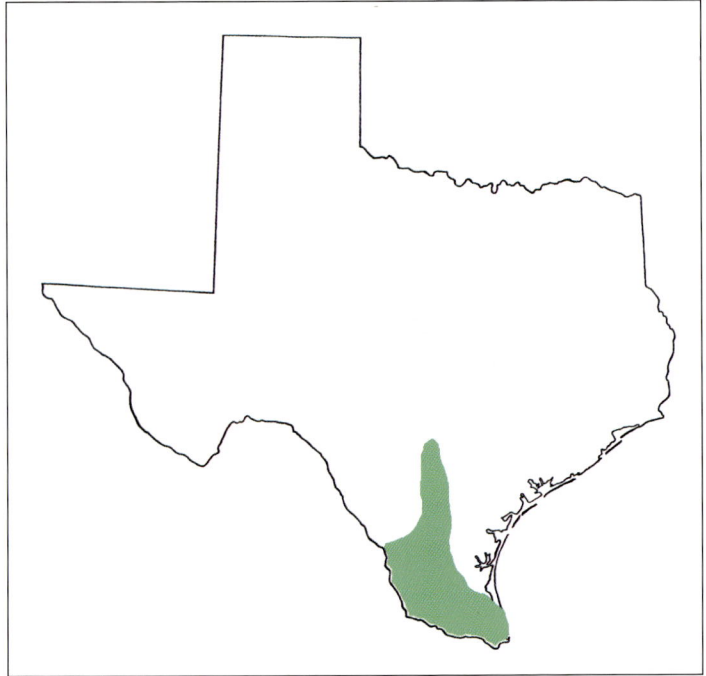

Map 61. *Echinocereus pentalophus* (Alicoche, lady-finger cactus).

ascending, yellowish to brown. Radial spines 3–7, 0.2–2 cm long, whitish to yellowish, upper ones very small.

Flowers. Flowering Mar–Apr. Flowers 8–10 cm long, 10–15 cm in diameter, bright pink to light magenta, the throat white to yellow.

Fruits. Fruit ovoid, green, to 1.9 cm long, covered by brownish bristlelike spines and white wool, irregularly dehiscent.

Full Name and Synonym. Echinocereus pentalophus (DC.) Lem. *Cereus pentalophus* DC.

Echinocereus berlandieri
Berlandier's Alicoche
PLATES 176, 177

Both *E. berlandieri* and *E. pentalophus* are similar in general appearance. When in flower, the species are easily distinguished from each other by the dark throat in *E. berlandieri* and the white to yellow throat in *E. pentalophus. Echinocereus berlandieri* is named after the French explorer Jean Louis Berlandier, who traveled in Mexico and settled in Texas in the early 1830s.

Distribution. Found in brushland, commonly mesquite thickets, and grassland, S TX, "near the Nueces River" and near the Rio Grande, Webb Co. SE to Cameron Co. (Map 62). Mexico: reportedly also in adjacent Tamaulipas and Nuevo León.

Vegetative Characters. Plants forming low clumps to ca. 1 cm in diameter. Stems branched, 5–65 cm long, 1.5–3 cm in diameter, dark green or purple-tinged, prostrate, apexes turned upward. Ribs 5–7, low, tubercles conspicuous. Central spines

Plate 176. *Echinocereus berlandieri* (Berlandier's alicoche), **S TX**; cultivated.

Plate 177. *Echinocereus berlandieri* (Berlandier's alicoche), **S TX**; nearly mature fruit; cultivated.

1–3, 2.5–5.1 cm long, porrect to deflexed, yellowish to brown; radial spines 6–9, to 1 cm or more long, whitish, the tips often brown.

Flowers. Flowers 6–9 cm long, 7–10 cm in diameter, purplish, the throat dark; anthers orange-yellow, filaments pale purplish; stigma lobes nine, green.

Fruits. Fruit obovoid, green (turning reddish; Taylor, 1988), 2–3 cm long, 1.5–1.8 cm in diameter, the areoles with spine clusters similar to those on the stems, deciduous.

Full Name and Synonym. *Echinocereus berlandieri* (Engelm.) Rümpler. *Cereus berlandieri* Engelm.

Map 62. *Echinocereus berlandieri* (Berlandier's alicoche).

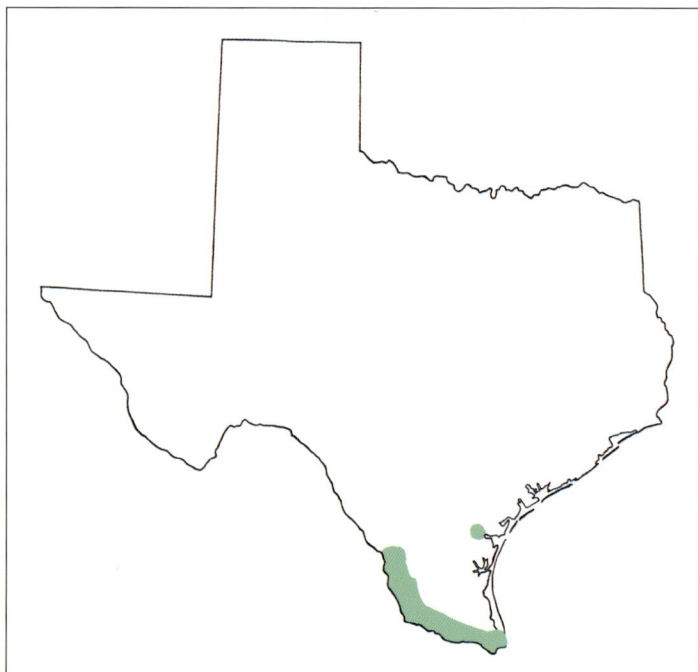

Echinocereus papillosus
Yellow-Flowered Alicoche

PLATES 178–80

Some authors have attempted to recognize an endemic variety, *E. papillosus* var. *angusticeps* (Clover) W. T. Marshall (Plates 179, 180), distinguished by smaller plants with 5–95 stems, these 4–8 cm long and 2–3 cm in diameter; restricted to Starr County and northern Hidalgo County in South Texas, growing under mesquite thickets.

Presumably, the specific epithet alludes to the conspicuous stem tubercles of *E. papillosus*. The varietal epithet *angusticeps* is a reference to the more slender stems of this taxon. *Echinocereus papillosus* is not cold-hardy enough to be propagated outside in most of the Trans-Pecos.

Distribution. Rio Grande plains, sandy loam, low elevations, from McMullen Co. S to near the Rio Grande (Map 63).

Vegetative Characters. Plants forming loose clumps of 2–95 stems. Stems mostly erect or leaning, to 4–20 cm long, 2–5 cm in diameter, deep green to brownish-green. Typical *E. papillosus* var. *papillosus* (Plate 178), is distinguished by clumps of up to 10–12 stems, to 20 cm long, 7 cm in diameter. Ribs 7–10, mostly divided into conspicuous tubercles to 9 mm long. One central spine, ca. 2 cm long, porrect, brownish to yellow, or with a brown base, yellow middle, and brown tip; radial spines 7–11, whitish to yellowish-brown, usually brown-based.

Flowers. Flowers 5–7 cm long, 6–10 cm in diameter, pale straw-yellow with orange-red throats.

Plate 178. *Echinocereus papillosus* var. *papillosus* (yellow-flowered alicoche), Jim Hogg Co., TX; cultivated.

Plate 179. *Echinocereus papillosus* var. *angusticeps* (yellow-flowered alicoche), S TX (photo by James H. Everitt).

Fruits. Fruit subglobose, greenish, covered with short bristles.

Full Name. *Echinocereus papillosus* Linke ex C. F. Först.

Plate 180. *Echinocereus papillosus* var. *angusticeps* (yellow-flowered alicoche), S TX.

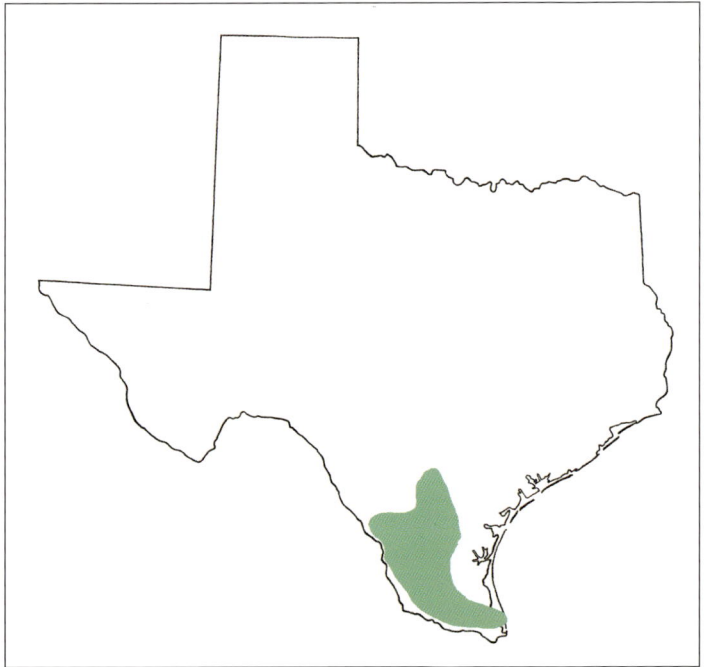

Map 63. *Echinocereus papillosus* (yellow-flowered alicoche).

Echinocereus poselgeri
Dahlia Cactus, Sacasil, Pencil Cactus

PLATES 181, 182

Outdoor plantings of *E. poselgeri* in 1996 at Alpine have survived the winter temperatures. Single large plants, properly supported, produce numerous large pink and white or magenta flowers. The specific epithet honors Dr. H. Poselger, the German cactus collector who conducted fieldwork in the southwestern United States and northern Mexico between 1849 and 1852.

Distribution. South TX, Cameron Co.

Plate 181. *Echinocereus poselgeri* (dahlia cactus); cultivated.

Plate 182. *Echinocereus poselgeri* (dahlia cactus); nearly mature fruits; cultivated.

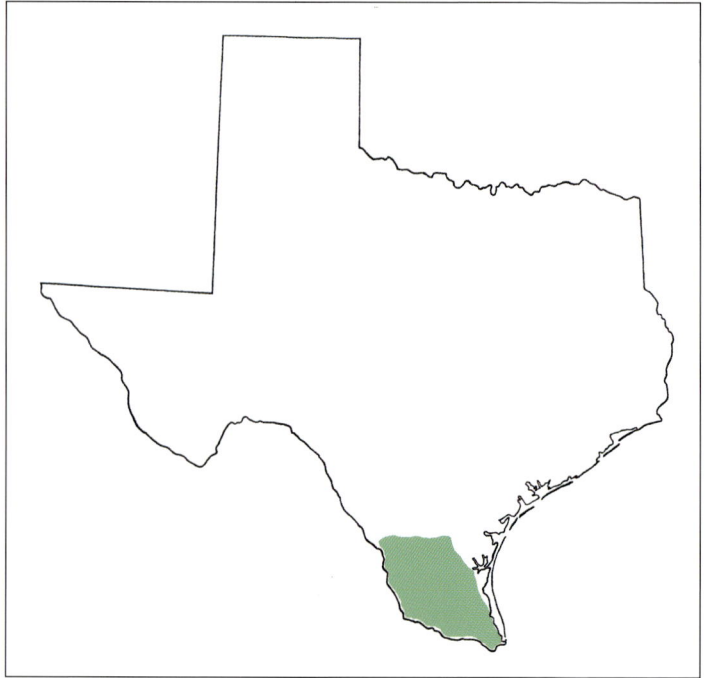

Map 64. *Echinocereus poselgeri* (dahlia cactus, sacasil, pencil cactus).

and western Hidalgo Co. N along the Rio Grande to Laredo (Map 64), sandy soils in brushlands. Mexico: Tamaulipas, Nuevo León, and Coahuila, extending into the eastern Chihuahuan Desert from eastern Coahuila to eastern San Luis Potosí, to 3,500 feet.

Vegetative Characters. Plants clambering through shrubs, arising from fascicled (one to several), tuberous roots 5–10 cm long, 2.5–5 cm in diameter. Stems branched, 30–120 cm long, 0.7–1.5 cm in diameter, terete, more slender and woody near the base. Ribs 8–10, inconspicuous. Spines appressed except near stem apex; central spine one, to 7–9 mm long, turned against the upper radials, dark brown to black, bases lighter; radial spines 8–16, 2–4.5 mm long, flat against the stem, whitish or gray, dark-tipped.

Flowers. Flowers to 6 cm long and 7 cm in diameter, pinkish to magenta, midregions of inner tepals darker especially near the base, margins sometimes white.

Fruits. Fruit ovoid, dark green to brownish, fleshy and juicy but drying soon after ripening, to 2 cm long and 1 cm in diameter, wool and spines persistent.

Full Name and Synonyms. Echinocereus poselgeri Lem. *Wilcoxia poselgeri* (Lem.) Britton & Rose; *Cereus poselgeri* (Lem.) J. M. Coult.

ECHINOCACTUS

Eagle-Claw Cactus, Horse-Crippler

Echinocactus Link & Otto is a genus of 6–7 species ranging from Texas west to California and south into central Mexico. Four species occur in the United States. Four species occur in the Chihuahuan Desert Region, two of them in the Trans-Pecos. The genus name is derived in part from the Greek *echinos,* "hedgehog."

Weniger (1984) included *Ferocactus, Glandulicactus, Ancistrocactus, Thelocactus, Echinomastus, Neolloydia, Astrophytum, Hamatocactus,* and *Sclerocactus* (sensu stricto, not found in Texas) in a broad concept of *Echinocactus.* Benson (1982) excluded all of these genera from *Echinocactus* except for *Astrophytum.*

Key to the Trans-Pecos Species

1. Ribs usually eight, their crests rounded; mature plants rarely depressed-globose but usually subglobose, subpyramidal, or short-cylindroid; spines glabrous; flowers magenta; fruit pink, drying quickly after ripening, indehiscent or dehiscent basally, at maturity completely obscured by a dense wool covering
   E. horizonthalonius, p. 200
1. Ribs 13 or more, at least after reaching sexual maturity, their crests relatively sharp; mature plants depressed or flattened and hemispheroidal; spines microscopically pubescent; flowers pink, rarely whitish; fruit bright red, succulent, drying slowly, not or only irregularly dehiscent, not obscured by wool
   E. texensis, p. 202

Echinocactus horizonthalonius

Eagle-Claw Cactus

PLATES 183, 184

This is one of the few cactus species in the Trans-Pecos, other than *Opuntia*, for which the characteristic flowering time is delayed until late spring or early summer. The meaning of the specific epithet, frequently misspelled *horizontalonius*, is obscure.

Distribution. Common in desert mountains and flats, of both igneous and sedimentary origin, including clay and gypsum, every county of the Trans-Pecos. 2,500–5,500 ft. East of the Trans-Pecos mostly in Val Verde Co.; rare in Crockett, Kinney, and Duval counties; N to central NM and W to S-central AZ. Mexico: Sonora (where disjunct and rare) and S through the Chihuahuan Desert to San Luis Potosí and possibly even farther S, where it displays obvious geographic variation. Map 65.

Vegetative Characters. Easily recognized by solitary eight-ribbed stems, orange- to grapefruit-sized (or larger), gray-green to gray-blue and glaucous, with rigid, subulate spines prominently annulate and glabrous. Ribs rounded and smooth or slightly constricted between areoles that are widely separated on immature plants and more closely spaced (1–2 cm apart) on adult plants. Spines prominently annulate or cross-ribbed, pale reddish to reddish-brown, or grayish-white to nearly black overlying reddish; largest (lower) of 1–3 central spines is 1.8–4.3 cm long, 1–2.5 mm thick, decurved; two smaller upper centrals are compressed against or near the stem, along with ca. 5 radial spines, typically curved toward the stem. Spines gray to pink, tan, brownish, or nearly black, with darker spine color sometimes evidently overlying pale color. The only other Trans-Pecos species with prominently annulate spines are *E. texensis* and *Ferocactus wislizeni*.

Flowers. Flowering Apr–Jun; after rains, in Jul–Sep. Magenta or bright rose-pink flowers 5–7 cm long and 5–6.5 cm in diameter when fully open, usually with more intensely pigmented (darker) throat. Flowers are produced from among dense, white, wool-like trichomes that cover the stem apex; flower buds are hidden by wool until near anthesis. Magenta or rose-pink inner tepals ca. 3 cm long and 1.5 cm wide, with entire to serrate margins; outer tepals aristate or attenuate-spinose. Filaments yellowish, anthers yellow. Pinkish style supports 6–8 stigma lobes, pinkish to olive.

Fruits. Fruits completely or mostly hidden by apical wool, globose to ovoid-cylindroid, 1–3 cm long, ca. 1 cm in diameter, with some scales on the surface. At maturity, fruits quickly drying from the top downward, the relatively thin walls turning tan. Fruits indehiscent or basally poricidal, in the latter case leaving a pile of seeds among apical trichomes of the stem after the dry "shell" has been removed. Seeds black or gray, subglobose to obovoid and angular, 2–3 mm long, the surface wrinkled and weakly papillate. The hilum appears to be basal or lateral.

Full Name. *Echinocactus horizonthalonius* Lem.

Other Common Names. Turk's head cactus; bisnagre; blue barrel.

Plate 184. *Echinocactus horizonthalonius* (eagle-claw cactus), S Brewster Co., TX; mature fruit.

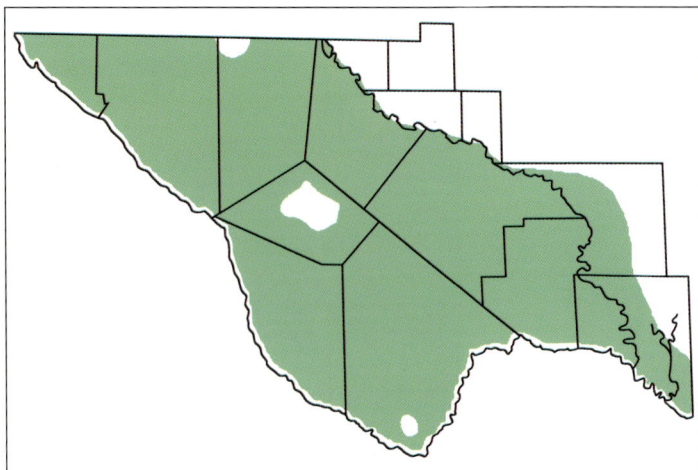

Map 65. *Echinocactus horizonthalonius* (eagle-claw cactus).

Echinocactus texensis
Horse-Crippler
PLATES 185, 186

Reportedly, *E. texensis* is more suitable to cultivation than many of the other barrel cactus types because it is more cold-hardy and capable of withstanding excessive ground moisture. Its wide distribution from sea level to desert floors and mountain slopes reveals its adaptability. In the Trans-Pecos *E. texensis* is particularly abundant in Pecos County and near Dog Canyon in Big Bend National Park. Numerous plants also have been located in alluvial basins near Alpine, Brewster County.

Distribution. Widespread in desert flats, substrates of igneous or sedimentary origin, desertscrub, grasslands, lower mountain slopes and basins, eastern two-thirds of the Trans-Pecos. 1,000–5,500 ft. Widespread in TX, from the S coastal plains at near sea level N, except in E TX, to SW OK, across to SE NM. Mexico: the Chihuahuan Desert (mostly in Coahuila and Nuevo León), and in Tamaulipas, S to vicinity of Tampico. Map 66.

Vegetative Characters. Depressed (flattened), solitary stems, deeply seated in the soil, provide one of the best traits for field identification. During dry periods the stems shrink to or slightly below ground level, and they may expand to several centimeters above the ground when moisture is plentiful. Stems are armed with long, rigid spines, inspiring the common name. Mature stems, with short white wool at apex, always have 13–27 ribs, and the ribs have sharper crests than those of *E. horizonthalonius*. Green to gray-green stems are as stiff as rubber tires and can be stepped on by heavy animals without harming the plant. Each areole has one central spine and usually 6–7 radials. The single, menacing central spine, mostly responsible for the "horse-crippler" image, usually arching downward, 4–6.5 cm long, 2–4 mm wide. Radial spines appressed or low-spreading, straight, or slightly decurved. Microscopically pubescent spines flattened, tan, light reddish-brown to pinkish or gray, prominently annulate or cross-ribbed, broad at base, tapering to point.

Plate 185. *Echinocactus texensis* (horse-crippler); cultivated.

Plate 186. *Echinocactus texensis* (horse-crippler); mature fruits; cultivated.

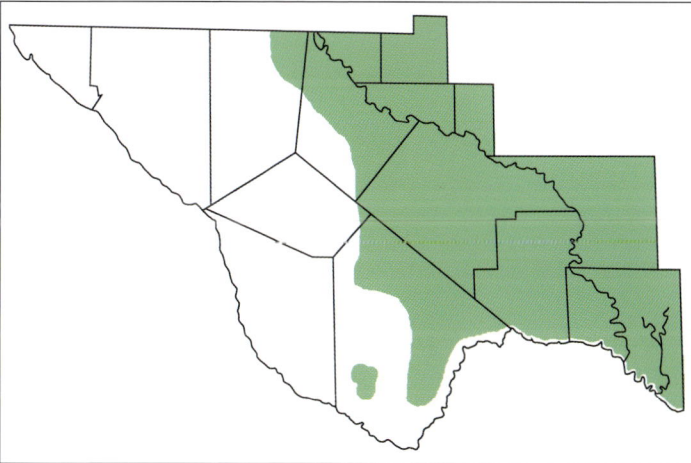

Map 66. *Echinocactus texensis* (horse-crippler).

Flowers. Flowering Apr–May. On warm, sunny days, flowers opening between 10:00 a.m. and 12:30 p.m., usually pink but varying to rose-pink or salmon, rarely whitish, with prominent reddish centers. Flowers 3–6 cm long and wide, usually smaller and paler than those of *E. horizonthalonius.* Delicate inner tepals usually erose-margined and apically aristate. Filaments pinkish to reddish; anthers yellow. Style pinkish to yellowish, supporting 7–14 stigma lobes ca. 3 mm long and pink to pinkish-white, sometimes red-striped underneath.

Fruits. Fruits ovoid to globose, 1.5–5 cm long, 1.5–4 cm in diameter. Floral remnant persistent. Crimson fruits fully exposed, but inconspicuously ornamented by wool from axils of bristlelike scales. Mature fruits persist on plants for months, if not removed by animals. Fruits mostly indehiscent, but some rupturing, forming longitudinal openings through which the bulging complement of black seeds is evident. Seeds 2.5–3 mm long, weakly glossy, subreniform or approaching globose or obovoid in shape, with a lateral, deeply concave hilum.

Full Name and Synonym. *Echinocactus texensis* Hopffer. *Homalocephala texensis* (Hopffer) Britton & Rose.

Other Common Names. Devils pincushion; devil's head; candy cactus; viznaga.

Barrel Cactus

Ferocactus Britton & Rose is a genus of 25–30 species distributed from Texas west to California but mostly in Mexico south to Oaxaca. Five species occur in the United States. Two are represented in Trans-Pecos Texas. Benson (1982) recognized three species of *Ferocactus* for Texas, with one of them, *F. setispinus* (Engelm.) L. D. Benson, mostly in South Texas. In the current treatment, *F. setispinus* is recognized as belonging to the separate genus *Hamatocactus*, as *H. bicolor*. The two true barrel cacti in Texas, *F. wislizeni* and *F. hamatacanthus*, are actually quite distinct. The genus name *Ferocactus* means "wild or fierce cactus" after the Latin *ferus*, "wild." The only really large barrel cacti in the Trans-Pecos are larger specimens of *F. wislizeni* remaining in the Franklin Mountains near El Paso.

Key to the Trans-Pecos Species

1. Mature stems, 20–200 cm tall (in Texas), 19–60 cm in diameter; spines, at least some of them in addition to the hooked main central, strongly cross-ribbed or annulate; central spines rigid, 1.5–3.5 mm wide at base; fruit yellow at maturity, the rind thick, pulp dry

    F. wislizeni, p. 206

1. Mature stems 10–40 cm tall, 7.5–17 cm in diameter; spines, even the hooked main central, not cross-ribbed but sometimes weakly annulate; central spines flexible, 1–2 mm wide at base; fruit green or reddish-tinged, the rind thin, pulp juicy

    F. hamatacanthus, p. 208

Ferocactus wislizeni
Arizona Barrel Cactus
PLATES 187–89

Ferocactus wislizeni is the largest barrel cactus in the United States, with mature plants reaching 2.5–3 m or more high and ca. 0.6 m in diameter, particularly in southwestern Arizona. One plant measured in the Franklin Mountains in 1996 was 1.8 m tall and 2 m in circumference.

A few plants of Arizona barrel cactus have been found in Big Bend National Park, Brewster County, near old houses along the Rio Grande, probably brought in by the early residents. A report of two supposedly naturally occurring plants of *F. wislizeni* near Glenn Spring in Big Bend National Park (in a mesquite thicket) has been verified by photographs. Glenn Spring was the site of early ranching activity and a general store, and plants of *F. wislizeni* could have been transported there. In 2001 G. G. Raun observed a plant of *F. wislizeni* in natural surroundings ca. 33 miles south of Marfa, Presidio County. Naturally occurring *F. wislizeni* in Brewster and Presidio counties, if verified, would be remarkable.

The specific epithet honors early botanist-explorer Frederick Adolphus Wislizenus, M.D. The correct spelling of the epithet is *wislizeni*, the genitive form of Wislizenus, and not *wislizenii*, as used in some treatments of the species.

Distribution. Rocky slopes and canyons, alluvial fans, and gravel soils, in TX apparently restricted to the Franklin Mts. 3,500–5,300 ft. Reported from the Pecos River drainage in SE NM, W across southern NM from SW Lincoln Co. to Hidalgo Co.; southern AZ, mostly in Pima Co. and N to Yapavai Co., W to Yuma Co.

Plate 187. *Ferocactus wislizeni* (**Arizona barrel cactus**), **Franklin Mts, El Paso Co., TX; cultivated.**

Plate 188. *Ferocactus wislizeni* (Arizona barrel cactus), Franklin Mts, El Paso Co., TX; cultivated.

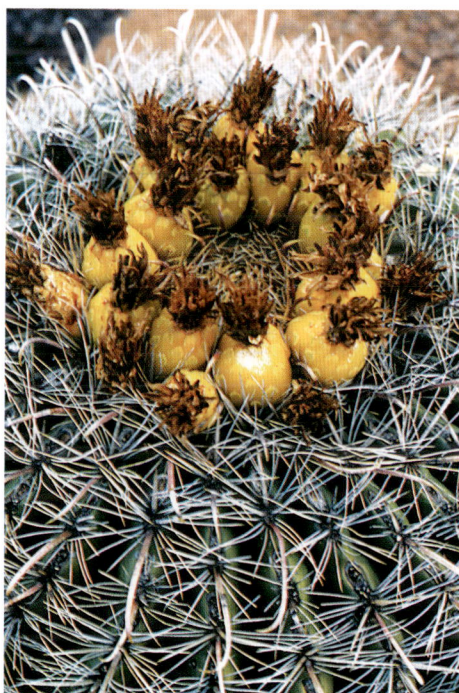

Plate 189. *Ferocactus wislizeni* (Arizona barrel cactus), Franklin Mts, El Paso Co., TX; mature fruits; cultivated.

Mexico: N Chihuahua and N Sonora. Map 67.

Vegetative Characters. Plants hemispheroidal, barrel-shaped, or short-cylindroid single stems. Usually over 20 ribs are shallowly notched above each areole and collectively support rather dense spines that partially obscure the stem. Each areole with ca. four central spines and 16–20 radial spines. Central spines arranged in a near cross, although the spines are angled outward, and the lateral centrals

Map 67. *Ferocactus wislizeni* (Arizona barrel cactus).

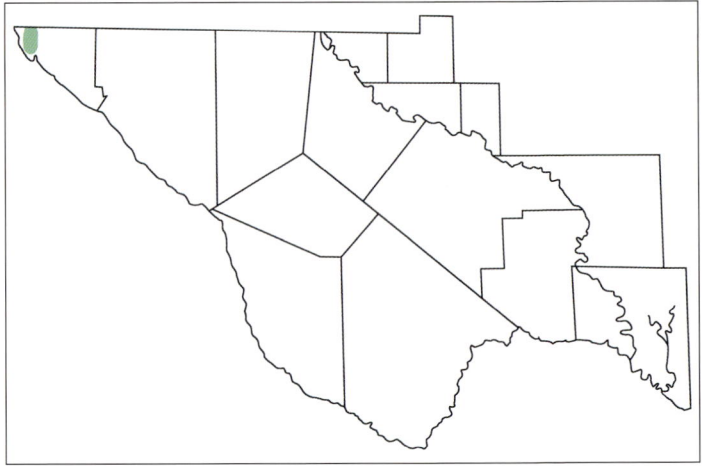

are directed upward. Lower central much larger and hooked downward or twisted to one side. Central spines basically reddish or salmon-colored but overlain with a gray surface and strongly cross-ribbed. Hooked central flattened and/or sulcate. Radial spines ashy-gray, to 3.5–5.4 cm long, the lateral ones mostly appressed against the stem, flexible, and wavy, the upper and lower radials often angled out from the stem and stiff, the lower three approaching the centrals in appearance but much smaller. Slender radials cross-ribbed or not.

Flowers. Flowering late summer. Usually several flowers produced at the stem apex, in good seasons ringing the apex. Showy flowers 4–7 cm long, 4–5 cm in diameter. Inner tepals, outer tepals, and staminal filaments typically orange-yellow but may range from pure yellow to red-striped or red. Anthers yellow. Stigma lobes 18–20, yellow to pale orange-yellow.

Fruits. Fruit yellow, barrel-shaped, with thick, fleshy wall but dry interior and with a persistent floral remnant. In good flower-ing years, apical rings of yellow fruits, each 4–5 cm long, ca. 2.5 cm in diameter, may persist for several months. Fruit walls support numerous semicircular, fringed scales. Mature fruits contain hundreds of edible black seeds. Some seeds escape the fruits through a basal opening during detachment of the fruit. Seeds about 2–2.5 mm long, shiny or dull, microscopically reticulate and warty, with a broad rim associated with the hilum.

Full Name and Synonym. *Ferocactus wislizeni* (Engelm.) Britton & Rose. *Echinocactus wislizeni* Engelm.

Other Common Names. Southwestern barrel cactus; compass barrel; candy barrel; fishhook barrel-cactus.

Ferocactus hamatacanthus
Giant Fishhook-Cactus

Two varieties of *F. hamatacanthus* are recognized in Texas, extreme southern New Mexico, and Mexico.

Full Name. *Ferocactus hamatacanthus* (Muehlenpf.) Britton & Rose.

Key to the Varieties of *Ferocactus hamatacanthus*

1. Ribs relatively thick, in younger plants with rounded tubercles; hooked central spine angled on one side, or terete, relatively stiff but still flexible, microscopically merely scabrous or sometimes short-pubescent; stigma lobes 11–14; fruits green at first, turning reddish-green with age

F. hamatacanthus var. *hamatacanthus*, p. 209

1. Ribs compressed and narrow, even in younger plants; hooked central spine flattened, very flexible, microscopically pubescent; stigma lobes 8–10; fruits remaining green

F. hamatacanthus var. *sinuatus*, p. 204

Ferocactus hamatacanthus var. hamatacanthus
Giant Fishhook-Cactus
PLATES 190–92

The epithet *hamatacanthus* is derived from the Latin *hamatus*, "hooked," and Greek *akantha*, "thorn," in reference to the hooked main central spine. *Ferocactus hamatacanthus* var. *hamatacanthus* is one of the most widely distributed cacti, but not most abundant, in the Big Bend and east-central regions of the Trans-Pecos. Plants occur from the lowest desert up to the wooded mountains (rarely 6,000 feet and higher), frequently under shrubs and in crevices of rocks. They are locally common

Plate 190. *Ferocactus hamatacanthus* var. *hamatacanthus* (giant fishhook-cactus), S Brewster Co., TX; cultivated.

Plate 191. *Ferocactus hamatacanthus* var. *hamatacanthus* (giant fishhook-cactus), S Brewster Co., TX; cultivated.

Plate 192. *Ferocactus hamatacanthus* var. *hamatacanthus* (giant fishhook-cactus), S Brewster Co., TX; mature fruits, cultivated.

in the limestone mesas and alluvial valleys of the eroded Stockton Plateau in Pecos County and the desert mountains and desert basins in the Big Bend region.

Distribution. Rock crevices and various soil types, igneous and limestone, wooded mountains, desert mountains, alluvial deposits in basins and desert. Widely distributed in the Trans-Pecos, except rare northward and westward; SE to the Devils River in Val Verde Co. 800–7,500 ft. South-central NM probably in Doña Ana

and Otero counties. Northern Mexico: W of the Sierra Madre Oriental, S to San Luis Potosí and Durango, W to Chihuahua, in the CDR. Map 68.

Vegetative Characters. Globular to short-cylindroid, usually dark green, solitary or branched stems commonly 10–30 cm high in the Trans-Pecos. Larger plants 40–45 cm high rarely encountered north of Mexico, although occasional plants 60–90 cm high have been reported in the Trans-Pecos. In overall aspect, stem tips of

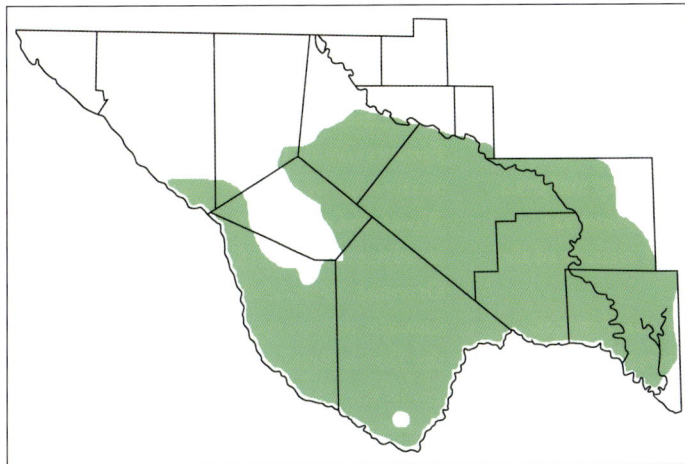

mature plants have a covering of red spines or a mixture of red spines and yellowish hooked central spines. Ribs 10–13, partially obscured, mostly by interlacing radial spines, large, 2.5–5 cm high and thick, partially divided into rounded tubercles.

In Texas, spine pattern characterized by four centrals and 10–14 radials. Hooked lower central potentially longest of the four central spines, typically reaching 9–11 cm. Partially flattened, flexible shaft of the lower central basically straight (not loosely spiraling), although usually twisted one or two turns. Two hooked central spines occur in the areoles of some plants. Radial spines up to 3.5–7 cm long, terete, somewhat compressed or angled, may be relatively stout and stiff or somewhat thin and flexible or even wavy. Longest radials are upper ones that project upward and away from the stem and lateral ones appressed against the horizontal plane of the stem, often interlacing with the lateral, appressed spines of the adjacent areoles. Spines in mature areoles may vary in color from reddish to gray.

Flowers. Flowering usually Jun–Aug. Fragrant flowers yellow, funnelform, 6–8 cm long, 6.5–8.5 cm in diameter, with yellow anthers, stigma lobes 4.5–6 mm long.

Fruits. Fruits obovoid to oblong, 3–5 cm long, ca. 2.5 cm in diameter, usually with a persistent floral remnant, with 20–40 widely spaced, fringed scales on the greenish surface. Fruits thin-walled, with white pulp, juicy and sweet. Juicy fruit interior markedly different from dry fruit of *F. wislizeni.* Seeds black, shiny, ca. 1.5 mm long, ca. 1 mm broad, ovate, microscopically pitted on the surface, with a basal-lateral hilum marked by a sharp, narrow hilum-micropylar rim.

Other Distinctions. Even in sterile condition adults of *F. hamatacanthus* var. *hamatacanthus* are not likely to be confused with any other species in Texas. The natural populations of *F. hamatacanthus* are not sympatric with *F. wislizeni,* the other large barrel cactus species in Texas. Smaller plants of *F. hamatacanthus* are superficially similar to *Glandulicactus uncinatus* var. *wrightii,* but the two species are readily distinguished by the all-straight radial spines in *F. hamatacanthus*, whereas in *Glandulicactus* the three lower radial spines in each areole are hooked. The Texas distribution of *F. hamatacanthus* var. *hamatacanthus* may overlap and even intergrade with *F. hamatacanthus* var. *sinuatus*, particularly in the vicinity of the Devils River.

Full Name and Synonyms. *Ferocactus hamatacanthus* var. *hamatacanthus*. *Echinocactus hamatacanthus* Muehlenpf.; *Hamatocactus hamatacanthus* (Muehlenpf.) F. M. Knuth; *Ferocactus hamatacanthus* var. *crassispinus* (Engelm.) L. D. Benson.

Other Common Names. Whiskered barrel cactus; turkshead echinocactus.

Ferocactus hamatacanthus var. sinuatus
Lower Rio Grande Valley Barrel Cactus
PLATE 193

Distribution. Not known to occur in the Trans-Pecos. This small eastern geographic race replaces var. *hamatacanthus* E of the desert, from the Devils River eastward; eventually some populations of *F.*

hamatacanthus W of the Pecos River might be interpreted as var. *sinuatus*. The main distributional range for var. *sinuatus* is in S TX, mostly in the brushlands from near Brownsville to Eagle Pass, extending NE to the edge of the Hill Country near Camp Wood and on W past the Devils River, and in adjacent Mexico E of the Sierra Madre Oriental. Varietal identification of *F. hamatacanthus* specimens is difficult near the Devils River and along the eastern margin of the Chihuahuan Desert in Coahuila and Nuevo León, where taxonomically ambiguous intermediates occur. Map 69.

Vegetative Characters. Variety *sinuatus* is distinguished by its smaller stems to 30 cm long and 20 cm in diameter; ribs ca. 13, narrow and more acute at the crest; central spines four, the lowermost one strongly flattened, pubescent, and somewhat flexuous; radial spines 8–12, at least some of them markedly flattened.

Fruits. Fruits sometimes globose and to 2.5 cm long, maturing green to dark brownish-red; seeds ca. 1 mm in diameter.

Full Name and Synonym. *Ferocactus hamatacanthus* var. *sinuatus* (A. Dietr.) L. D. Benson. *Echinocactus sinuatus* A. Dietr.

Plate 193. *Ferocactus hamatacanthus* var. *sinuatus* (lower Rio Grande Valley barrel cactus), S TX (photo by James H. Everitt).

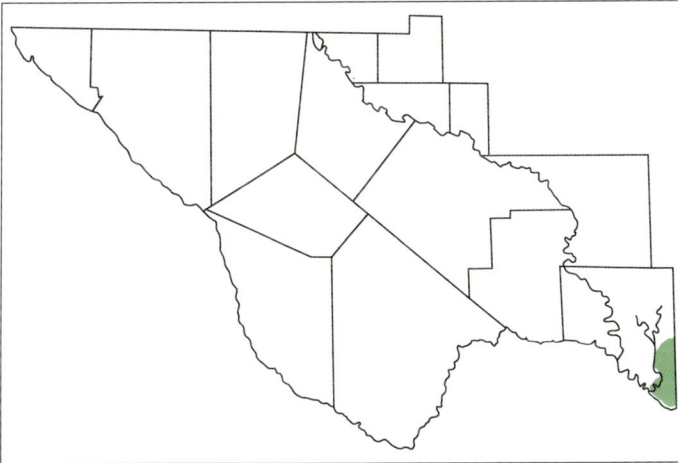

Map 69. *Ferocactus hamatacanthus* var. *sinuatus* (lower Rio Grande Valley barrel cactus).

Hamatocactus Britton & Rose is a genus of one species, as interpreted here, distributed from Central Texas to South Texas and adjacent northeastern Mexico. The taxon has been aligned with at least four different genera. The genus name *Hamatocactus* is rooted in the Latin *hamatus*, "hooked," and the Greek *akantha*, "thorn," a reference to the hooked central spines.

Hamatocactus bicolor
Twisted-Rib Cactus
PLATES 194, 195

Hamatocactus bicolor is not documented to occur in the Trans-Pecos. The known occurrence of *H. bicolor* near the mouth of the Devils River suggests that it could exist in sheltered microhabitats up the Rio Grande at least as far as Terrell County. The specific epithet, *bicolor,* presumably refers to the two colors in the flowers, yellow with red centers.

Distribution. Heavy soils in grassland and shrubland habitats, particularly in mesquite thickets, from Brown and San Saba counties of Central TX SE to Travis Co., S to Cameron Co. at the S tip of TX, and W to Val Verde Co. in limestone habitats near the mouth of the Devils River. Reported from S Terrell Co. From near sea level to ca. 1,000 ft. Mexico: Adjacent Tamaulipas, to be expected in Nuevo León and Coahuila. Map 70.

Vegetative Characters. Plants unbranched or branched at the base. Roots diffuse. Stems green, hemispheric when young, then becoming ovoid to cylindroid, 3.6–12 cm long, 4.5–12 cm in diameter. Ribs 13, spiraling or straight, slender, wavy, sharp at the crests; narrow tubercles raised on the ribs. Areoles 2.5–3 mm in diameter, becoming elliptic or ovate, 4.5–5 mm long; after sexual maturation a felted groove expands adaxially for a few millimeters; usually one or a few golden (dark with age), cylindrical glands adjacent to the spines. Spines not obscuring the stem. Central spine 1, porrect, hooked, 1–3.8 cm long, acicular or rarely flattened, minutely scabrous, yellowish, becoming ashy-gray or reddish-brown, not as long as in other relatively large cacti with hooked central spines and rather stiff and enlarged at the base. Radial spines 10–19, thin and flexible, yellowish, whitish, or reddish-brown, acicular, slightly diffuse, straight or slightly curved toward the stem, the longest (upper ones) 1.2–3.2 cm. Number of radial spines appears to be lower in northern and eastern parts of the range, with a norm of 10–15 spines, whereas in Cameron,

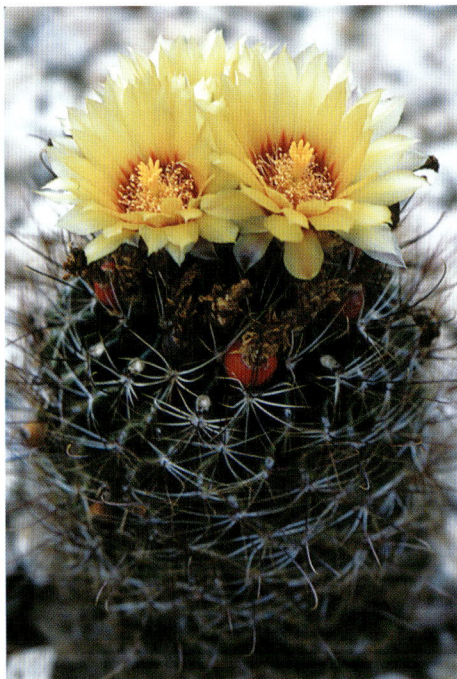

Plate 194. *Hamatocactus bicolor* (twisted-rib cactus), S TX; cultivated.

Plate 195. *Hamatocactus bicolor* (twisted-rib cactus), S TX; mature fruits; cultivated.

Starr, and Hidalgo counties of South Texas 12–19 radial spines are common.

Flowers. Flowering spring–fall. Flower buds produced in a felty adaxially extended groove between the stem and the gland-/spine-bearing portion of the areole, leaving a circular pit after flower or fruit abscission. Flowers yellow with red centers, 3.7–7 cm long, 4–7 cm in diameter, smaller than the pure yellow flowers of *F. hamatacanthus*. Inner tepals yellow (to cream or ivory) with red bases. Stamens with weak,

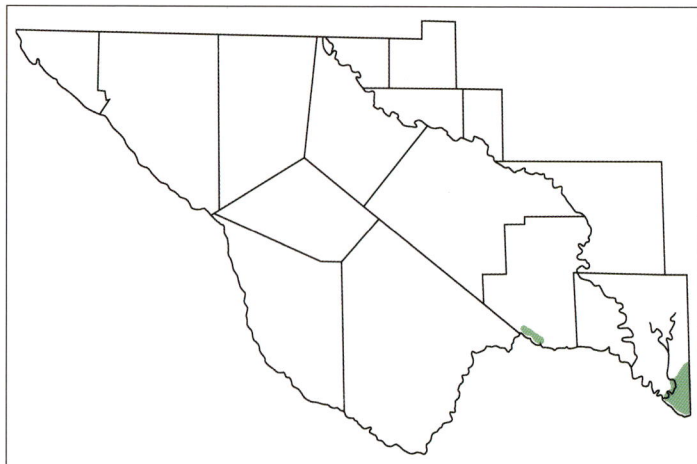

Map 70. *Hamatocactus bicolor* (twisted-rib cactus).

swirled filaments, reddish, pale yellow, or whitish, supporting pale yellow anthers. Style yellow to greenish-yellow, with 5–11 stigma lobes, 3–7 mm long and pale yellow to orangish. Ovary ca. 4.5 mm long.

Fruits. Fruit bright red, fleshy, spherical or nearly so, 8–13 mm wide, with a maximum of about 10–15 scales, usually fewer, whitish, with fringed margins. By the time of ripening, often positioned on the upper stem some distance away from the apex. Fruits indehiscent for a long period but eventually may form vertical slits. Floral remnant persistent. Seeds black, minutely papillate, somewhat obovoid but with a larger, somewhat globose, minutely papillate end and a smaller, somewhat cylindroid smooth end, usually 1–1.4 mm long, 0.8–1 mm wide, expanded around the micropyle, hilum basal.

Other Distinctions. In the field, most likely to be confused with *Ferocactus hamatacanthus, Ancistrocactus* spp., and also possibly *Glandulicactus uncinatus.* Adult plants of *H. bicolor* not in flower usually are easily recognized by the 13 ribs, typically wavy and spiraling around hemispheric to columnar stems. Ribs have shallow to rather deep recesses between the areoles. In some plants, the ribs are straight; for example, in the population recognized by Weniger (1984) as var. *setaceus* Engelm. (*Echinocactus setispinus* var. *setaceus*). Flowers of *Glandulicactus* are relatively small, red-maroon, and usually in clusters at the stem apex. In flowers of *H. bicolor,* the red centers are the collective manifestation of red midregions coloring the basal-central portion of each yellow inner tepal, with the basal midregions tapering to an attenuated point from the base to about midtepal.

Full Name and Synonyms. Hamatocactus bicolor (Terán & Berland.) I. M. Johnst., non *Thelocactus bicolor* (Galeotti) Britton & Rose. *Echinocactus setispinus* Engelm.; *Hamatocactus setispinus* (Engelm.) Britton & Rose; *Ferocactus setispinus* (Engelm.) L. D. Benson; *Thelocactus setispinus* (Engelm.) E. F. Anderson.

Other Common Name. Fishhook cactus.

THELOCACTUS

Glory of Texas

Thelocactus (K. Schum.) Britton & Rose is a genus of 11 species located mostly in the CDR of Texas and northeastern Mexico, but also occurring as far south as Tamaulipas, Querétaro, and Hidalgo. Only one species of *Thelocactus* occurs in Texas. The genus name is taken from the Greek *thele*, "nipple," in reference to the tubercles. *Thelocactus* has been recognized as a separate genus by all twentieth-century cactus specialists except Weniger (1984), who combined it and many other genera with *Echinocactus*.

Thelocactus bicolor
Glory of Texas, Marathon Basin Thelocactus

Thelocactus bicolor primarily is a species of northeastern Mexico that extends into Texas in the lower Big Bend region of the Trans-Pecos and on the Rio Grande plains in South Texas. One or more additional varieties of *T. bicolor* occur in Mexico. The specific epithet presumably alludes either to the color-banded (stramineous and reddish) spines or the red-centered "target" pattern of the flowers.

Full Name. *Thelocactus bicolor* (Galeotti) Britton & Rose.

Key to the Varieties

1. Stems 7.5–18 cm long, 5–9 cm in diameter; ribs notched by well-defined tubercles; central spines 4 (upper 3 like the radials but slightly larger), tinged with pink, longest one to ca. 4.5 cm, keeled; radial spines 15–17, mostly pinkish-tinged; flattened upper spines 5–10 cm long, the main one ca. 1.5 mm broad

T. bicolor var. *bicolor*, p. 218

1. Stems 5–9 cm long, 3.5–6 cm in diameter; ribs almost completely divided into tubercles; central spines 1 or 3 (upper 2 like the radials), stramineous (or slightly pink-tinged), ca. 2 cm long, not keeled; radial spines 12–14, stramineous or slightly pinkish-tinged; flattened upper spines 2.5–3.8 cm long, the main one 0.5–1.5 mm broad

T. bicolor var. *flavidispinus*, p. 220

Thelocactus bicolor var. bicolor

Glory of Texas

PLATES 196, 197

Thelocactus bicolor var. *bicolor* is infrequent almost throughout its range in southern Presidio and Brewster counties, but plants are locally common in certain sites. Collectors highly prize the plants, mostly because of their attractive flowers.

Distribution. Infrequent and/or localized, in silty alluvial and gravelly soils of both sedimentary and igneous origin, desert flats, hills, and bajadas with *Larrea* Cav., *Agave lechuguilla* Torr., and *Prosopis* L., southern Presidio and Brewster counties. 2,300–4,000 ft. Disjunct to S TX, on Rio Grande plains in Starr Co., ca. 50–300 ft, in Tamaulipan scrub. Mexico: Chihuahua and Coahuila S to San Luis Potosí, E through Nuevo León to northern Tamaulipas. Map 71.

Vegetative Characters. Mature stems ovoid to short-cylindroid or subconical, mostly tuberculate when young, or with rows of tubercles either vertical or spiraling on 8–13 poorly developed ribs when older. Most prominent spine characters are wood-shaving-like upper spines in each areole and bi- or tricoloration of central and certain radial spines. Flattened upper spines usually whitish throughout, not bicolored. The bi- or tricolored spines pinkish to reddish in central part of each spine, if not all the way or nearly all the way to the base, with gray to tan bases and stramineous or tan tips. Flattened "main" central spine, centrally located in the areole, 2–6 cm long.

Flowers. Flowering Mar–Sep, after

Plate 196. *Thelocactus bicolor* var. *bicolor* (glory of Texas), S Brewster Co., TX; cultivated.

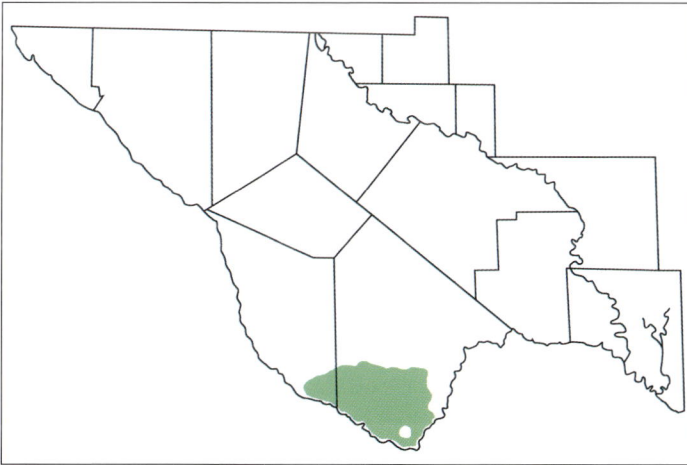

Map 71. *Thelocactus bicolor* var. *bicolor* (glory of Texas).

rains. Flowers among the most spectacular of the Texas cacti, bright magenta to rose-pink, 4–8 cm in length and diameter, with dark red centers. When fully open, the tepals reflexed. Filaments cream to whitish at the base, reddish distally, ca. 1.0 cm long, anthers yellow; style whitish, 2–3 cm long, stigma lobes 7–13, ca. 4.5 mm long, reddish to orange or yellowish.

Fruits. Fruits ovoid to nearly globular, green to brownish-red, 0.7–1.8 cm long, 0.6–1.2 cm in diameter; pericarpel thin, not juicy, the fruit rapidly drying after ripening, dehiscent through a large basal pore, the pericarpel wall with 15–20 fringed scales. Seeds black, obovoid, truncate at the base, 1.5–2.5 mm long, ca. 1.75 mm broad, finely reticulate-papillate. Floral remnant persistent.

Other Distinctions. No other Texas cactus species has spine characters identical to those of *T. bicolor.* See Powell and Weedin (2004) for details. Variety *bicolor* is distinguished from var. *flavidispinus* most conspicuously by its lack of a yellow spine-color phase and by its larger ovoid to cylindrical or subconical stems. In var. *bicolor* there are four central spines, three of them borne in the upper part of the areole near the radials and one positioned near the center of the areole. The lower central is longer, usually to 4.5–6 cm, porrect, and flat on the upper side. The upper flattened spines are longer (5–10 cm), and ashy-gray in mature areoles. The two varieties also are separated geographically and ecologically. Variety *bicolor* occurs in miscellaneous substrates in southern Brewster and Presidio counties, whereas var. *flavidispinus* is mostly restricted to novaculite exposures in the Marathon Basin of northeast Brewster County. Plants of var. *flavidispinus* are spheroidal to short-cylindroid, typically 3–8 cm high, with 1–3 central spines, one in the areole center and two arising in the upper areole margin almost among the radials.

Full Name and Synonyms. *Thelocactus bicolor* var. *bicolor.* *Echinocactus bicolor* Galeotti var. *schottii* Engelm.; *E. bicolor* var. *tricolor* (K. Schum.) Knuth; *T. bicolor* (Galeotti) Britton & Rose var. *schottii* (Engelm.) Krainz.

Other Common Names. Texas pride; bicolor cactus.

Thelocactus bicolor var. flavidispinus
Marathon Basin Thelocactus
PLATES 198, 199

The varietal epithet *flavidispinus* is derived from the Latin *flavidus,* "somewhat yellow," and *spina,* "spine." Plants of var. *flavidispinus* are abundant at some sites in the Marathon Basin.

Distribution. Common in soils in and near novaculite exposures, hillsides, and basins, often in alluvium associated with or derived at least in part from novaculite, occasionally in soils of other derivation, Marathon Basin, Brewster Co. 4,000–4,500 ft. The reports of *T. bicolor* var. *flavidispinus* in Starr Co. of S TX (Benson, 1982) are based upon misidentified *T. bicolor* var. *bicolor.* Map 72.

Vegetative Characters. Any *Thelocactus* located in the Marathon Basin probably is var. *flavidispinus.* Sexually mature plants typically are smaller than those of var. *bicolor.* Plants of var. *flavidispinus* may produce flowers when stems reach only ca. 3 cm high. Porrect or downcurved, essentially needlelike "main" central spine is rather short (1.5–2.4 cm). In areoles of mature plants, there may be only one central spine. Typically, however, there are what appear to be 1–2 additional central spines at the upper margin of the areole, their position essentially in line with the radials, but they are slightly larger at the base and slightly longer than the radials. One or two flattened upper spines to 3.8 cm long occur typically in areoles of mature plants, these always borne among or under normal radials at the upper edge of the usually circular areoles; they typically are yellowish in the upper areoles, but weathering to grayish in areoles down the stem.

Plate 198. *Thelocactus bicolor* var. *flavidispinus* (Marathon Basin thelocactus), Marathon Basin, Brewster Co., TX; cultivated.

Plate 199. *Thelocactus bicolor* var. *flavidispinus* (Marathon Basin thelocactus), Marathon Basin, Brewster Co., TX; mature fruit; cultivated.

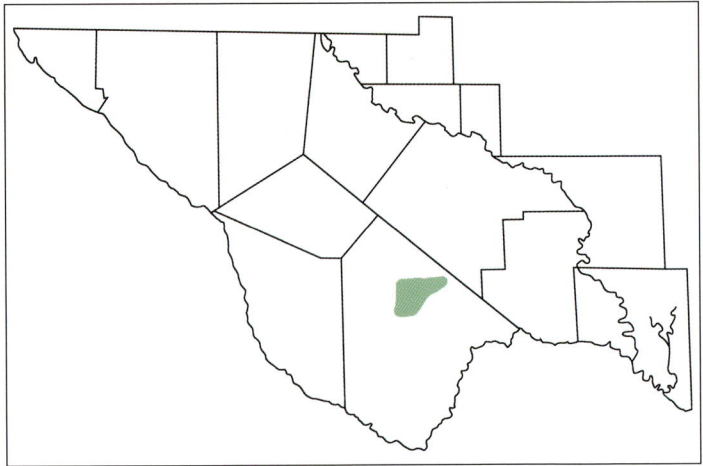

Map 72. *Thelocactus bicolor* var. *flavidispinus* (Marathon Basin thelocactus).

Radial spines number 15–17. The majority of radial spines around the usually circular areoles are 1.5–2 cm long, basically needlelike except typically laterally compressed near the base, and many curved back toward the stem. In color, "typical" radials yellowish (rarely reddish) in upper areoles and rose to grayish or bicolored with reddish central zones, gray bases, and yellowish tips in areoles down the stem. The 1–2 flat upper spines at the top of the areole may be poorly defined on young plants or flat on only one side; on older plants, more like pointed wood shavings and becoming grayish, as in var. *bicolor*.

Flowers. Flowering Mar–Sep, opportunistic. Flowers like those of var. *bicolor*.

Fruits. Fruits and seeds, so far as known, like those of var. *bicolor*.

Full Name and Synonyms. *Thelocactus bicolor* var. *flavidispinus* Backeb. *Thelocactus flavidispinus* (Backeb.) Backeb.; *Echinocactus flavidispinus* (Backeb.) Weniger.

Anderson (1980, 1996) recognized two closely related species for *Lophophora* J. M. Coult. One species is *L. williamsii,* the wide-ranging entity that occurs in Mexico and in Texas, and the other is *L. diffusa* (Croizat) Bravo, which is restricted to near Vizarrón in the state of Querétaro, Mexico. Historically, *Lophophora* has been placed in several different genera, *Echinocactus, Anhalonium* Lem., *Mammillaria,* and *Ariocarpus,* but among leading cactologists today there appears to be virtual unanimity in accepting *Lophophora.* All populations of *L. williamsii* in Mexico and Texas produce essentially the same alkaloids. *Lophophora diffusa* has a slightly different alkaloid content. Both *Lophophora* species have received extensive cultural attention (Anderson), and the common name, peyote, is applied to both taxa of *Lophophora.* The origin of the genus name, *Lophophora,* is from the Greek *lophos,* "the crest," and *phoreus,* "a bearer," in reference to the tufts of hairs borne in each areole. The common name, peyote, originated from peyotl, the Aztec name for the taxon.

Note: It is a violation of U.S. federal law (the Comprehensive Drug Abuse Prevention and Control Act, Public Law 91-513) to possess or use either peyote or its alkaloid mescaline. A special exception to the law is the use of peyote in bona fide religious ceremonies of the Native American Church. Under this law, peyote is considered to be a Schedule I controlled substance. Thus, the law prohibits even simple possession by cactus collectors, horticulturists, and scientists, although scientists can apply for a permit to use peyote in teaching or research. Conviction for illegal possession of peyote can result in a fine of up to $5,000 or imprisonment for up to one year or both.

Lophophora williamsii
Peyote

PLATES 200-202

Weniger (1984) recognized two varieties of *L. williamsii* for Texas. We follow Anderson in not recognizing varieties for the Texas *Lophophora*.

Distribution. Mostly in rocky limestone substrates or in alluvium derived from calcareous parent material, among and under shrubs and other Chihuahuan Desert vegetation or in rock crevices. Presidio Co., at least two localities in southern part of county, nearly extirpated from a significant portion of its range in one population; Brewster Co., one documented locality, on an igneous mountain; Val Verde Co., at least two known sites near the Pecos River. 1,000–4,500 ft. South TX in Hidalgo, Starr, Jim Hogg, Zapata, and Webb counties and then NE near the Rio Grande to the Pecos River. Mexico: from Tamaulipas and Nuevo León SW to San Luis Potosí and NE Zacatecas, and N through eastern Chihuahua and much of Coahuila to TX . Map 73.

Vegetative Characters. Plants spineless, low, flat to the ground or rounded, often gray, blue-green to darker green, rarely reddish-green; taproot fleshy, broadly carrot-shaped, 6–12 cm long. Stems solitary to numerous (up to 50), flesh soft, extending well below the ground as fleshy basal stems or roots, depressed-globose or flattened and depressed in the centers, 2–7.5 cm high, 5–10 cm in diameter. Ribs typically eight, low and demarked by narrow grooves, straight or spiraled, each rib composed of usually low, fused tubercles, more distinct at the stem apex, the tubercles imperfectly hexagonal and often with slight wrinkles when desiccated, to 2.5 cm in diameter. Areoles essentially round, 0.5–1.5 cm apart at the tips of the humplike tubercles, each bearing a compact, cylindroid tuft of soft whitish or yellowish trichomes, the trichomes 7–10 mm long, with age turning grayish and eventually broken or worn away; stem surface dull, microscopically papillate; pith and cortex more or less flaccid, not mucilaginous. Spines absent in adults; rudimentary and weak in seedlings.

Plate 200. *Lophophora williamsii* (peyote), **Presidio Co., TX.**

Plate 201. *Lophophora williamsii* (peyote), Presidio Co., TX; cultivated.

Plate 202. *Lophophora williamsii* (peyote), Presidio Co., TX; mature fruits; cultivated.

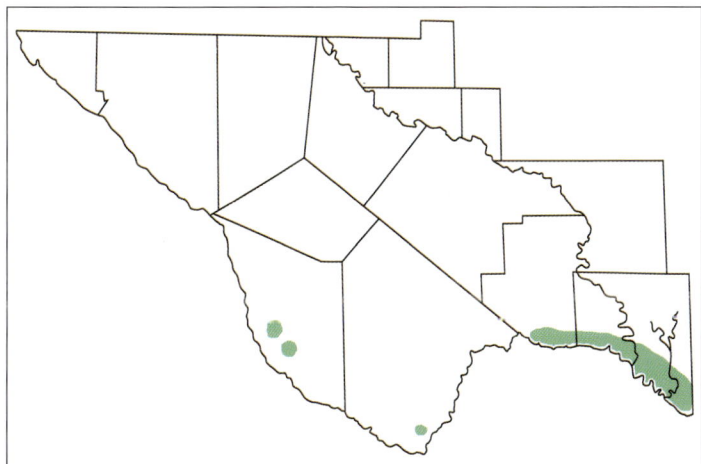

Map 73. *Lophophora williamsii* (peyote).

Flowers. Flowering Mar–Sep, commonly Jun–Aug in the Trans-Pecos. Usually pinkish flowers morphologically similar to those of a *Mammillaria*, a *Coryphantha*, or *Ariocarpus fissuratus*, 1.5–2.5 cm in diameter, 1–3 cm long, in the woolly depressed centers of stems. Inner tepals mostly pink in ours, also pinkish-red to nearly white, usually darker near midribs, rarely yellowish-white; stamens with filaments white, rarely magenta, 1–2 cm long, anthers yellow, to 1.4 mm long; style white, rarely pinkish, 0.5–1.4 cm long, stigma lobes 4–8, 1–3 mm long, white, rarely pinkish.

Fruits. Fruits partially hidden by long trichomes ("wool") of the stem apex. Fruit naked, clavate or nearly cylindroid, 1.1–2.5 cm long, 2–4.5 mm in diameter distally, whitish, pinkish in most of our specimens, or pinkish-red, weakly succulent, quickly becoming dry, translucent, and brownish-white or membranous-white after ripening, having first grown rapidly and exserted above the apical hairs. Seeds black, 1–1.5 mm long, 1–1.2 mm wide, somewhat pear-shaped. Cells of testa strongly convex. Hilum basal, large and flattened. Usually only upper half of the fruit contains seeds. Seeds are almost identical to those of *Ariocarpus*. After the thin fruit walls rupture, not by patterned dehiscence, the seeds are disseminated by rainwater.

Other Distinctions. In the Trans-Pecos *L. williamsii* is easily distinguished from all other species, but it has been confused by some inexperienced observers with *Ariocarpus fissuratus* and juveniles of *Echinocactus horizonthalonius*, both of which are widely distributed in the same type of limestone habitats that appear to be capable of supporting *L. williamsii*. Like *L. williamsii*, *A. fissuratus* is completely spineless and characterized by flattened stems extending just above ground level, but *A. fissuratus* is easily recognized by its rigid, copiously fissured, and wrinkled tubercles. In aspect the flat or weakly hemispheric stems of *A. fissuratus* support a rosette of roughened, usually pointed tubercles, often a starlike habit. Immature stems of *E. horizonthalonius* often are depressed-globular and have relatively broad, smooth tubercles and may resemble *L. williamsii* in these traits, but *E. horizonthalonius* has stout spines in the areoles. Juvenile plants of *E. horizonthalonius* with few spines or, rarely, with almost no spines are most likely to be confused with *L. williamsii*. See Anderson (1996) and Powell and Weedin (2004) for biosystematic and ethnobotanical discussion of *L. williamsii*.

Full Name and Synonyms. *Lophophora williamsii* (Lem.) J. M. Coult. *Anhalonium williamsii* Lem.; *Echinocactus williamsii* Lem.; *Mammillaria williamsii* J. M. Coult.; *Ariocarpus williamsii* Voss. For additional synonymy, see Anderson (1996).

Other Common Names. Dry whisky; mescal buttons; divine cactus; for other common names, see Anderson (1996).

ARIOCARPUS
Living Rock Cactus

Ariocarpus Scheidw. is a mostly Chihuahuan Desert genus of six species. All six occur in northeastern Mexico, extending south to San Luis Potosí, disjunct to one taxon in Querétaro. Four species occur in the CDR, where three species and two varieties are endemic. Only one of the species, *A. fissuratus*, reaches the United States, where its primary distribution is in limestone habitats of the Big Bend region of the Trans-Pecos along and near the Rio Grande. The genus name is derived from the intent by Scheidweiler, who described the genus, to depict the fruit as shaped like a small pear (the Latin *aria*, usually a suffix denoting similarity, but in this case used as a prefix meaning "pear," or *pyrus*, and the Greek *karpos*, "fruit"). [*Anhalonium* Lem.; *Neogomesia* Castañeda; *Roseocactus* A. Berger].

Ariocarpus fissuratus var. fissuratus
Living Rock Cactus
PLATES 203–5

Ariocarpus fissuratus is unusual among Trans-Pecos cacti in that it flowers only in the fall. The epithet *fissuratus* refers to the fissured upper surfaces of the tubercles. See Powell and Weedin (2004) for a biosystematic and ethnobotanic review of *A. fissuratus*.

Plate 203. *Ariocarpus fissuratus* var. *fissuratus* (living rock cactus), Black Gap Wildlife Management Area, Brewster Co., TX.

Plate 204. *Ariocarpus fissuratus* var. *fissuratus* (living rock cactus), Rosillos Mts, Brewster Co., TX.

Plate 205. *Ariocarpus fissuratus* var. *fissuratus* (living rock cactus), Big Bend National Park, TX; mature fruits.

Distribution. Rocky limestone habitats, rarely in gypseous soils, Hudspeth, Presidio, Brewster, and Pecos counties, S and E to Terrell and Val Verde counties, usually with lechuguilla and other desertscrub; rarely with *Pinus remota* in the Del Norte Mts ca. 10 mi SE of Alpine. 1,500–4,500 ft. Mexico: NE Chihuahua, N Coahuila (from Cuatro Ciénegas northward), NE Durango. Map 74.

Vegetative Characters. Plants spineless, usually unbranched, deeply seated, taprooted, usually barely extending aboveground, often drawn below ground level. Stems turnip-shaped, flattened, concave, or weakly hemispheric, usually 0–3 cm high (aboveground), 5–10 cm in diameter, gray-green, becoming yellowish and horny with age. Tubercles crowded and overlapping, flattened or slightly convex, the exposed part deltoid (pointed or slightly rounded at the tip), 1–2 cm long, 1.5–2.5 cm wide, tough-skinned and rigid, prominently fissured and tuberculate. Areoles 10–15 mm

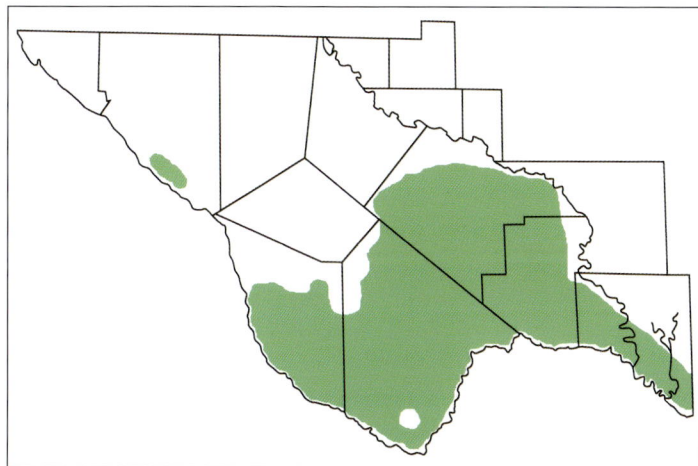

Map 74. *Ariocarpus fissuratus* var. *fissuratus* (living rock cactus).

long, 3–4 mm wide, evident as densely woolly (matted hairs or trichomes) short grooves to 3 mm wide in the centers of the flattened tubercles (the tubercles with naked lateral grooves as well), the areolar groove rarely absent. In younger plants tubercles may be suberect, or peripheral tubercles may extend beyond sphere of stem and yield a star-shaped habit, inspiration for one of the popular names of the species.

Flowers. Flowering Sep–Nov. Flowers apical, usually rose-pink to magenta, less commonly pale pink or rarely nearly white, 1.5–3.5 cm long, 2.5–4.5 cm wide. Inner tepals dark to light magenta or pink; filaments whitish; anthers deep yellow, ca. 0.7 mm long. Style whitish, 1.5–1.9 cm long, ca. 1 mm thick. Stigma lobes white, 5–10, 1.2–5 mm long, exserted 1–3 mm above anthers.

Fruits. Fruit maturing in the summer or fall. Fruit naked, oval to clavate or cylindroid, whitish to greenish at maturity, 1–2.4 cm long, 0.5–1 cm wide, fleshy at first, drying brownish. Mature dry fruits with thin walls remain firmly attached to the stem, often remaining hidden in the apical wool, ultimately disintegrating, leaving numerous seeds in the woolly center. Seeds black, 1–2 mm long, shiny, globose to obovoid; cells of the testa strongly convex, with a large, pale, basal hilum.

Full Name and Synonyms. *Ariocarpus fissuratus* (Engelm.) K. Schum. var. *fissuratus*. *Mammillaria fissurata* Engelm.; *Anhalonium fissuratum* (Engelm.) Engelm.; *Roseocactus fissuratus* (Engelm.) A. Berger.

Other Common Name. Star cactus.

Neolloydia Britton & Rose is a genus of two species in southwest Texas and Mexico. Neolloydia has been associated systematically with a number of other genera, including the Mexican *Turbinicarpus* Buxb. & Backeb., *Thelocactus,* and *Echinomastus* (see Powell and Weedin, 2004, for a taxonomic review). In the Trans-Pecos *N. conoidea* is easily distinguished from *Thelocactus, Echinomastus,* and all other taxa by numerous vegetative and floral characters. The genus name is taken from the Greek *neos,* "new," and *lloydia,* after Francis Ernest Lloyd (1868–1947), professor of botany at McGill University in Montreal.

Neolloydia conoidea var. conoidea
Texas Cone Cactus
PLATES 206, 207

This species is exceedingly abundant at certain sites, for example, west of Sanderson, in the Dead Horse Mountains, and on certain stable limestone formations elsewhere, particularly in central and southern Brewster County. The specific epithet is derived from the Greek *konos,* "cone," and *oideos,* suffix meaning "a form or type of," presumably in reference to the somewhat cone-shaped stems or tubercles.

Distribution. Stable rocky limestone habitats, mostly in desert mountains, S, central, and E Brewster Co., SE Presidio Co., E to S Pecos Co., Terrell Co. into Val Verde Co. near Del Rio. Outliers reported in El Paso and Culberson counties. 1,500–4,000 ft. Edwards Co., TX. Mexico: Coahuila, Nuevo León, Durango, Zacatecas, San Luis Potosí, Querétaro, S to Hidalgo, and E to Tamaulipas. Map 75.

Vegetative Characters. Plants branched or not. Roots diffuse. Stems globose to cylindroid, 5–10 cm high, 2.5–6.5 cm in diameter, gray-green to rather yellowish-green, typically with a white-woolly apex; pith and cortex not mucilaginous; a yellow viscid layer present in the bark beneath old epidermal tissue, unique to this genus; pith relatively small, without vascular strands. Ribs none, but tubercles in 8–13 spiral rows. Tubercles prominent, 7–12 mm long, 8–18 mm broad, conical-compressed, each with a felted areolar groove from the apex to the axil. Areoles circular, 2–5 mm in diameter. Spines rather dense; central spines stout with bulbous bases, black or dark brown, mostly 2–4 per areole, straight, the lower one longest, 1.7–2.4 cm, 0.6–1 mm thick, porrect or angled slightly downward, the upper 1–3 angled upward, 0.9–1.7 cm long, ca. 0.5 mm thick; radial

Plate 206. *Neolloydia conoidea* var. *conoidea* (Texas cone cactus), S Brewster Co., TX; cultivated.

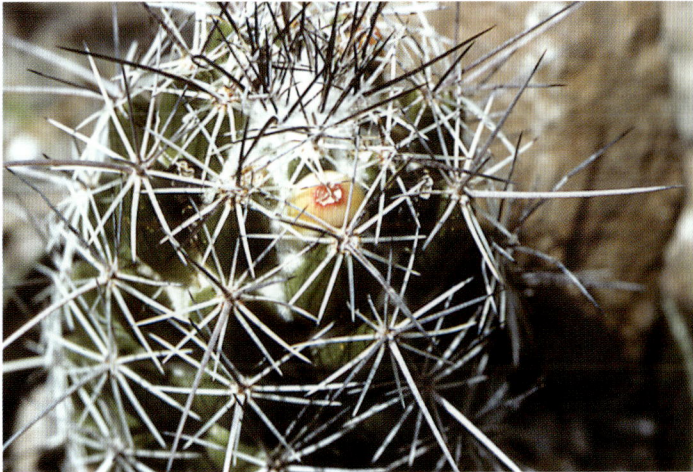

Plate 207. *Neolloydia conoidea* var. *conoidea* (Texas cone cactus), near Sanderson, Terrell Co., TX; cultivated.

spines 14–17, appressed, whitish, often with a dark tip, 0.6–1.2 cm long, straight, needlelike, 0.3 mm in diameter, with bulbous bases.

Flowers. Flowering Mar–Jul. Flowers 2.5–3.2 cm long, 3–5.5 cm in diameter; inner tepals magenta or bright rose-pink, filaments whitish, 3–6 mm long, anthers deep yellow, 1 mm long. Style whitish, 0.7–1.1 cm long, ca. 1 mm or less wide. Stigma lobes 4–7, white to cream, 2–3 mm long, slender. Magenta or bright rose-pink

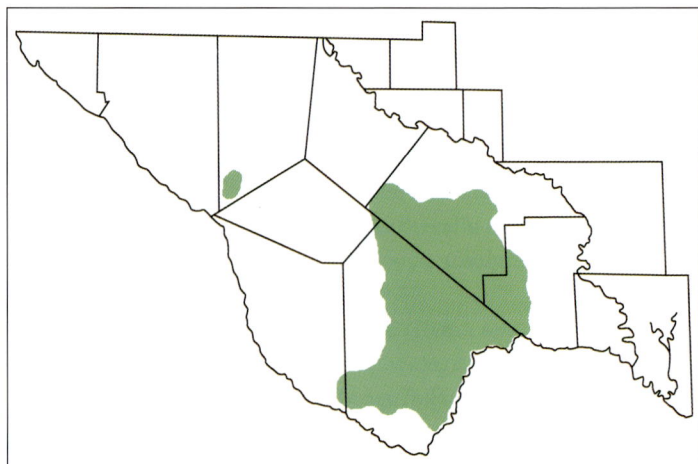

Map 75. *Neolloydia conoidea var. conoidea* (Texas cone cactus).

flowers allow easy distinction from *Echinomastus*, which has white, tan, greenish, or pale pink-striped flowers. Flowers of *N. conoidea* are similar to magenta flowers of *Coryphantha macromeris* and *C. ramillosa*, except that *Coryphantha* flowers are larger at 5.5–6 cm long and 4.5–6 cm across; *C. macromeris* has fringed, not entire, outer tepals.

Fruits. Fruit green then greenish-brown or tan at maturity, spheroidal, 4–8 mm long, 4–8 mm in diameter, dry and papery at maturity, dehiscing by vertical slits, the fruits persistent. Floral remnant deciduous. Seeds black to gray, strongly papillate,

1–1.6 mm long, 0.8–1.2 mm in diameter, pear-shaped. Hilum large and basal, with a lip over part of the hilum. Fruits of *N. conoidea* are inconspicuous even when fully ripe, being relatively small and hidden in copious white wool at the stem apex. As the fruits enlarge, they become evident, in that the apical portion extends slightly above the wool.

Full Name and Synonyms. Neolloydia conoidea (DC.) Britton & Rose var. *conoidea. Mammillaria conoidea* DC.; *Echinocactus conoideus* (DC.) Poselger; *Coryphantha conoidea* (DC.) Orcutt ex A. Berger.

GLANDULICACTUS

Eagle-Claw Cactus

Glandulicactus Backeb. is a genus of two species: *Glandulicactus uncinatus* occurs in Texas, New Mexico, and Mexico; *G. crassihamatus* (Weber) Backeb. is restricted to Querétaro, Guanajuato, and San Luis Potosí, Mexico. Historically *Glandulicactus* has been associated taxonomically with several other genera (reviewed in Powell and Weedin, 2004). The genus name is derived from Latin *glandula,* "a gland," and cactus; there are several yellowish glands in the narrow areolar grooves.

Glandulicactus uncinatus var. wrightii

Eagle-Claw Cactus

PLATES 208–10

Glandulicactus uncinatus var. *wrightii* is one of the most widespread and common cacti in the Trans-Pecos, and it extends like *Thelocactus bicolor* disjunctly into South

Plate 208. *Glandulicactus uncinatus* var. *wrightii* (eagle-claw cactus), S Brewster Co., TX; cultivated.

Plate 209. *Glandulicactus uncinatus* var. *wrightii* (eagle-claw cactus), S Brewster Co., TX; cultivated.

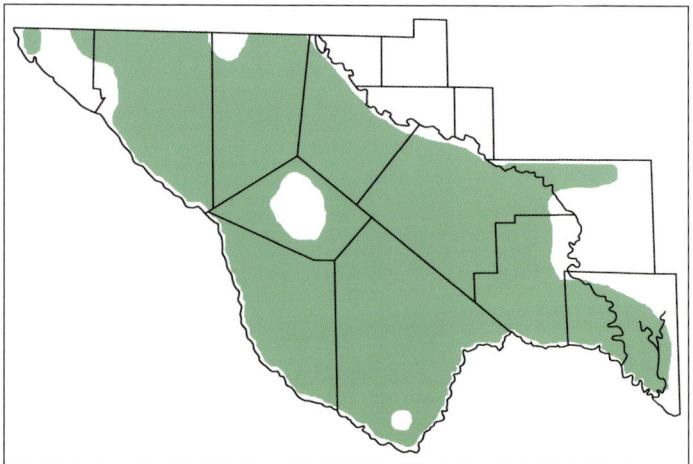

Plate 210. *Glandulicactus uncinatus* var. *wrightii* (eagle-claw cactus), Rosillos Mts, Brewster Co., TX; mature fruits.

Map 76. *Glandulicactus uncinatus* var. *wrightii* (eagle-claw cactus).

Texas. The descriptive specific epithet is taken after the Latin *uncus*, "hook," in reference to the hooked spines of this species. The varietal epithet honors the early plant collector Charles Wright (see *Mammillaria wrightii*).

Distribution. Desert hills, flats, in desertscrub and grasslands, igneous and limestone derivation, commonly growing in or adjacent to grass clumps. In TX mostly restricted to the Chihuahuan Desert, almost throughout the Trans-Pecos, with a few outliers in Crockett and Val Verde counties. 1,000–5,000 ft. Rare in Starr and Victoria counties of S TX. Chihuahuan Desert of southern NM, Sierra, Doña Ana, Otero, and Eddy counties. Mexico: E Chihuahua, Coahuila, Nuevo León, Zacatecas, Durango, and NW Sonora. Map 76.

Vegetative Characters. Plants typically with solitary stems, occasionally branched at the base. Stems globose, short-cylindroid, or ovoid, green to bluish-green, glaucous, 7.5–15 cm long, 5–7.5 cm wide. Ribs 9–13, somewhat prominent but deeply notched, 6–9 mm broad, protruding 9–15 mm. Areoles with several yellowish glands in a narrow area of the short groove connecting the upper spine-bearing portion of the areole and the lower flower-bearing portion. Spines abundant but not totally obscuring the stem; central spines tannish-white to stramineous and pinkish, 1–4 per areole (on mature stems), principal central spine turned upward, prominently hooked, 5–10 cm long, 1–1.5 mm wide at the base, somewhat flattened; radial spines 5–8 per areole, including 2–3 upper ones in a quasi-central spine position; three lower radials subterete, reddish or reddish-tan, hooked, slightly flattened; one straight lower radial may be present underneath the hooked ones; lateral and upper radials subulate, tan, stramineous, or cyanic, not hooked, somewhat flattened.

Stems dominated by presence of a rather long (to 10 cm), dull yellow, hooked central spine in each areole. In the field, *G. uncinatus* var. *wrightii* is most likely to be confused with *Ancistrocactus brevihamatus* and young plants of *Ferocactus hamatacanthus*, which occupy much the same range and share the presence of a single long, hooked central spine in each areole. Older plants of *F. hamatacanthus*, a barrel cactus, are not likely to be confused with *G. uncinatus* var. *wrightii* because of much larger stems and tubercles and a very different spine combination. The presence of three hooked lower radial spines in *G. uncinatus* var. *wrightii*, evident even on immature plants, is a unique character in Texas cacti.

Flowers. Flowering Mar–May. Distinctive flowers cylindrical-funnelform, 2–4 cm long, 2–3 cm wide, several borne near the stem apex at adaxial ends of short areolar grooves. Inner and outer tepals brick-red (or maroon or garnet), much the color of *Mammillaria pottsii* flowers. Filaments yellow (sometimes brownish or maroon), ca. 6 mm long, anthers yellow. Style reddish, to 1.2 cm long, stigma lobes 10–14, 5–6 mm long, pale yellow to dull orange. Flowers of *F. hamatacanthus* are considerably larger and yellow in color.

Fruits. Fruit indehiscent, fleshy, red, ovate or globose, 1.5–2.5 cm long, 1–2 cm thick, with numerous (13–35) conspicuous white-fringed scales, deltoid-auriculate, ca. 4.5 mm long, scarious-margined with naked axils; fruit pulp white (red?). Floral remnant persistent. Seeds black, basically oblong-obovoid and curving, 1.3–1.5 mm long, ca. 1 mm broad, ca. 0.8 mm thick, upper portion minutely papillate, base smooth; hilum basal, with a conspicuous rim. In *F. hamatacanthus* the fleshy fruits are much larger, to 4.5 cm long and 3.5 cm in diameter, green to reddish-green, and with numerous conspicuous scales.

Full Name and Synonyms. *Glandulicactus uncinatus* (Galeotti ex Pfeiff.) Backeb. var. *wrightii* (Engelm.) Backeb. *Echinocactus uncinatus* Galeotti ex Pfeiff. var. *wrightii* Engelm.; *E. wrightii* J. M. Coult.; *Hamatocactus wrightii* (Engelm.) Orcutt; *Echinomastus uncinatus* (Galeotti) F. M. Knuth var. *wrightii* Engelm. ex F. M. Knuth; *Ancistrocactus uncinatus* (Galeotti ex Pfeiff.) L. D. Benson var. *wrightii* (Engelm.) L. D. Benson; *Sclerocactus uncinatus* (Galeotti) N. P. Taylor var. *wrightii* (Engelm.) N. P. Taylor; *Glandulicactus wrightii* D. J. Ferguson.

Other Common Name. Catclaw cactus.

Fishhook Cactus

Ancistrocactus Britton & Rose is a genus of two or three poorly defined species distributed from the southern Trans-Pecos to South Texas mostly along and near the Rio Grande and into northern Mexico. All three of the "species" occur in Texas. One species with three varieties occurs in the CDR. Two of the taxa, both varieties of *A. brevihamatus*, occur in and near the Trans-Pecos. The genus name is derived from the Greek *ankistron*, "fishhook," in reference to the hooked central spines and hence "fishhook cactus."

Ancistrocactus brevihamatus
Short-Spined Fishhook Cactus, Snipe Cactus

Ancistrocactus brevihamatus was treated by Benson (1982) as a synonym of *A. scheeri* (Salm-Dyck) Britton & Rose. Weniger (1984) and Zimmerman and Parfitt (2003a) recognized *A. brevihamatus* as a distinct species. Benson (1982) included *Glandulicactus uncinatus* in *Ancistrocactus*. The specific epithet is derived from the Latin *brevis*, "short," and *hamatus*, "hooked," a reference to the relatively short, hooked central spine.

Full Name. *Ancistrocactus brevihamatus* (Engelm.) Britton & Rose.

Key to the Varieties of *Ancistrocactus brevihamatus*

1. Hooked central spine 0.8–1.5 mm in diameter; flowers brown, pink and brown, green and brown, or rose-pink with pale margins
   A. brevihamatus var. *brevihamatus*, p. 238
1. Hooked central spine 0.3–0.8 mm in diameter; flowers off-white or cream-colored
   A. brevihamatus var. *pallidus*, p. 243

Ancistrocactus brevihamatus var. brevihamatus

Short-Spined Fishhook Cactus

PLATES 211, 212

The distribution of var. *brevihamatus* is mostly east of the Pecos River in Val Verde County, east to Kinney and Uvalde counties, but it has been documented as far west as southern Terrell County. At the northeast edge of its range it inter-grades with very weakly distinguished *A. tobuschii*. *Ancistrocactus tobuschii* is listed federally and by the state of Texas as an Endangered species. *Ancistrocactus brevihamatus* is more sharply distinct from *A. scheeri*, which in general has a more southerly distribution in Texas (and adja-cent Mexico), extending north to southern Kinney County, where it is sympatric with *A. brevihamatus* var. *brevihamatus*.

Distribution. Various limestone habi-tats, rocky or alluvial, mostly in Tamau-lipan scrub. Val Verde Co., Devils River W to near Comstock. Terrell Co., near Rio Grande. 1,000–2,500 ft. Kinney and Uvalde counties. Mexico: Coahuila W to near Cuatro Ciénegas, and Nuevo León. Map 77.

Vegetative Characters. Stems usually globular and unbranched. In very old specimens, stems may be short cylindroid. Stems vary in size, 3–6 cm in length and diameter in most sexually mature plants, to ca. 13 cm long and 8–9 cm in diameter in old plants. Stems tuberculate in younger plants; older specimens have 8–13 ribs. Vegetatively, the most conspicuous feature of *Ancistrocactus* is the single porrect, hooked, lower central spine, 1.5–5 cm long. Usually there are three additional straight central spines, which are erect, conspicuously dorsiventrally flattened, and may be slightly curved but never hooked. Typically, 7–14 appressed, all-straight radial spines, 6–12 mm long, longer in old plants. Both central and radial spines usu-ally whitish-gray to gray-white in color, or centrals gray and radials whitish.

Flowers. Flowering Jan–Mar. Incon-spicuous flowers funnelform, 2–4 cm or more long, 2–3 cm in diameter, borne on new growth near stem apex, like those of

Plate 211. *Ancistrocactus brevihamatus* var. *brevihamatus* (short-spined fishhook cactus), Val Verde Co., TX; cultivated.

Plate 212. *Ancistrocactus brevihamatus* var. *brevihamatus* (short-spined fishhook cactus), Val Verde Co., TX; cultivated.

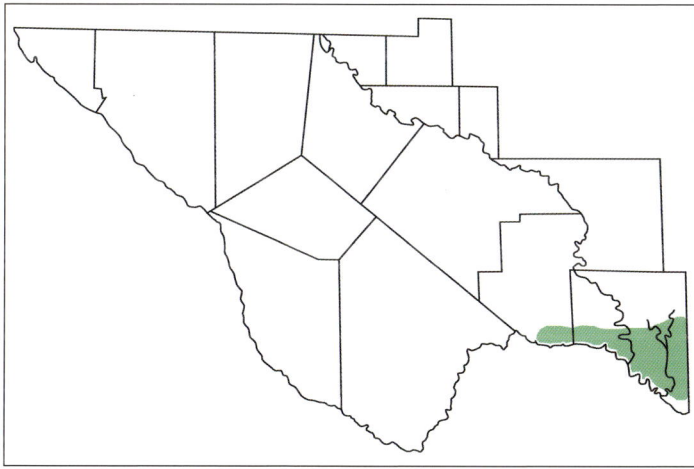

Map 77. *Ancistrocactus brevihamatus* var. *brevihamatus* (short-spined fishhook cactus).

all other *Ancistrocactus* taxa, differing only in color. Inner tepals usually pinkish-brown to olive-green with reddish-brown midregions, may be brownish, pink and brown, green and brown, or dull rose-pink with pale margins, varying to "green to rose or salmon-orange," and "darkened or muddy looking." Filaments pale or reddish, anthers yellow; style green to greenish-white, stigma lobes 4–10, 1.5–2.5 mm long, green to yellowish. Apical spines typically restrict opening of flowers. Flowers of

A. tobuschii are bright yellow, turning golden-yellow with age (rarely cream-yellow or yellowish-green), and those of *A. scheeri* are green (even bright green) or greenish-yellow.

Fruits. Fruits cylindroid-ovoid, 1.5–2.5 cm long, 0.6–1.3 cm in diameter, green at maturity, or often with a tinge of pink or rose when very ripe. Green fruits of the closely related *A. tobuschii* also may develop a rose tinge when fully mature, and when fully ripe the fruit walls become papery thin and turn tan to papery, at least in cultivated specimens. Fruits of var. *brevihamatus* are indehiscent and weakly succulent until after maturity, when the pericarpel wall dries. Floral remnant persistent. Helmet-shaped or reniform seeds 1.7–2 mm across in maximum dimension and dark reddish-brown to nearly black.

Other Distinctions. Other species with hooked spines that may be sympatric and potentially confused with *Ancistrocactus* in vegetative condition are *Glandulicactus uncinatus* var. *wrightii*, *Ferocactus hamatacanthus*, and *Hamatocactus bicolor*. *Glandulicactus* is by far the most similar to *Ancistrocactus*, but it is easily distinguished by its longer, yellowish, often upturned hooked central spine, and especially by its hooked lower radial spines. In *Ancistrocactus* the radial spines are never hooked. Mature plants of both *Ferocactus* and *Hamatocactus* are larger than those of *Ancistrocactus*, and even in immature plants, the ribs and spine patterns are different from those of *Ancistrocactus*.

At immaturity the globular, often flat-topped stems of var. *brevihamatus* are different from the cylindroid, relatively tall stems of *A. scheeri* but identical to those of *A. tobuschii*. The single-stemmed plants of var. *brevihamatus* often are deep-seated. The roots of var. *brevihamatus* are diffuse and without tuberous enlargements, or there may be a short, conical taproot. In contrast, the root system of *A. scheeri* features a rather long, fleshy taproot, ca. 1 cm in diameter, sometimes with bulbous swellings, and tuberous secondary roots as well. Both var. *brevihamatus* and *A. tobuschii* have similar root systems.

Ancistrocactus tobuschii (Plates 213, 214), most similar morphologically to var. *brevihamatus*, was described from a single locality near Vanderpool in Bandera County. Very few additional populations were known when it was federally listed as Endangered in 1983. The status of *A. tobuschii* as a rare plant generated considerable interest in its distribution, and presently it is known from eight Hill Country counties: Bandera, Edwards, Kerr, Kimble, Kinney, Real, Uvalde, and Val Verde (Map 78). At present it is not clear whether *A. tobuschii* and *A. brevihamatus* should be

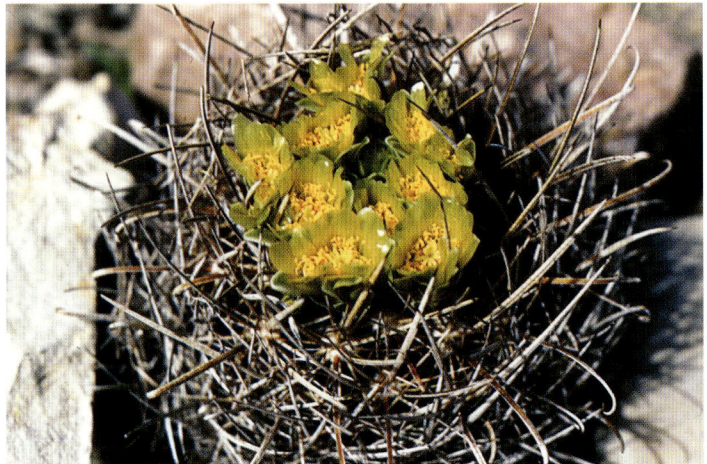

Plate 213. *Ancistrocactus tobuschii* (Tobusch fishhook cactus), McKinney Co., TX; cultivated.

Plate 214. *Ancistrocactus tobuschii* (Tobusch fishhook cactus), Edwards Co., TX; mature fruits (photo by Mark Lockwood).

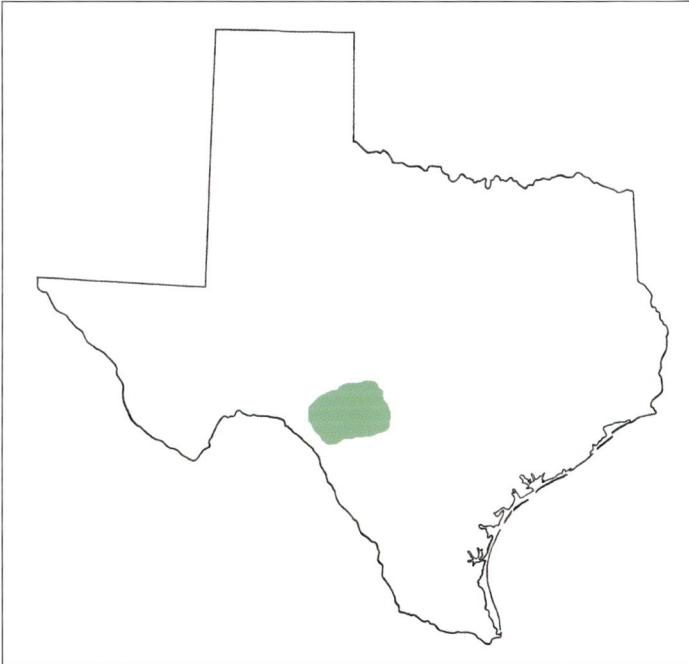

Map 78. *Ancistrocactus tobuschii* (Tobusch fishhook cactus), interranging with *A. brevihamatus*, (short-spined fishhook cactus).

regarded as distinct species, two intergrading varieties, or merely integrating flower-color morphs of the same taxon.

Ancistrocactus scheeri (Plates 215, 216), commonly known as Scheer's fishhook cactus, is regarded by most workers as a distinct species distributed in South Texas and adjacent Mexico. This Tamaulipan taxon extends northwest in Texas to the Anacacho Mountains in Kinney County (Map 79), where it is sympatric with *A. brevihamatus*. *Ancistrocactus scheeri* also approaches *A. brevihamatus* in distribution and spine number in eastern

Plate 215. *Ancistrocactus scheeri* (Scheer's fishhook cactus), S TX; cultivated.

Plate 216. *Ancistrocactus scheeri* (Scheer's fishhook cactus), S TX; mature fruit; cultivated.

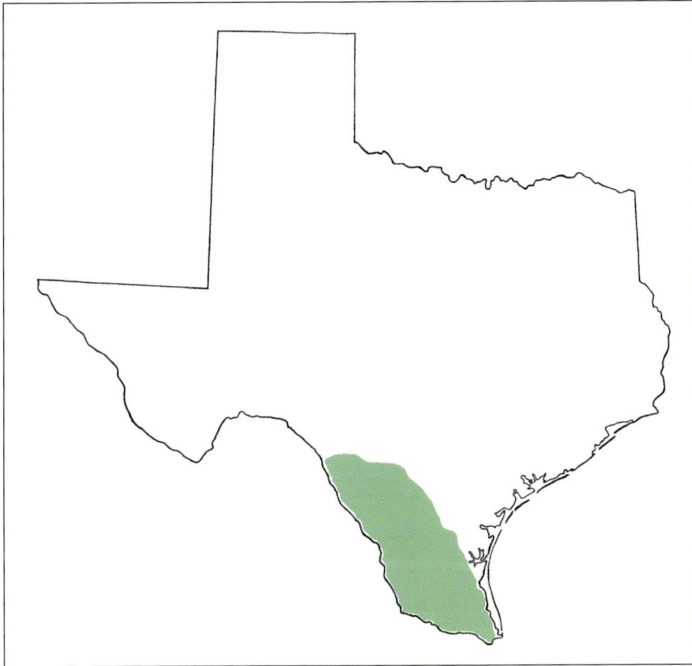

Map 79. *Ancistrocactus scheeri* (Scheer's fishhook cactus).

Coahuila. Other morphological, distributional, and biosystematic considerations of *Ancistrocatus* in Texas are reviewed in Powell and Weedin (2004).

Full Name and Synonyms. Ancistrocactus brevihamatus var. *brevihamatus*. *Echinocactus brevihamatus* Engelm.; *E. scheeri* Salm-Dyck var. *brevihamatus* (Engelm.) F. A. C. Weber.

Other Common Name. Fishhook cactus.

Ancistrocactus brevihamatus var. pallidus
Snipe Cactus
PLATES 217–19

Variety *pallidus* often is sympatric with *Glandulicactus uncinatus* var. *wrightii*, and small plants of *G. uncinatus* are similar to those of var. *pallidus*. The two taxa are easily distinguished in vegetative condition by the straight radial spines in

var. *pallidus* and hooked lower radials in *Glandulicactus*. Variety *pallidus* has not been widely collected probably because it is rare and also because individual plants are difficult to see in their habitats. Repeated expeditions over a period of several years in search for the exceedingly cryptic plants of var. *pallidus* were labeled "snipe hunts" by one of the regular participants, the inspiration for the affectionate common name. The varietal epithet is after the Latin *pallidus*, "pale," descriptive of the white to cream-colored flowers.

Distribution. Rounded limestone hills with *Parthenium argentatum* and other desertscrub, mixed novaculite alluvium in degraded grassland, bentonite in desertscrub, or Del Rio clay exposed in and among limestone mesas. Central and SW Brewster Co. and Terrell Co. S and E of Sanderson. 1,400–2,000 ft. Endemic. Map 80.

Plate 217. *Ancistrocactus brevihamatus* var. *pallidus* (snipe cactus), NE Brewster Co., TX (photo by Shirley A. Powell).

Plate 218. *Ancistrocactus brevihamatus* var. *pallidus* (snipe cactus), Terrell Co., TX; cultivated.

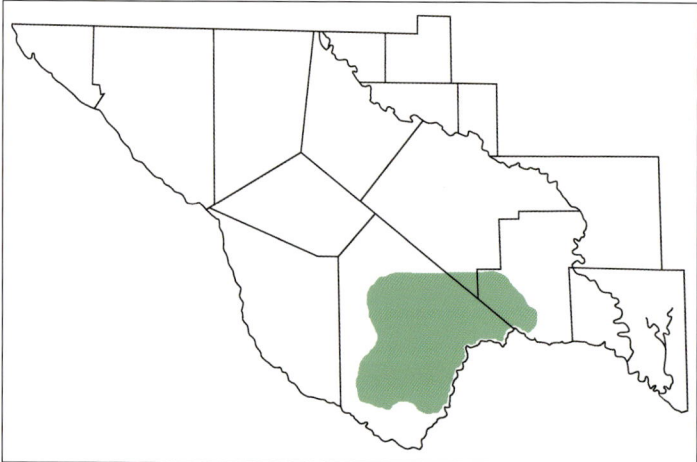

Map 80. *Ancistrocactus brevihamatus* var. *pallidus* (snipe cactus)

Plate 219. *Ancistrocactus brevihamatus* var. *pallidus* (snipe cactus), NE Brewster Co., TX; immature fruit; cultivated.

Vegetative Characters. Stems appear always to be unbranched and globular or flat-topped. Largest stems we have observed are 6–7 cm long and 6.5 cm in diameter. One large stem extended only 3 cm aboveground. Most plants are smaller and seem to be sexually mature when 1.5–2.5 cm in diameter. Plants of var. *pallidus* closely resemble those of var. *brevihamatus* except that the hooked central spines in most plants are more slender. In addition, in var. *pallidus* the central spines, 1.8–4.5 cm long, are yellowish or tan, especially in mid- and upper areoles, in Brewster County populations, and light brown to varnish brown on upper spine surfaces in the Terrell County population. In lower areoles the hooked spines usually are gray.

Flowers. Flowering Feb–Mar. Flowers are like those of var. *brevihamatus* except that the inner tepals are off-white or cream-colored, sometimes with pale brown midregions. Flowers 2.5–3.8 cm long, basically tubular, opening wider, to 2.5 cm, only where the spines permit. Sensitive stamens with colorless or pale green filaments. Stigma lobes 5–9, cream-colored or pale green, ca. 2 mm long.

Fruits. Fruits of var. *pallidus* green in development and in maturity. Mature fruits 0.8–1.4 cm long, 0.6–1 cm wide, may turn light brown or tannish. Upon drying, pericarpel wall tan or paper-white in color, thin and easily torn, with seeds barely visible through the wall. Seeds black, helmet-shaped or subreniform, 1.7–2 mm long. Floral remnant persistent.

Full Name. *Ancistrocactus brevihamatus* var. *pallidus* A. D. Zimmerman ex A. M. Powell.

Toumeya Britton & Rose is a genus of one species distributed in east-central Arizona, western New Mexico, and into Trans-Pecos Texas in northeastern Hudspeth County, possibly in adjacent Culberson County. *Toumeya papyracantha* has an unsettled taxonomic history, having been placed in several different genera: *Mammillaria*, *Pediocactus*, *Sclerocactus*, and *Toumeya*. The genus name was proposed by Britton and Rose in honor of Dean James W. Toumey, whose work with certain cacti was significant.

Toumeya papyracantha
Grama-Grass Cactus
PLATES 220, 221

The plants of *T. papyracantha* are small and commonly overlooked, particularly because they typically grow in or near clumps of grama grass. In Hudspeth County plants near 10 cm or more tall were found associated with a rather large, clumped grass species, alkali sacaton, but still well camouflaged within or beside the grass. In Texas the taxon surely ranks with *Ancistrocactus brevihamatus* var. *pallidus* as being either rare or so well camouflaged that it appears to be rare. Modern data strongly suggest that *T. papyracantha* belongs with *Sclerocactus*.

Distribution. Salt flats and gypsum habitats, northeastern Hudspeth Co., associated with alkali sacaton and possibly blue grama, as in NM and AZ. Northeastern Hudspeth Co. near the NM state line N of Dell City and about 18 mi or more S into TX. 3,600–4,000 ft. Arizona, in S Navajo Co., western and central NM (at 5,000–7,200 ft). Map 81.

Vegetative Characters. Plants usually with solitary stems. Stems cylindroid or slightly obconic, 2.5–7.5 cm long, 1.2–2 cm thick. Tubercles elongate, 1.5–5 mm long, dark gray-green, rather soft; glands pale reddish to tan. Stem surface mostly obscured by spines. Central spines 1–3 per areole, whitish or pale brown to gray, erect or ascending, often curled, flattened, thin, 1.2–2.7 cm long, 0.4–2 mm wide; radial spines 6–9 per areole, silvery-white to grayish, flexible, flattened, straight, appressed, 2–5 mm long, 0.3–0.6 mm wide.

Throughout most of its range, stems extend aboveground for only a few centimeters, but often they are taller in the salt flats population in Hudspeth County. The most identifiable feature is flattened, papery, twisted, and curved central spines that resemble dried blades of grass or wood shavings. In upper areoles the central spines tend to stand or curve upward, col-

lectively overarching the stem in a disorganized fashion. Centrals in low areoles extend in any direction.

Flowers. Flowering Mar–May. White flowers produced at the stem apex, narrowly bell-shaped or tubular-funnelform, 2–2.5 cm wide, ca. 2.5 cm long. Filaments whitish; anthers cream-colored. Style greenish to cream-colored, to 2 cm long. Stigma lobes 4–5, cream-colored or pale green, 1.5 mm long.

Fruits. Subglobose fruits are 4.5–7 mm

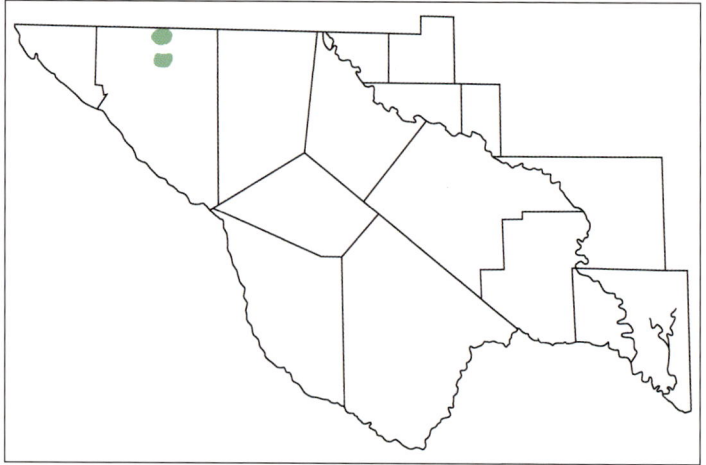

Map 81. *Toumeya papyracantha* (grama-grass cactus).

long, to ca. 4.5 mm wide, green, usually indehiscent, at full maturity tan and dry. Floral remnant persistent. Seeds shiny, black, ca. 3 mm in largest dimension, ca. 1 mm thick, surprisingly large for a small cactus species, broadly obovoid, finely papillate-checkered. Hilum relatively small.

Full Name and Synonyms. *Toumeya papyracantha* (Engelm.) Britton & Rose. [= *Sclerocactus papyracanthus* (Engelm.) N. P. Taylor]. *Mammillaria papyracantha* Engelm.; *Pediocactus papyracanthus* (Engelm.) L. D. Benson.

Other Common Name. Paperspine cactus.

ECHINOMASTUS

Pineapple Cactus

Echinomastus Britton & Rose is a genus of 5–9 species distributed from the southwestern United States to central Mexico. Four species occur in the CDR. About five species occur in the United States. Three species are found in Texas, all of them in the Trans-Pecos. The genus *Echinomastus* has been the subject of considerable taxonomic repositioning, by several cactus specialists, in *Neolloydia*, *Echinocactus*, and *Sclerocactus*, probably its closest relative (Powell and Weedin, 2004). The genus name is derived from the Greek *echinos*, "hedgehog" (for "spiny"), and *mastos*, "breast," in reference to the spiny tubercle apexes.

Key to the Trans-Pecos Species

1. Stems green, usually visible through the spines; central spines stramineous, gray-ish, or reddish, usually the distal part if not the whole spine, pink or dark reddish; anthers cream-colored; stigma red

E. intertextus, p. 250

1. Stems blue-green or blue-gray, tinted by powdered wax on the surface, visible through the spines or not; central spines stramineous or tan, the distal parts, if not nearly the whole spines, chalky-blue, blue-gray, or blue-brown; anthers yellow; stigma green (2).

2(1). Stems usually visible through spreading radial spines; radial spines tan or stramineous and chalky-blue, at least at the tips, 11–15 per areole, 1.5–2.4 cm long; central spines straight or slightly curved, similar to radials but slightly larger in diameter, upper three (usually) angled upward, blue-gray distally, and slightly longer (1.8–2 cm) than the lower central one that is angled upward (or porrect)

E. warnockii, p. 256

2. Stems usually not visible through somewhat appressed and overlapping radial spines; radial spines ashy-white, 26–32 per areole, 0.4–0.7 cm long; central spines (upper ones) markedly larger in diameter (0.4–0.9 mm basally) and longer than radials (upper centrals to 1.5–2.1 cm long), the upper three upswept, curved, and chalky-blue for most of the length, particularly those of the upper areoles in mature plants, lower central usually much shorter, 0.5–0.7 cm long, slightly turned downward or curving downward

E. mariposensis, p. 258

Echinomastus intertextus
Woven-Spine Pineapple Cactus,
Longcentral Woven-Spine
Pineapple Cactus

Two varieties of *E. intertextus* have been recognized by most authors who have dealt with cacti of Texas. The specific epithet *intertextus* presumably refers to the interwoven spines, particularly at the plant apex, and is from the Latin *inter*, "between," and *textilis*, "woven."

Full Name. *Echinomastus intertextus* (Engelm.) Britton & Rose.

Key to the Varieties of *Echinomastus intertextus*

1. Stems spherical in young plants, becoming ovoid to ovoid-cylindroid in age, 5–12 cm high; radial spines and upper centrals markedly appressed, the longest 0.9–1.9 cm; lower central spine 0.05–0.4 cm long

E. intertextus var. *intertextus*, p. 251

1. Stems ovoid to cylindroid, 7–15 cm long; radial spines and upper centrals not appressed, somewhat diffuse, the longest 1.2–2.0 cm; lower central spine 0.4–1.5 cm long

E. intertextus var. *dasyacanthus*, p. 253

Echinomastus intertextus var. intertextus
Woven-Spine Pineapple Cactus
PLATES 222, 223

Echinomastus intertextus var. *intertextus* is a common variety of *Echinomastus* in the Trans-Pecos, being most abundant in the grama grasslands and woodlands of the central mountain region (the Davis Mountains). *Echinomastus intertextus* var. *dasyacanthus* in the Trans-Pecos seemingly is most common in the Franklin Mountains of El Paso County. Most published records of var. *intertextus* from central New Mexico, and many reports from southern New Mexico, probably pertain to *E. intertextus* var. *dasyacanthus* or intermediates between the two varieties. In general the two varieties of *E. intertextus* replace each other geographically in New Mexico. See Powell and Weedin (2004) for further discussion of *E. intertextus*.

Distribution. Mostly in the central mountains, Jeff Davis, N Brewster, N Presidio counties, also Chinati Mts, Del Norte Mts, E to SW Pecos Co., W to E Hudspeth Co. and El Paso Co. (Franklin Mts). 4,000–5,800 ft. West to SW NM (Hidalgo, Luna, and Socorro counties) and SE AZ. Mexico: NE Sonora, Chihuahua, possibly N Coahuila. Map 82.

Vegetative Characters. In the field, var. *intertextus* is most easily distinguished from var. *dasyacanthus* by its appressed spines (radials and upper centrals) lying flat against the stem and by the short, porrect lower central spine, usually only 1–3 mm long in the Davis Mountains populations. In aspect, plants of var. *intertextus* appear as if they could be safely handled. Plants

Plate 222. *Echinomastus intertextus* var. *intertextus* (woven-spine pineapple cactus), N Brewster Co., TX.

Plate 223. *Echinomastus intertextus* var. *intertextus* (woven-spine pineapple cactus), N Brewster Co., TX; mature fruits and seeds.

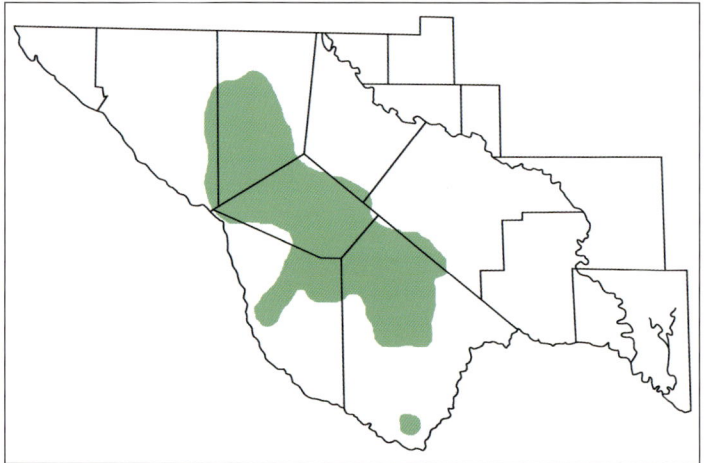

Map 82. *Echinomastus intertextus* var. *intertextus* (woven-spine pineapple cactus).

are often globular to ovoid in shape, compared to more ovoid to cylindroid stems of var. *dasyacanthus*. In var. *dasyacanthus*, radial and upper central spines somewhat diffuse, obviously not flattened against the stem as in var. *intertextus*, and the porrect lower central spine is noticeably longer.

Among Trans-Pecos cacti *E. intertextus* might be confused in the field mostly with *E. warnockii*, although both of these taxa in their pure forms are ecologically separated. Putative hybrids involving *E. intertextus* and *E. warnockii* may be

found in adjacent or mixed populations with the parental species in some localities at or near the desert margin, such as the southeastern Davis Mountains and Chisos Mountains (Plate 224).

Flowers. Flowering Feb–Apr. Flowers of var. *intertextus*, 2.5–3 cm long, 2.3–3 cm in diameter, are borne on new growth at the stem apex. Closely spaced or interlocking apical spines often interfere with full opening of the flowers. Inner tepals bright white to pale pink, or with pale pink midstripe. Outer tepals dull white to pink-

ish, with pink midstripes. Filaments pale greenish and anthers cream-yellow. Greenish style supports 6–12 stigma lobes that are bright red, pink, or rarely white.

When in flower, both varieties of *E. intertextus* are easily distinguished from *E. warnockii* and *E. mariposensis*. Typically, the stigma is red in *E. intertextus* and green in *E. warnockii* and *E. mariposensis*. Also, the tepals in *E. intertextus* are white, often with a pinkish tinge and pink midstripes, whereas the tepals of *E. warnockii* typically are white with greenish, greenish-brown, or tan midstripes. The tepal midstripes in flowers of some *E. warnockii* are pinkish to pinkish-brown.

Fruits. Fruits of *E. intertextus* are green, globose to ovoid, and 0.8–1.5 cm long, at maturity drying to tan, brownish, or dull reddish, with persistent floral remnant, circumscissily dehiscent at the base

in a manner similar to that of *Thelocactus*. After dehiscence the fruit wall is easily dislodged, leaving a small pile of black seeds on the basal fruit remains and the stem surface. The ovate-veniform seeds, 1.5–2.2 mm across, are readily dry and not at all adherent.

Full Name and Synonyms. Echinomastus intertextus var. *intertextus*. *Echinocactus intertextus* Engelm.; *Neolloydia intertexta* (Engelm.) L. D. Benson; *Sclerocactus intertextus* (Engelm.) N. P. Taylor.

Echinomastus intertextus var. dasyacanthus
Longcentral Woven-Spine Pineapple Cactus
PLATES 225, 226
Variety *dasyacanthus* is considered here to be a taxon of central and southwestern

Plate 225. *Echinomastus intertextus* var. *dasyacanthus* (longcentral woven-spine pineapple cactus), Franklin Mts, El Paso Co., TX (photo by Richard D. Worthington).

Plate 226. *Echinomastus intertextus* var. *dasyacanthus* (longcentral woven-spine pineapple cactus), Doña Ana Co., NM.

New Mexico that barely enters the Trans-Pecos in El Paso County. The varietal epithet *dasyacanthus* refers to the shaggy aspect created by the relatively long, protruding spines and is from the Greek *dasy,* "shaggy," and *akantha,* "spine," which contrasts with the smooth aspect of typical *E. intertextus.*

Distribution. Desert mountains, slopes, and grasslands. In the Trans-Pecos restricted to the Franklin Mts, El Paso Co.; the isolated population keying to

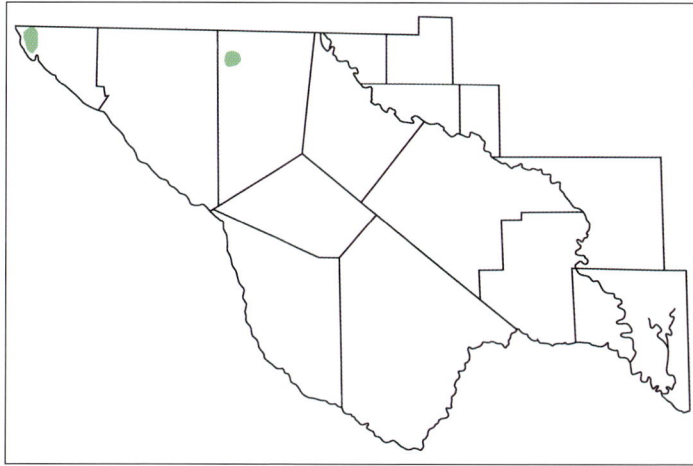

var. *dasyacanthus* in Big Bend National Park may owe its long spines to introgressive hybridization from *E. warnockii* into ordinary short-spined var. *intertextus* and may not be a genuine var. *dasyacanthus* (Powell, 2002; Powell and Weedin, 2004). 3,800–5,600 ft. New Mexico, Otero Co. (Jarilla Mts) N to Bernalillo Co., W to Luna Co. (Florida Mts) and Hidalgo Co. (Little Hatchet Mts). Map 83.

Vegetative Characters. Stems ovoid to cylindroid, usually 7–17 cm long when sexually mature. In habit, plants bristly and uncomfortable to handle, compared to the smooth-looking var. *intertextus.* "Bristly" or "shaggy" look results from slightly appressed to spreading radial and upper central spines, along with a porrect lower central spine usually 0.7–1.5 cm long. Radial spines of var. *dasyacanthus* number 16–24, and usually are longer (0.8–2.4 cm) than those of var. *intertextus.* Dominant spine color in both varieties varies from gray or brown to dull reddish; especially on the distal portions of the spines, reddish inner layers and spine tips may lend whole stems a dull and dark reddish hue, especially when the spines are wet.

Flowers. Flowering Feb–Mar. Flowers appear to be identical to those of var. *intertextus.*

Fruits. Fruits and seeds are virtually identical to those of var. *intertextus.* In some specimens of var. *dasyacanthus* from El Paso County, fruits tend not to be as crowded at the stem apex as in var. *intertextus,* where they are physically compressed into a low cone by restricting appressed spines. Fruits in var. *dasyacanthus* are not as tightly trapped because the spines are more divergent.

Full Name and Synonyms. *Echinomastus intertextus* var. *dasyacanthus* (Engelm.) Backeb. *Echinocactus intertextus* Engelm. var. *dasyacanthus* Engelm.; *Neolloydia intertexta* (Engelm.) L. D. Benson var. *dasyacantha* (Engelm.) L. D. Benson; *Sclerocactus intertextus* (Engelm.) N. P. Taylor var. *dasyacanthus* (Engelm.) N. P. Taylor; *Echinomastus dasyacanthus* (Engelm.) Britton & Rose; *Thelocactus dasyacanthus* W. T. Marshall ex Kelsey & Dayton.

Echinomastus warnockii
Warnock's Cactus

PLATES 227, 228

Echinomastus warnockii is one of the most common cactus species in the desert southern Big Bend region and adjacent areas. The specific epithet honors Barton H. Warnock, longtime professor of biology at Sul Ross State University.

Distribution. Common on desert gravel hills and benches, limestone hills, alluvial flats, occasional in gypsum flats with desertscrub vegetation, most abundant in southern Presidio and Brewster counties, but also in desert habitats NW to Jeff Davis, Culberson, and southern Hudspeth counties. 1,900–4,500 ft. Mexico: adjacent to the Big Bend region in NE Chihuahua (W of Ojinaga) and expected in NE Coahuila. Map 84.

Vegetative Characters. Solitary, globose to oblong blue-green stems with tubercles coalescent into low ribs, and spreading, divergent spines not hiding the stem in turgid, healthy plants. Most plants in the Big Bend region approximately 3–8 cm tall. Much larger stems, to 20 cm long, occasionally encountered in populations near the Rio Grande. Radial spines, 11–15, sufficiently divergent, stiff, and overlapping that it is difficult to measure their length with an ordinary ruler, although some radial spines may be somewhat appressed. Radial spines acicular and of slightly smaller caliber than the usually three (in adult plants) upper central spines that merge upward with the radials. In adult plants a single porrect or upward-directed lower central spine is evident. Radial spines tan to stramineous, gray at the base and

Plate 227. *Echinomastus warnockii* (**Warnock's cactus**), **S Brewster Co., TX; cultivated.**

Plate 228. *Echinomastus warnockii* (Warnock's cactus), Mariscal Mt, S Brewster Co., TX; mature fruits; cultivated.

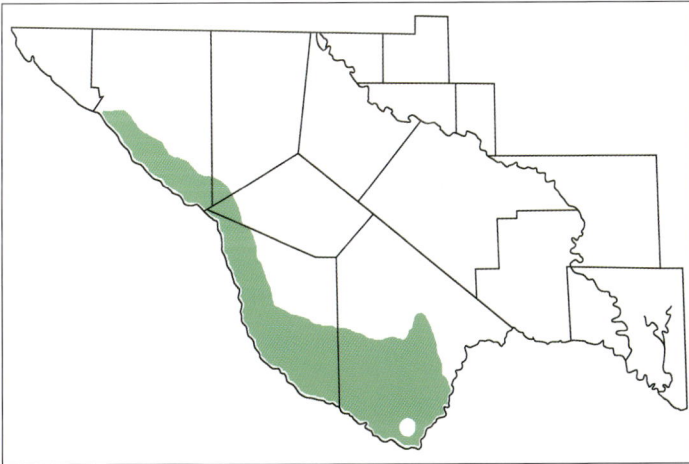

Map 84. *Echinomastus warnockii* (Warnock's cactus).

"chalky-blue" or blue-gray (rarely reddish-purple) on the distal portion, or nearly the whole spines may be cyanic. Color of central spines similar to that of radials, or the centrals, particularly the lower central, may be blue-gray (rarely dark reddish-purple) for more that half of their length. *Echinomastus warnockii* is so different from the related species *E. mariposensis* that the two taxa are not likely to be confused in the field.

Flowers. Flowering usually Feb–Mar,

Big Bend National Park and vicinity. White flowers 2.5–3 cm long and 2.3–3 cm in diameter when widely open, with reflexed tepals. Size and general appearance the same as in *E. intertextus* and *E. mariposensis*. Anthers of *E. warnockii* bright yellow; style and 6–10 stigma lobes green. Green stigma lobes and relatively dark yellow anthers are reliable floral differences between *E. warnockii* and *E. intertextus*, which has red stigma lobes and cream-yellow anthers.

Fruits. Light green fruits globular, 0.7–1 cm in diameter, thin-walled, mostly indehiscent. Thin-walled fruits become dry and fragile after ripening, in some cases may split longitudinally. Ripe fruits usually green but may develop pinkish tinge on exposed surfaces. Floral remnant persistent.

Full Name and Synonyms. Echinomastus warnockii (L. D. Benson) Glass & R. Foster. *Neolloydia warnockii* L. D. Benson; *Sclerocactus warnockii* (L. D. Benson) N. P. Taylor; *Echinocactus warnockii* (L. D. Benson) Weniger, nom. nud.

Other Common Name. Warnock biznagita.

Echinomastus mariposensis
Mariposa Cactus
PLATES 229, 230

Echinomastus mariposensis is one of the early-blooming cactus species in the southern Big Bend region. The relatively small plants, covered with ashy-white spines and with white flowers, are difficult to locate against the pale background of limestone rock. *Echinomastus mariposensis* was listed federally as a Threatened species in 1979, and in 1983 it was accorded the same status by Texas. *Echinomastus mariposensis* is restricted to specific habitats within its geographic range, but it is widespread and abundant across that range. The specific epithet is taken from the name of a cinnabar mine and mining town near the type locality on private land northwest of Terlingua in Brewster County.

Distribution. Locally common in thin Boquillas limestone soil, frequently on

Plate 229. *Echinomastus mariposensis* (Mariposa cactus), S Brewster Co., TX.

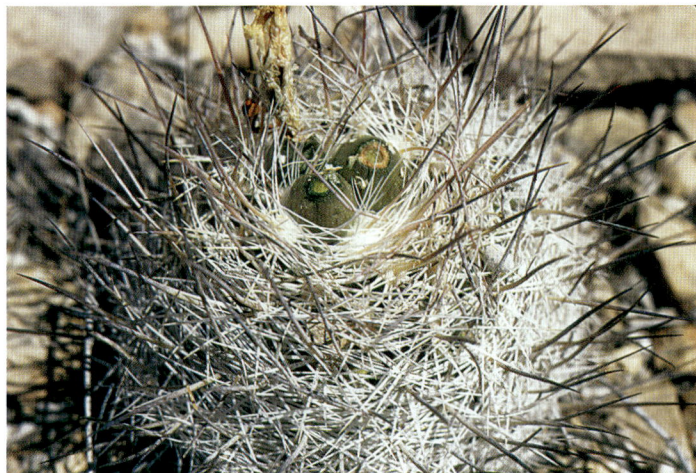

Plate 230. *Echinomastus mariposensis* (Mariposa cactus), S Brewster Co., TX; mature fruits.

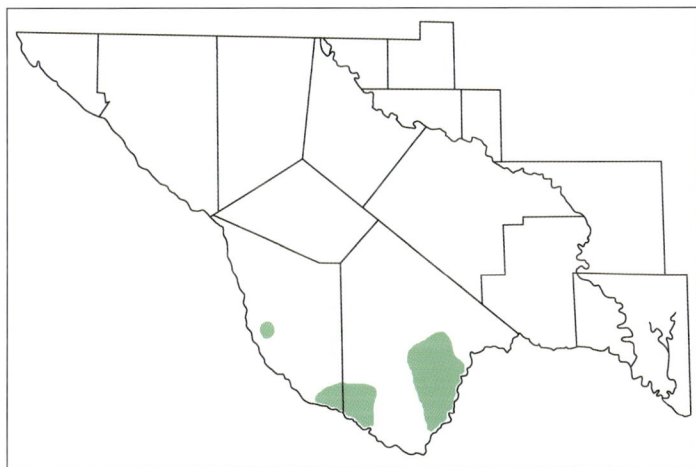

Map 85. *Echinomastus mariposensis* (Mariposa cactus).

mesas and ridgetops, but also in saddles and on slopes, often in gravel-like rubble or in relatively stable rocky substrates, southern Presidio and Brewster counties, most common from S and NE of Lajitas E beyond Black Gap Wildlife Management Area; one collection from near Ruidosa in W-central Presidio Co. 2,500–3,700 ft. Mexico: well documented in eastern Coahuila. Map 85.

Vegetative Characters. Sexually mature plants the general size and shape of golf balls or tennis balls. Thus, unbranched stems typically subglobose, broadly ellipsoid, or cylindroid. Larger plants approaching 10–12 cm tall are not common in Texas. Ashy-white habit is the result of overlapping white radial spines. Plants often exhibit slightly darker hue because central spines are cyanic on distal portion. Blue-green stem surfaces are obscured by dense spine cover.

Radial spines slender, 25–32 in number, 0.3–1.2 cm long. In older areoles most radial spines are oriented horizontally. Usually four central spines per areole (rarely six) in mature plants: one lower central, three upper centrals. Central spines markedly heavier than slender radials and with small-bulbous bases. Porrect or down-curved lower central spine 0.6–1.2 cm long, 0.3–0.6 mm thick at the base. Tentative field identification of *E. mariposensis* comes from observing subglobose, small, ashy-white plants with upswept cyanic central spines, often forming a diffuse tuft of spines at the apex.

Flowers. Flowering Feb–Mar. White flowers of *E. mariposensis* 2–3 cm long and 2–3 cm in diameter, borne at stem apex. Inner tepals white or very pale pink except for flesh-pink, pale yellowish-tan, or pale green midregions. Outer tepals either white or pale pink at margins and with greenish or brownish midregions. Anthers cream-yellow to slightly darker yellow. Style pale green, supporting 5–8 green to yellow-green stigma lobes.

Fruits. Green to yellowish-green fruits globose to oblong, ca. 1 cm long, ca. 0.8 cm thick. After ripening, fruits remain green, usually indehiscent, but may split longitudinally on one side. Seeds black, subreniform or subglobular, 1.6–2 mm long, and surfaces finely papillate.

Other Distinctions. Echinomastus mariposensis is easily distinguished from the closely related *E. warnockii* by the spine characters discussed above. In the field some nonflowering plants of *E. mariposensis* resemble single-stemmed plants of *Mammillaria pottsii*. Plants of both species are whitish in overall appearance; both have an abundance of appressed,

slender, ashy-white radial spines; and both display up-curved, cyanic central spines in the upper areoles. In addition, both species occur in similar limestone habitats in the western portion of the range of *E. mariposensis*. Plants of *M. pottsii*, however, typically are multistemmed, and at least in larger plants, the stems are cylindroid. *Mammillaria pottsii* has a very different spine pattern, with eight centrals and usually over 40 radials. When in flower in late spring, *M. pottsii* is distinctive with a row of small reddish-maroon flowers circling the stem below the apex.

In the field *E. mariposensis* also could be confused with *Coryphantha sneedii* var. *albicolumnaria, C. tuberculosa* var. *tuberculosa, C. tuberculosa* var. *varicolor,* or *C. dasyacantha.* All of these taxa can be distinguished from *E. mariposensis* by their central spine morphology, which includes usually several spreading, central spines oriented as inner and outer centrals.

Full Name and Synonyms. Echinomastus mariposensis Hester. *Neolloydia mariposensis* (Hester) L. D. Benson; *Echinocactus mariposensis* (Hester) Weniger, nom. nud.; *Sclerocactus mariposensis* (Hester) N. P. Taylor.

Other Common Names. Mariposa viszagita; Mariposa Lloyd's cactus; Lloyd's Mariposa cactus; golfball cactus. The U.S. Fish and Wildlife Service and the Texas Parks and Wildlife Department list this taxon as Threatened under the common name Lloyd's Mariposa cactus. The application of "Lloyd's" to a common name for *Echinomastus mariposensis* is enigmatic, because Lloyd (presumably F. E. Lloyd) apparently had nothing to do with this species.

EPITHELANTHA

Button-Cactus

Epithelantha F. A. C. Weber ex Britton & Rose is a genus of 2–5 species distributed from western Texas to Arizona and in northeastern Mexico. Two taxa occur in the United States and in Texas. Five or six taxa of *E. micromeris* are recognized for Mexico. The genus name is derived from the Greek *epi*, "on"; *thel*, "nipple"; and *anthos*, "flower," in reference to the position of flowers near the apex of tubercles.

Key to the Trans-Pecos Species

1. Stems whitish-gray, relatively rough in appearance; areoles and their spines 4–5 mm across (on sides of stems); spines of each areole in 1–3 series, 20–30 in all; flowers 6–8.5 mm long, 3–4.6 mm wide, only the tips exposed above the spines and wool of the stem apex; stamens 10–16

E. micromeris var. *micromeris*, p. 262

1. Stems whitish (or cream-colored), smooth and shiny in appearance; areoles and their spines 2–2.5 mm across (on sides of stems); spines of each areole in many series, 33–40 or more in all; flowers 10–17 mm long and wide, conspicuously exserted above the spines at stem apex; stamens 20–40

E. bokei, p. 264

Epithelantha micromeris var. micromeris
Common Button-Cactus
PLATES 231, 232

In the Trans-Pecos the plants of *E. micromeris* var. *micromeris* usually are unbranched, and they have been described aptly as fuzzy white balls. The whitish-gray spines completely cover the wartlike greenish tubercles. Small, usually pinkish flowers are borne at the stem apex in an upswept apical tuft of spines. In late spring slender, bright red fruits stand prominently on the "button-size" whitish or gray plants. The specific epithet is taken after the Greek *mikros*, "small," and *meros*, "part," for the small characters of the species.

Distribution. Infrequent to locally common in rocky substrates of limestone and igneous origin, hills and ridges in the desert and grasslands, to be expected in every county, but currently known in the eastern Trans-Pecos in eastern Presidio Co., Brewster Co., and Pecos Co. E to Val Verde Co., and in western Culberson Co. and El Paso Co. 2,000–5,000 ft. East in Texas to Medina, Real, and Bandera counties, and to Upton Co., N to Howard Co., W across S-central NM (rare except in Otero and Eddy counties), N to near Belen on limestone to AZ, rare in S Cochise Co. Presumably northern Mexico, but identifiable Mexican specimens thus far are *E. micromeris* var. *greggii* (Engelm.) Borg. Map 86.

Vegetative Characters. In the field, potentially confused with the closely related *E. bokei*, *Mammillaria lasiacantha*, and seedlings of *Coryphantha vivipara*. All of these have small stems hidden by whitish spines. In *E. micromeris* the stems are solitary or in clumps, each stem 1.5–6 cm long, 1–4 cm across. Spines are 2–6 cm long. In sexually mature *Epithelantha* the spines at the stem apex are about twice as long as the spines in areoles on the sides of the stem. The reason is that during the course of one season or more the distal halves of the longer spines break away, leaving shorter spine clusters that are appressed against the stem in older, lateral areoles.

Flowers. Flowering late winter to early spring. In *E. micromeris*, 5–8 pink or pale yellow inner tepals per flower, with the largest of these only 2.5–3.5 mm long and 1–2 mm wide. Only 10–16 stamens, with pink to cream-yellow anthers. The 3–4 stigma lobes are white. Reduced, somewhat funnelform flowers of *E. micromeris* are the smallest (to 4.6 mm across) of any cactus species in the Trans-Pecos. When in flower, *Epithelantha* and *M. lasiacantha* are easily distinguished. The flowers of *M. lasiacantha* are on the shoulders of the stem, and both inner and outer tepals usually are "candy-striped" with sharply defined red to brownish midregions or midstripes and prominent whitish margins.

Fruits. Red fruits naked, cylindroid, usually 1–2 cm long, 2–3 mm thick, hollow, few-seeded, with thin but modestly juicy walls. Floral remnant caducous. Seeds black, glossy, 1–1.4 mm long, somewhat comma- or helmet-shaped, with a deep cavity associated with the hilum.

Full Name and Synonym. *Epithelantha micromeris* (Engelm.) F. A. C. Weber ex Britton & Rose var. *micromeris*. *Mammillaria micromeris* Engelm.

Plate 231. *Epithelantha micromeris* var. *micromeris* (common button cactus), S Brewster Co., TX; cultivated.

Plate 232. *Epithelantha micromeris* var. *micromeris* (common button cactus), S Brewster Co., TX; mature fruits; cultivated.

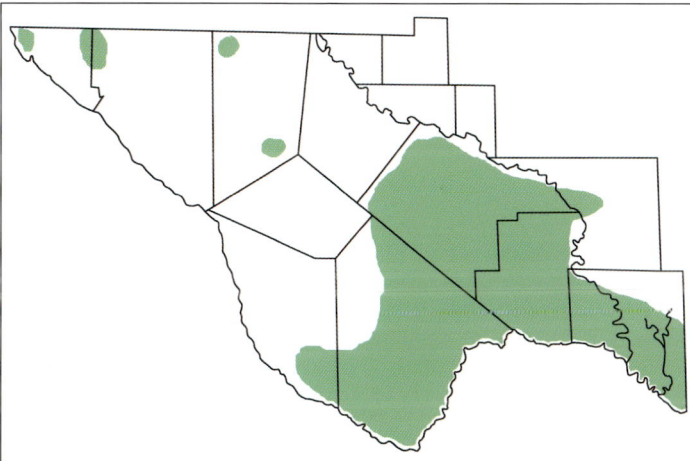

Map 86. *Epithelantha micromeris* var. *micromeris* (common button cactus).

Epithelantha bokei
Boke's Button-Cactus
PLATES 233–35

Epithelantha bokei often appears as one of the more common small cacti in the limestone hills of southern Brewster County, particularly southeast of the Chisos Mountains. In this area *E. bokei* is much more common than *E. micromeris.* The two small cactus species are easily distinguished in the field, even from a standing position. The plants of *E. bokei* are usually seen as short-cylindroid stems, solitary or in clusters, flat or concave at the apex, with a shining, smooth coat of close-set spines covering the sides of the stems. Characteristically, the stems of *E. micromeris* are more rounded at the apex (but immatures of both species have concave apexes), the close-set coat of spines is ashy-white or gray, and the texture is more coarse in appearance and feel (but immatures of both species are smooth and shiny). The species *E. bokei* is named after Norman H. Boke, plant anatomist, who at the University of Oklahoma studied this species and other cacti. Boke also super-vised several students whose investigations contributed to knowledge about the Cactaceae.

Distribution. Rocky limestone soils, ridges, tops, and slopes of desert mountains, near the Rio Grande in southern Brewster and Presidio counties. 2,300–4,400 ft. Mexico: Coahuila, SE to near Saltillo. Map 87.

Vegetative Characters. Stems 1.5–5 cm long and wide, with small tubercles ca. 1.5 mm long. Adjacent spine clusters are merely touching or barely overlapping in *E. bokei,* whereas spine tips from adjacent clusters in *E. micromeris* usually are overlapping. The tiny spines themselves in *E. bokei* have margins that appear to be whiter than the central portion of the spines. The ashy-white spines of *E. micromeris* do not have margins that are distinctive in color.

Flowers. Flowering May–Jun. Pale pink to silvery-white flowers always larger at 1–1.7 cm in diameter than those of *E. micromeris.* In each flower, 13–21 inner tepals, translucently pale, usually with very diffuse pink, pinkish-tan, or yellow-green

Plate 233. *Epithelantha bokei* (Boke's button-cactus), S Brewster Co., TX; cultivated.

Plate 234. *Epithelantha bokei* (Boke's button-cactus), S Brewster Co., TX; cultivated.

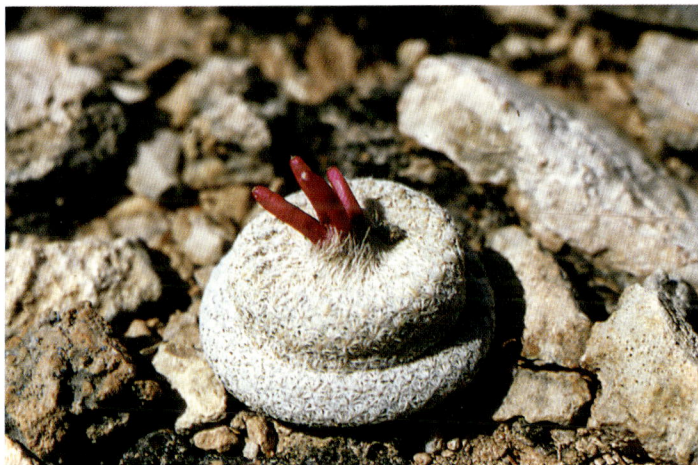

Plate 235. *Epithelantha bokei* (Boke's button-cactus), Solitario Uplift, Presidio Co., TX; mature fruits.

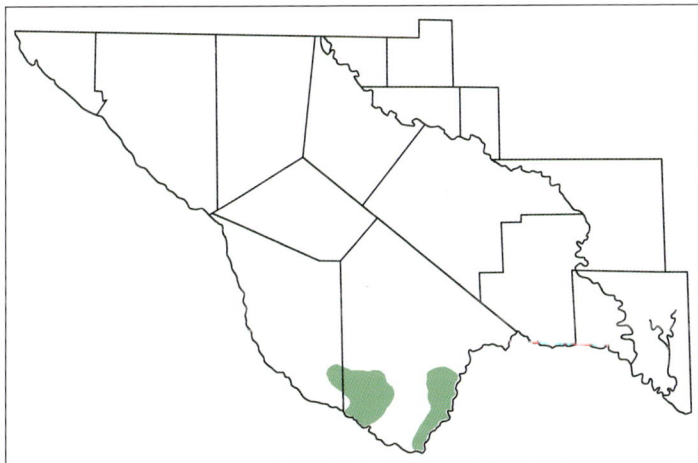

Map 87. *Epithelantha bokei* (Boke's button-cactus).

midstripes. Stamens number 20–40 in each flower, with pinkish to cream-yellow anthers. Stigma lobes white.

Fruits. Red fruits naked, cylindroid when elongated, 0.8–1.3 cm long, 1.5–2.3 mm thick. Fruits elongate, turning red only under favorable conditions; under unfavorable conditions remaining small and green. Seeds black, glossy, 1–1.4 mm long, comma- or helmet-shaped, and somewhat bowl-like with a deep cavity associated with the hilum.

Full Name and Synonym. *Epithelantha bokei* L. D. Benson. *Epithelantha micromeris* var. *bokei* (L. D. Benson) Glass & R. Foster.

Other Common Name. Boquillas button-cactus.

Pincushion Cactus

Mammillaria Haw. is a genus of ca. 164 species occurring in the southwestern United States and Mexico, with a few species in South America (Colombia and Venezuela), Central America, and the West Indies. Eight species occur in Texas, all but *M. sphaerica* in the Trans-Pecos. All twentieth-century authors except for Weniger (1970, 1984) have maintained *Mammillaria* as distinct from *Coryphantha* while pointing out that the most consistent difference between the two genera is the lack of an adaxial areolar groove on the tubercles in *Mammillaria*. *Mammillaria* usually also differs by producing flowers on older growth in a ring 1 cm or more below the apex (typically between tubercles), whereas in *Coryphantha* flowers usually are produced on current growth at the stem apex, although some coryphanthas exhibit intermediate flower position. In general, the Trans-Pecos mammillarias are a group of dissimilar species. *Mammillaria grahamii, M. wrightii, M. lasiacantha, M. pottsii,* and *M. prolifera* are all distinctive, but *M. meiacantha–M. heyderi* is a species-pair. The genus name is from the Latin *mammilla,* "breast" or "teat," in reference to the usually prominent tubercles.

Key to the Trans-Pecos Species

1. Hooked central spine(s) present; flowers rose-pink, rose-purple, or magenta (2).
1. Hooked central spine(s) absent; flowers white to pink with prominent-colored midstripes, or flowers maroon-red or yellow (3).

2(1). Central spines 3–4, only one (the lower one) prominent, porrect, hooked, dark red-brown or blackish, the others inconspicuous, appressed-erect, straight, relatively slender; radial spines 19–35; pith and cortex not mucilaginous; outer tepals minutely fimbriate; fruit when fully ripe bright red, clavate; Franklin Mts, El Paso Co., to NW Presidio Co.

M. grahamii var. *grahamii*, p. 269

2. Central spines 1–4, one or usually all hooked, all protruding, brownish; radial spines usually ca. 13; pith and cortex mucilaginous; outer tepals prominently fimbriate; fruit when fully ripe green to dull purplish, globose to ovoid; Franklin Mts, El Paso Co.

M. wrightii var. *wrightii*, p. 270

3(1). Plants when mature in low, matlike clumps, with 10–20 or more subglobose to short columnar stems of different sizes; radial spines numerous, the ones closest to the stem white, hairlike, flexible, usually curved and twisted; distributed in Val Verde Co.

M. prolifera var. *texana*, p. 283

3. Plants when mature with solitary stems, flat-topped, hemispheric, to depressed-globose, or with usually branched stems, cylindroid or clavate; radial spines few to numerous, white, tan, yellowish, brown, reddish-brown or gray, not hairlike, all straight; distributed throughout Trans-Pecos (4).

4(3). Stems narrowly cylindroid or clavate, 6–15 cm or more tall, usually branched at the base; flowers maroon-red, the inner tepals usually with darker midregions

M. pottsii, p. 275

4. Stems depressed globose, flat-topped or hemispheric (aboveground), unbranched; flowers white to pink, the inner tepals usually with conspicuous midstripes of pink or purple to brown or tan (5).

5(4). Stems depressed globose to short cylindroid, usually 1.5–4 cm in diameter, stem surface completely hidden by very numerous, thin, interlaced whitish spines

M. lasiacantha, p. 272

5. Stems flat-topped or hemispheric (the part aboveground), usually 7–15 cm or more in diameter, stem surface and its prominent tubercles evident through slender or thick, spreading, stramineous, gray, to red-brown radial spines (6).

6(5). Radial spines usually 5–7 per areole, to 0.35–0.7 mm thick; fruits dull red to purplish-pink

M. meiacantha, p. 277

6. Radial spines usually 13–17, to 0.15–0.45 mm thick; fruits bright red

M. heyderi, p. 279

Mammillaria grahamii var. grahamii
Graham's Fishhook Cactus
PLATES 236, 237

The occurrence of *M. grahamii* in northwestern Presidio County marks the southeastern distributional limits of hooked-spined mammillarias, so far as known. Benson (1982) recognized another variety of this species, *M. grahamii* var. *oliviae* (Orcutt) L. D. Benson, from southeast Arizona, with shorter, straight central spines. The specific epithet is after Colonel James Duncan Graham (1799–1865), topographical engineer, astronomer, and surveyor with the eastern portion of the U.S.-Mexican border survey. Mount Graham in Arizona was named after him.

Distribution. Desert mountains, igneous or limestone, desertscrub or grasslands. Northwest Presidio Co. near Candelaria, below the rim of the Sierra Vieja, to El Paso Co., Franklin Mts. 3,200–4,600 ft. Southern NM, S and central AZ, and extreme E CA. Mexico: Sonora and Chihuahua. Map 88.

Vegetative Characters. The plants of *M. grahamii* are unbranched or branched at the base. Plants in Presidio County often are clumped with 5–10 or more stems of different sizes. The stems are globose to short-cylindroid, 5–10 cm long, 4–6.5 cm in diameter, and whitish to gray in aspect because of a rather dense covering of radial spines that typically number 19–25 in each areole. Upon close examination the single, porrect, hooked central spine is evident. The slender hooked spine is 1.1–2 cm long, and its dark red-brown to blackish color is in contrast to the whitish radial spines.

Plate 236. *Mammillaria grahamii var. grahamii* (Graham's fishhook cactus), W Presidio Co., TX.

Plate 237. *Mammillaria grahamii* var. *grahamii* (Graham's fishhook cactus), S Presidio Co., TX; mature fruits; cultivated.

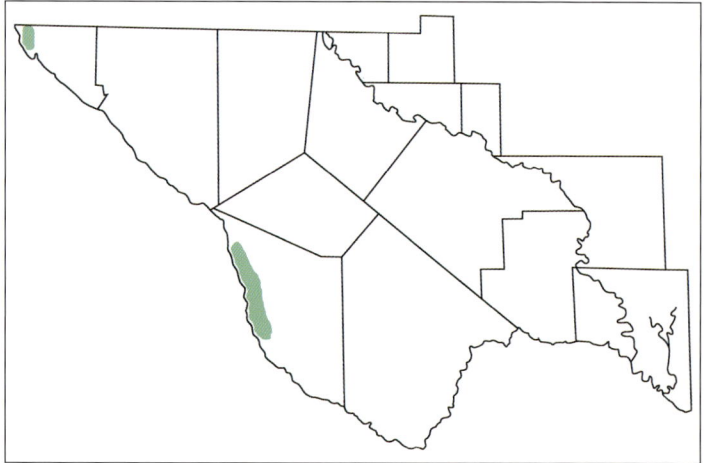

Map 88. *Mammillaria grahamii* var. *grahamii* (Graham's fishhook cactus).

Flowers. Flowering May–Jun, sporadically in Sep. The flowers of *M. grahamii* are bright rose-pink or magenta, 2–4 cm in diameter when fully open with reflexed tepals. The stamens have pinkish filaments and yellow to pale orange anthers. The relatively long style extends ca. 6 mm or more above the anthers and supports 6–10 slender, green stigma lobes, 3–7 mm long.

Fruits. The fruits are bright red at maturity. Before maturation the fruits are green and spheroidal but usually gradually elongate, becoming somewhat club-shaped,

1.2–2.7 cm long and 5–8 mm in diameter, as they turn red. The seeds are black, globose to pear-shaped, and ca. 1 mm long.

Full Name. *Mammillaria grahamii* Engelm. var. *grahamii*.

Mammillaria wrightii var. wrightii
Wright's Fishhook Cactus
PLATES 238, 239
Mammillaria wrightii barely enters Texas in the Franklin Mountains of western El Paso County. It was collected from

Plate 238. *Mammillaria wrightii* var. *wrightii* (Wright's fishhook cactus), W slope, Organ Mts, Doña Ana Co., NM (photo by Dale and Marian Zimmerman).

Plate 239. *Mammillaria wrightii* var. *wrightii* (Wright's fishhook cactus) near Santa Rita, Grant Co., NM (photo by Dale and Marian Zimmerman).

the Franklin Mountains in 1909, but since then it has not been rediscovered. Its major distribution is in New Mexico, where it is part of a complex of hooked-spined mammillarias. Another variety of this species, *M. wrightii* var. *wilcoxii* (Toumey ex K. Schum.) W. T. Marshall, occurs in New Mexico, Arizona, and Sonora and Chihuahua, Mexico. The specific epithet honors Charles Wright (1811–85), early plant collector in Texas and a plant collector with the early U.S.-Mexican border survey.

Distribution. Rare in the Franklin Mts, El Paso Co. ca. 5,000 ft? Grassy slopes, Great Plains grasslands, pinyon-juniper woodland, most of NM except the E plains and the extreme N part, into E-central AZ. Mexico: reportedly Chihuahua (Laguna de Santa Maria). Map 89.

Vegetative Characters. In habit, usually unbranched, with the tapering base of the stem deeply seated in the ground. During long dry periods and in winter, the whole stem may retract to near ground level.

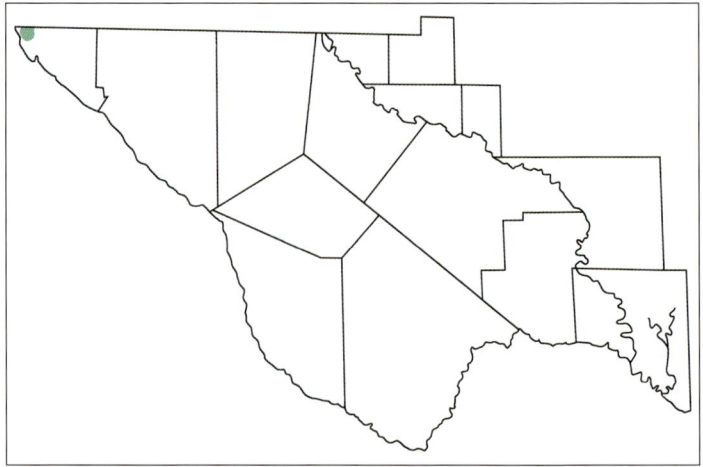

Stems flat-topped or globose, 4–8 cm in diameter, with tubercles ca. 1.2 cm long. Both *M. wrightii* and the superficially similar *M. grahamii* occur in the Franklin Mountains, if *M. wrightii* still is extant there. *Mammillaria wrightii* generally is easily distinguished by most of the characters listed above under *M. grahamii* and usually by 2–3 (less often 1–4 or more) hooked central spines. *Mammillaria wrightii* also differs vegetatively in having usually 8–15 radial spines, compared to 20 or more radials in *M. grahamii*.

Flowers. Flowering after summer rains. Rose-purple or magenta flowers typically 2.5–4 cm long and 2.5–7.5 cm in diameter. Anthers bright yellow. Style greenish, pinkish above, supporting 7–11 stigma lobes that are yellow or pale green, or rarely reddish.

Fruits. Globose or ovoid fruits, previously described as "grapelike," usually 1.3–2 cm long, 1.2–2.5 cm in diameter. Fruits green or dull purple, very juicy, and differ conspicuously from the elongate, bright red mature fruits of *M. grahamii*.

Seeds of *M. wrightii* black, pitted, 1.3–1.5 mm long.

Full Name. *Mammillaria wrightii* Engelm. var. *wrightii*.

Mammillaria lasiacantha
Golf Ball Cactus
PLATES 240–42

Mammillaria lasiacantha is one of the most widespread small desert cacti in the Trans-Pecos. Usually its common name, golf ball cactus, is an accurate reference to size, but the plants are mature at the size of ping-pong balls or smaller, and plants the size of tennis balls are found occasionally, especially near Candelaria, Presidio County, and near the Old Ore Road, Big Bend National Park. In the Trans-Pecos *M. lasiacantha* is most common in rocky limestone hills and in alluvium derived from limestone rather than in igneous substrates. Some of the largest plants of this species occur in igneous-derived soils in northwest Presidio County below the Sierra Vieja rim. The specific epithet is after the Greek *lasios*, "hairy," "woolly," or "shaggy," and

Plate 240. *Mammillaria lasiacantha* (golf ball cactus), Fresno Creek, Presidio Co., TX; pinkish midregions.

Plate 241. *Mammillaria lasiacantha* (golf ball cactus), Rosillos Mts, S Brewster Co., TX; brownish midregions.

akantha, "thorn" or "spine," a reference to the whitish spines that clothe the plants.

Distribution. Desert mountains, hills, and alluvial flats, usually limestone but also igneous substrates. Often with lechuguilla scrub or other desertscrub. Throughout much of the Trans-Pecos except the central Davis Mts. 1,500–4,500 ft. Southeast and S-central NM. Mexico: NE portion S to San Luis Potosí, NW Sonora. Map 90.

Vegetative Characters. Unbranched stems usually depressed-globose, 1–3 cm tall, 1.5–3.5 cm in diameter, deeply seated in the substrate and inconspicuous, with the stem completely obscured by whitish spines. Larger plants found at some sites are subglobose to short cylindroid, to 7–8 cm long, 6–7 cm in diameter. Usually 40–60 white or ashy-gray radial spines per areole, 2.5–5 mm long, thin, mostly appressed, those of adjacent areoles often

Plate 242. *Mammillaria lasiacantha* (golf ball cactus), Candelaria, Presidio Co., TX; mature fruits.

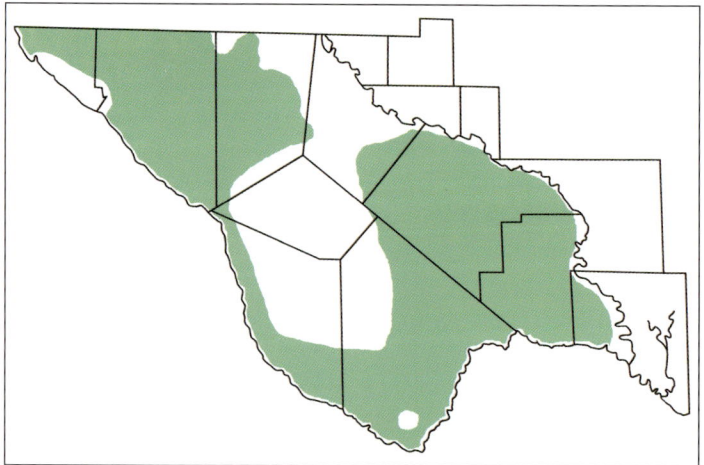

Map 90. *Mammillaria lasiacantha* (golf ball cactus).

overlapping. All the spines are interpreted as radial spines, although 1–6 short spines, 0.6–2 mm long, resembling true radials, occupy the central part of the areole. The plants can be handled without spine pricks. The spines are glabrous or plumose, with the short, fine hairs easily detected on the spines with the aid of a hand lens. *Mammillaria lasiacantha* in habit resembles *Epithelantha*, particularly *E. micromeris*. Spine clusters in *M. lasiacantha* usually are 9–10 mm across, but in *Epithelantha* the spine clusters generally are 2–5 mm in diameter. Both *M. lasiacantha* and *E. micromeris* exhibit a relatively coarse stem surface texture, whereas *E. bokei* has a satinlike surface texture.

Flowers. Flowering Feb–Mar. Flowers white or cream-colored with conspicuous midstripes of pink, purplish, reddish-brown, salmon-red, or greenish-tan colors. Flowers 0.9–1.5 cm long, 0.8–1.2 cm in diameter, and rotate when fully open. Filaments pale yellow, anthers yellow. Style greenish, slightly longer than the stamens, supporting usually 4–5 green or yellow-green stigma lobes, 0.3–1 mm long.

Fruits. Fruits bright red, cylindroid or clavate, 1–2.3 cm long, 4–5 mm in diameter. Fruit surface naked. A floral remnant is persistent at the fruit apex. Seeds black, pitted, globose or somewhat comma-shaped, 1–1.2 mm long. Elongated red fruits of *M. lasiacantha* are similar in appearance to those of *Epithelantha.*

Full Name and Synonym. Mammillaria *lasiacantha* Engelm. *Mammillaria lasiacantha* var. *denudata* Engelm.

Mammillaria pottsii
Potts' Mammillaria
PLATES 243, 244

Mammillaria pottsii is distinctive among other mammillarias of the Trans-Pecos. It is mostly restricted to the southern Big Bend region, in assorted sedimentary habitats such as those in the vicinity of Terlingua. The major distribution of *M. pottsii* is in Mexico, where its relatives are located. Presumably, the specific epithet honors F. H. Potts (1824–88), a mining engineer in the Sierra Madre, Mexico. Reportedly, Potts was a source of cactus plants for the Kew Royal Botanic Gardens

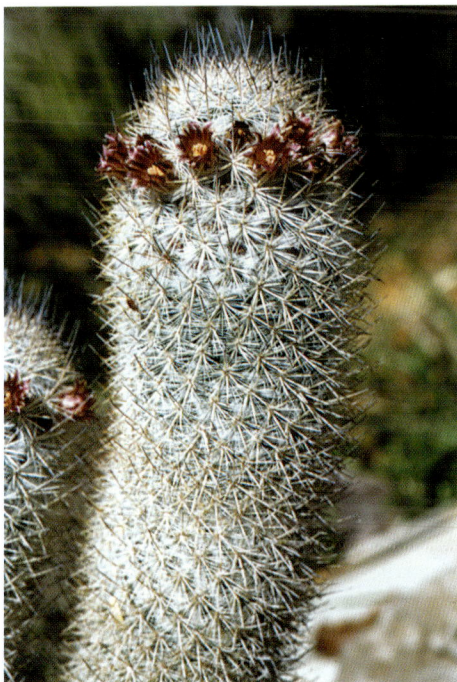

Plate 243. *Mammillaria pottsii* (Potts' mammillaria), S Brewster Co., TX; cultivated.

Plate 244. *Mammillaria pottsii* (Potts' mammillaria), S Brewster Co., TX; cultivated.

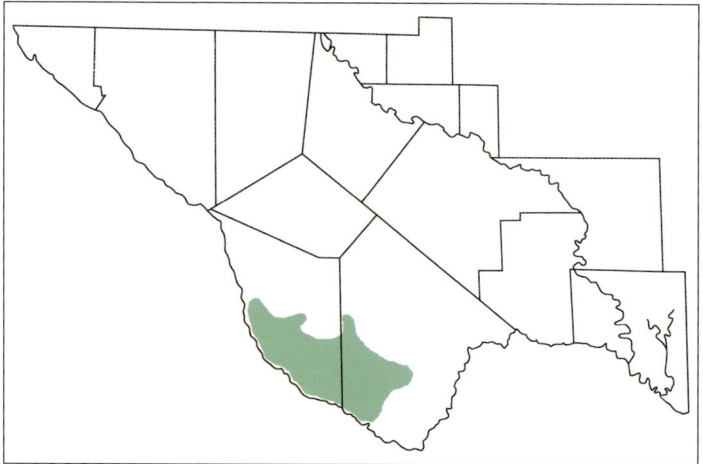

Map 91. *Mammillaria pottsii* (Potts' mammillaria).

in London and ultimately for the foremost German cactologist, Salm-Dyck. Instead, the epithet may be for John Potts.

Distribution. Mostly if not entirely limestone hills, slopes, mesas, flats, and desert habitats with lechuguilla and other desertscrub. South Presidio Co., N to near Shafter; SW Brewster Co., N to near Nine-Point Mesa. 2,800–4,000 ft. Mexico: NE Mexico S to Durango and Zacatecas. Map 91.

Vegetative Characters. Plants usually

branched from the base, although single-stemmed, presumably young plants are not uncommon. Stems narrowly cylindroid or clavate, 6–15 cm long, 2–3.5 cm in diameter, completely obscured by whitish radial spines. Central spines, 6–12 (usually 8 in the Trans-Pecos), mostly 5–8 mm long, pale reddish to gray or nearly white, with darker, usually dark gray, reddish, brown, or chalky-blue distal portions. Relatively heavy central spines contrast sharply with slender, numerous (36–49), whitish radial spines, 3–5 mm long. The upper central spine, particularly in areoles near the stem apex, characteristically is curved upward and blue-gray on the distal half. In this feature young single-stemmed plants of *M. pottsii* may resemble *Echinomastus mariposensis*, with which *M. pottsii* is sympatric in the vicinity of Terlingua.

Flowers. Flowering late Feb–Mar. Flowers maroon-red, deep red, rusty-red, or reddish-purple. Small bell-shaped flowers 0.9–1.3 cm long, 0.6–1.3 cm in diameter. Typically, a ring of flowers may encircle the stem one to several centimeters below the apex. The flowers usually do not open widely, but the inner tepals are reflexed near the tips. Inner tepals usually have darker midregions and paler margins. Stamens cream-colored to pale yellow. Style reddish, with 4–5 narrow stigma lobes, reddish-purple to orange-yellow.

Fruits. Fruits bright red at maturity, clavate, ca. 1.5 cm long. Seeds dark reddish-brown or brownish-black, pitted, nearly oval, ca. 1 mm long.

Full Name. Mammillaria pottsii Scheer ex Salm-Dyck.

Other Common Names. Foxtail cactus; rat-tail cactus.

Mammillaria meiacantha
Nipple Cactus
PLATES 245, 246

These flat-topped or hemispheric, circular plants with prominent, rather closely spaced tubercles are abundant in the central mountains of the Trans-Pecos and in desert habitats of the southwestern Big Bend. Only a few other cacti in the Trans-Pecos have such broad ecological tolerances. The specific epithet denotes the Greek *meion*, "fewer," and *akantha*,

Plate 245. *Mammillaria meiacantha* (nipple cactus), Davis Mts, Jeff Davis Co., TX; cultivated.

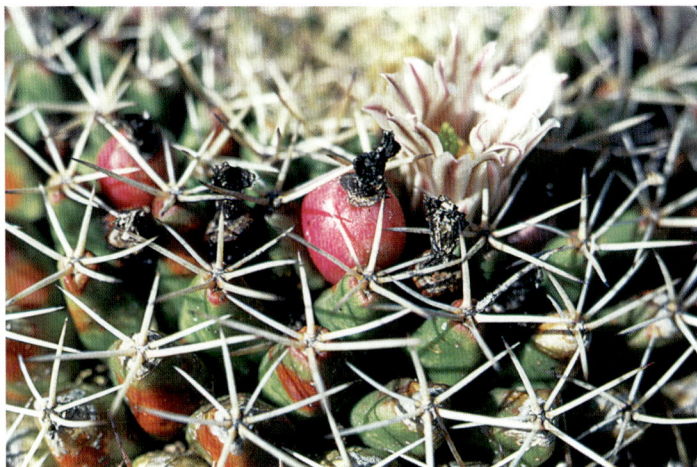

Plate 246. *Mammillaria meiacantha* (nipple cactus), Davis Mts, Brewster Co., TX; mature fruit; cultivated.

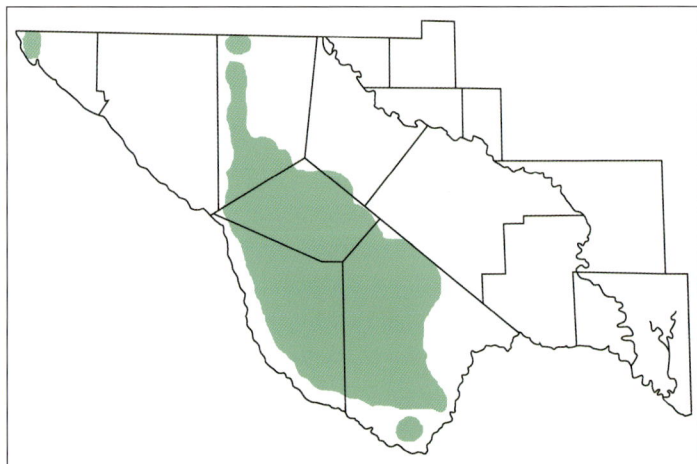

Map 92. *Mammillaria meiacantha* (nipple cactus).

"spine," in reference to the smaller number of spines per areole compared to related taxa.

Distribution. Mountains to desert, oak-juniper-pinyon woodland, grasslands, desertscrub. Widespread in the Trans-Pecos, most common in Culberson, Jeff Davis, Presidio, and Brewster counties, also Terrell Co., expected in adjacent counties as well. 3,000–7,300 ft. Central and S-central NM, probably to SE AZ. Mexico: mostly Coahuila, probably in E Chihuahua and N Zacatecas. Map 92.

Vegetative Characters. Similar in habit and several other characters to the regionally sympatric *M. heyderi*. Both species are unbranched, deep-seated (with a thick but basal stem), circular in outline, usually 4–16 cm in diameter, flat-topped or hemispheric, and produce a milky latex. The aboveground portion of the *M. meiacantha* stem when turgid is 2.5–5 cm high. When desiccated, the stem may shrink to near ground level. *Mammillaria meiacantha* is readily distinguished from *M. heyderi* by its typically 5–7 relatively thick radial spines, the longest 0.9–1.3 cm long. *Mammillaria heyderi* usually has 13–17 radial spines, 0.6–1.2 cm long. Both species produce 0–1 central spine, 0.5–1.5 cm long.

Flowers. Flowering Mar–May. Flowers white to pale pink, often with pink or reddish-brown tepal midstripes. Typically, a ring of blooms outside the newer growth in the center of the stem. Flowers 2–3 cm long, 2–3 cm or more in diameter. When fully open, tepals usually reflexed. Stamen filaments, ca. 8 mm long, pinkish to nearly white, and anthers are cream-colored to yellowish. Style 1.2–1.6 cm long, longer than the stamens, and supports 6–9 light green stigma lobes, 3–5 mm long.

Fruits. Fruits broadly clavate, 2–3.2 cm long, rose-pink to dull red at maturity. Seeds reddish-brown, pitted, 1.1–1.2 mm long.

Full Name and Synonyms. Mammillaria meiacantha Engelm. *M. runyonii* (Britton & Rose) Cory; *M. gummifera* Engelm. var. *meiacantha* (Engelm.) L. D. Benson; *M. heyderi* Muehlenpf. var. *meiacantha* (Engelm.) L. D. Benson.

Other Common Names. Little chilis; bisnaga de chilitos.

Mammillaria heyderi
Heyder's Pincushion Cactus, Western Heyder Pincushion Cactus

What might be termed the *M. heyderi* complex has been interpreted by previous workers as either a group of closely related species or as a species with several varieties, in some cases including *M. meiacantha*. Here we regard *M. heyderi* as a species of two varieties, both of which occur in the Trans-Pecos. The specific epithet honors Edward Heyder (1808–84).

Full Name. Mammillaria heyderi Muehlenpf.

Key to the Varieties of *Mammillaria heyderi*

1. Radial spines usually 13–17 or more, the longest ones usually 6–11 mm; central spines usually 0.35 mm in diameter or less; distribution widespread in the southeastern Trans-Pecos

    M. heyderi var. *heyderi*, p. 280

1. Radial spines usually 10–14, the longest ones 9–15 mm; central spines usually 0.35–0.45 mm in diameter; distribution in El Paso Co.

    M. heyderi var. *bullingtoniana*, p. 281

Mammillaria heyderi var. heyderi
Heyder's Pincushion Cactus
PLATES 247, 248

Mammillaria heyderi var. *heyderi* is by far the more common of the two varieties of *M. heyderi* in the Trans-Pecos, particularly in the southeastern portion. Variety *heyderi* extends southeast along and near the Rio Grande to deep South Texas in Cameron County, where it occurs at near sea level, up the coast to near Corpus Christi, and inland northwest to near Austin and beyond to the Edwards Plateau.

Distribution. Mostly in limestone substrates, desertscrub, or semidesert habitats. Sporadic in occurrence from near Sierra Blanca to near Indian Hot Springs, Hudspeth Co., SE to where more common in SE Brewster, Pecos, Terrell, and Val Verde counties. 1,000–4,600 ft. South TX and S-central TX N to SE OK and SE NM, excluding the TX Panhandle, where absent or rare; reports from SE AZ and SW NM

Plate 247. *Mammillaria heyderi* var. *heyderi* (Heyder's pincushion cactus), S Brewster Co., TX.

Plate 248. *Mammillaria heyderi* var. *heyderi* (Heyder's pincushion cactus), S Brewster Co., TX; mature fruits; cultivated.

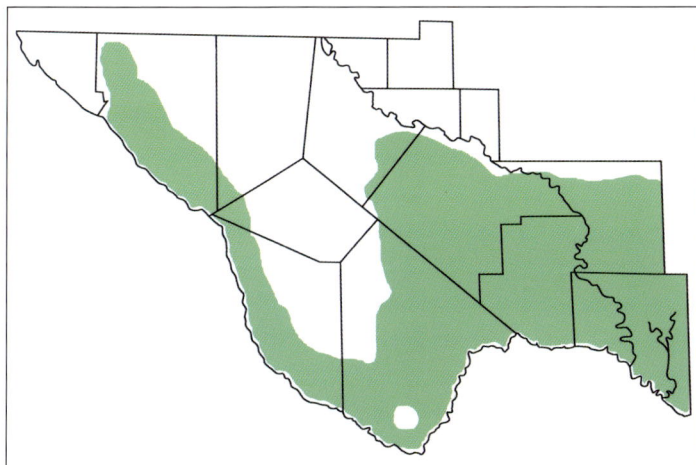

Map 93. *Mammillaria heyderi* var. *heyderi* (Heyder's pincushion cactus).

are based on *M. heyderi* var. *bullingtoniana*. Mexico: northern Tamaulipas, Nuevo León, and probably Coahuila. Map 93.

Vegetative Characters. Plants similar in appearance to those of *M. meiacantha:* flattened hemispheres 7–15 cm in diameter, extending 2–5 cm above the ground. Prominent tubercles, 9–15 mm long, 3–4 mm in diameter above the bases, arranged in spiral rows, are evident through the relatively thin spines. Usually one central spine. In *M. meiacantha* the spines are fewer and heavier in appearance, distinguishing features that in the field are evident at a glance.

Flowers. Flowering Mar–Apr. Flowers white to cream-colored or slightly pinkish, usually with prominent pink or greenish-brown midregions, typically appear in a circle outside new growth at the flattened apex. Flowers 2–3.8 cm in length and diameter. When fully open, the tepals are reflexed. Stamen filaments whitish to pink, with yellow anthers. Stigma lobes 5–10, light green, cream-colored, or pinkish-tan and elevated slightly above the anthers.

Fruits. Fruits bright red when ripe, broadly clavate, 1–3.5 cm long, usually easily distinguishable from the pinkish to dull red fruits of *M. meiacantha*. Seeds reddish-brown, pitted, 1–1.2 mm long.

Full Name and Synonyms. Mammillaria heyderi var. heyderi. Mammillaria applanata Engelm.; *M. heyderi* var. *applanata* (Engelm.) Engelm.; *M. gummifera* var. *applanata* (Engelm.) L. D. Benson; *M. heyderi* var. *hemispherica* Engelm.; *M. heyderi* var. *gummifera* (Engelm.) L. D. Benson; *M. gummifera* var. *hemispherica* (Engelm.) L. D. Benson.

Other Common Names. Heyder mammillaria; nipple cactus; biznaga de chilitos.

Mammillaria heyderi var. bullingtoniana
Western Heyder Pincushion Cactus
PLATE 249

The distribution of var. *bullingtoniana* in Arizona and New Mexico is reasonably well established, but its status in El Paso County remains to be determined.

Plate 249. *Mammillaria heyderi* var. *bullingtoniana* (western Heyder pincushion cactus), Peloncillo Mts, Hidalgo Co., NM; flowers and mature fruits (photo by Dale and Marian Zimmerman).

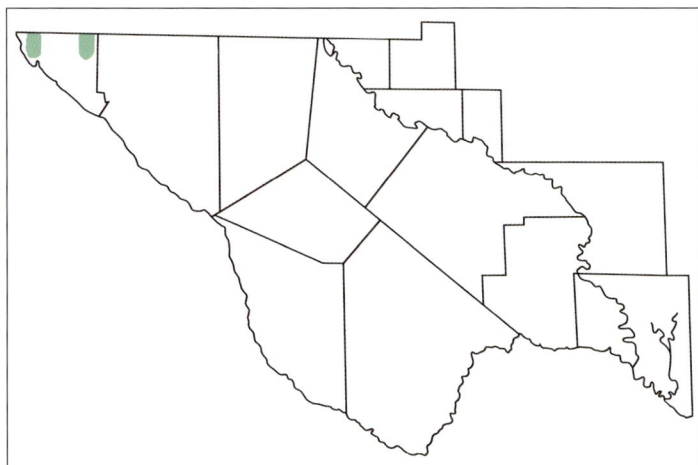

Map 94. *Mammillaria heyderi* var. *bullingtoniana* (western Heyder pincushion cactus).

Worthington (1995) has collected plants identified as *M. heyderi* in the Franklin and Hueco mountains.

Distribution. Reportedly in El Paso Co., desert grassland, and desertscrub. 3,800–4,800 ft. Southwestern NM from Doña Ana Co. W to SE AZ. Probably Chihuahua, Mexico. Map 94.

Vegetative Characters. Similar to var. *heyderi*, except that the stem and tubercle measurements are about one-fourth larger in var. *bullingtoniana*. Also, in var. *bullingtoniana* there are fewer radial spines than typical for var. *heyderi*, and the longest (lower) radials are heavier than the radials in var. *heyderi*. Central spine of var. *bullingtoniana* is slightly larger in diameter, usually 0.35–0.45 mm, than in var. *heyderi*.

Flowers and Fruits. Flowering Mar–Apr. Presumably, the flowers and fruits of var. *bullingtoniana* are like those of var. *heyderi*.

Full Name. Mammillaria heyderi var. *bullingtoniana* Castetter, P. Pierce, & K. H. Schwer.

Mammillaria prolifera var. texana
Hair-Covered Cactus
PLATES 250, 251

In North America, *M. prolifera* essentially is a species of south and south-central Texas and adjacent Mexico. It extends west of the Pecos River only a few miles to its only known locality near Langtry in Val Verde County. *Mammillaria prolifera* might be susceptible to freezing, a possible limiting factor for the species in the Trans-Pecos, except that cultivated *M. prolifera* has survived (–12°C [10°F]) in a cactus garden in Alpine. The epithets "prolifera" and "multiceps" (from the Latin *ceps*, "head") presumably are in reference to the prolific, many-stemmed plants (see synonyms).

Distribution. Base of limestone cliffs near the Rio Grande, Langtry, Val Verde Co. ca. 1,500 ft. East to Edwards and Bandera counties, S to near the Rio Grande valley in Hidalgo and Brooks counties, in

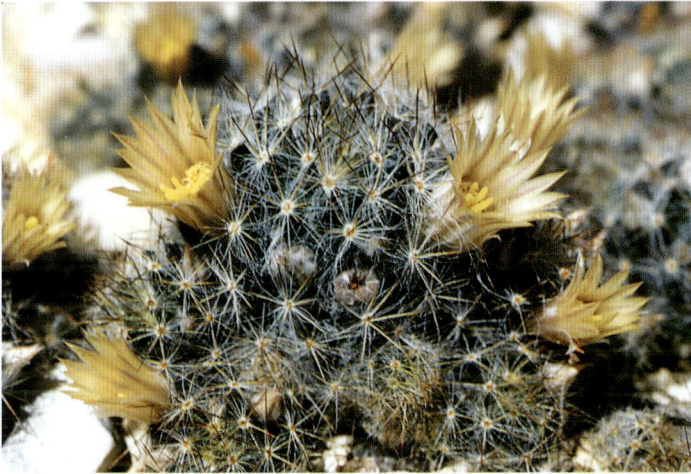

Plate 250. *Mammillaria prolifera* var. *texana* (hair-covered cactus), Terrell Co., TX; cultivated.

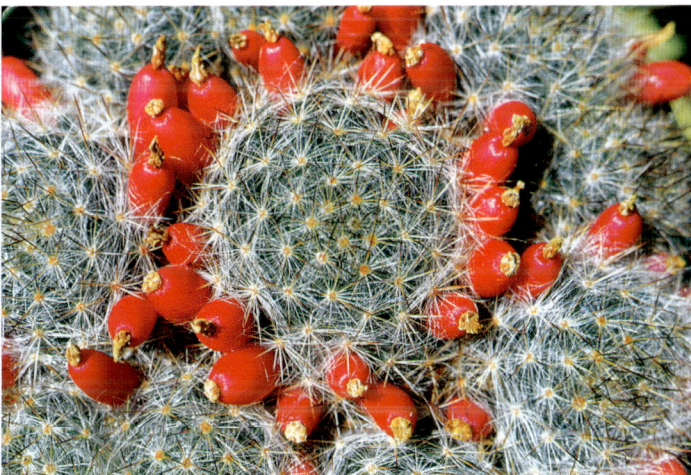

Plate 251. *Mammillaria prolifera* var. *texana* (hair-covered cactus), Terrell Co., TX; mature fruits; cultivated.

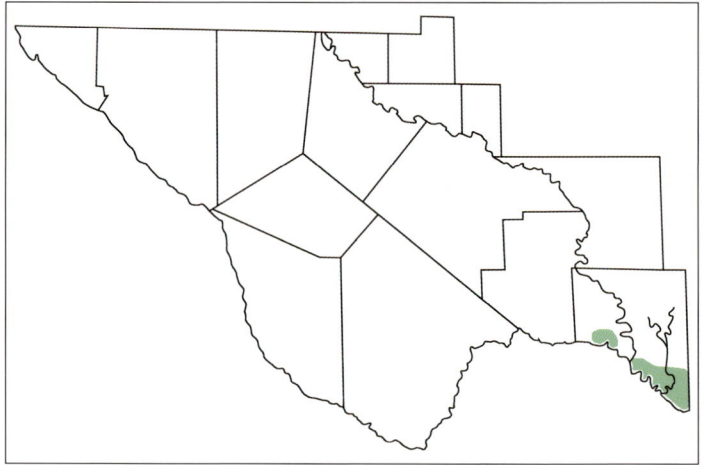

Map 95. *Mammillaria prolifera* var. *texana* (hair-covered cactus).

deep soil with shrubs and grasses, crests of rocky hills, and limestone ledges, E to Bexar Co. and Aransas Co. Mexico: adjacent Coahuila, Nuevo León, and Tamaulipas. Also West Indies. Map 95.

Vegetative Characters. Mature plants usually form dense, low mounds or mats, with several to many stems representing a range of sizes. Clumps of 12–20 or more stems are not unusual, forming mats to near 30 cm in diameter. Flowering stems are up to 5 cm or more long, 1.2–5 cm in diameter, spherical to ovate or short columnar. The most identifiable vegetative characters involve the spines. In each areole are 8–11 relatively slender, spreading, central spines, 7–8 mm long, with bulbous bases. Numerous slender radial spines (30–60) are in several series grading from flexible spines inside near the centrals to longer, undulating, hairlike spines 5–15 mm long at the outside of the areole. Central spines are translucent to pale yellow at the base and reddish-brown to nearly black on the distal half. Radial spines are white to pale yellow. Twisting, curved, hair-like radial spines overlapping and mostly obscuring the stem, also present in areoles of juveniles, provide the most unusual and most distinctive identifiable feature for the "hair-covered cactus."

Flowers. Flowering in Mar. Flowers usually are dirty yellow. Inner tepals also may be described as tannish-yellow or light yellow, with pale rose to rose-purple midregions. Cylindrical to funnelform flowers 1.5–2 cm long, 1–1.5 cm in open diameter. Inner tepals may be reflexed apically. Stamens 4.5–7 mm long with pale yellow or whitish filaments and yellow anthers. Style ca. 1 cm long, cream-colored, supporting 4–8 yellowish stigma lobes, 1.5–4 mm long. Flowers are subtended by some white wool and several hairlike bristles.

Fruits. Mature fruits bright red, juicy, clavate or narrowly obovoid, 1.3–2 cm long, 4.5–5 mm in diameter. Fruit surface smooth, lacking scales, and the floral remnant is persistent. Seeds black, pitted, asymmetrically obovate to nearly round, 1–1.3 mm long.

Full Name and Synonyms. Mammil-

laria prolifera (Mill.) Haw. var. *texana* (Poselger) Borg. *Mammillaria multiceps* Salm-Dyck; *M. multiceps* Salm-Dyck var. *texana* Engelm. ex F. M. Knuth; *M. pusilla* DC. var. *texana* Engelm.; *M. prolifera* var. *multiceps* Borg.

Other Common Name. Grape cactus.

Another Mammillaria Species

Another Texas species of *Mammillaria* not discussed elsewhere in the Trans-Pecos treatment of the genus is summarized below.

Mammillaria sphaerica
Pale Mammillaria

PLATE 252

Benson (1982) treated *M. sphaerica* as a variety of *M. longimamma*, a species otherwise restricted to central Mexico and often segregated from *Mammillaria* as the genus *Dolichothele*. Plants of *M. sphaerica* are readily distinguished from other mammillarias by their light green, almost yellow-green, stems. The specific epithet alludes to the characteristic spherical stems.

Distribution. Distribution in deep S TX (Map 96), where it is rather common under shade in the brush country, and into Tamaulipas and Nuevo León, Mexico.

Vegetative Characters. Plants with numerous stems, forming mounds or clumps to ca. 5 cm high and ca. 30 cm in diameter. Roots fleshy, to 2.5 cm thick. Stems light green, spherical, or depressed-spherical, ca. 5 cm long, 2.5–6 cm in diameter; tubercles soft, turgid, spreading, mammiform-cylindroid, 1.2–2.5 cm long, usually tapering toward apex, when desiccated becoming much smaller, firm, and closely arranged. Areoles circular, 1.5 mm in diameter. Spines not obscuring the stem; usually one central spine, 1–1.3 cm long, slightly thicker than radials; radial spines 12–15, slender, the longest ones 1–1.6 cm, smooth, brownish, yellow to gray distally.

Flowers. Flowering in spring and summer. Flowers fragrant, lemon-yellow, 5–6.5 cm in diameter, funnelform; filaments and anthers yellow, the stamens swirled around

Plate 252. *Mammillaria sphaerica* (pale mammillaria), McMullen Co., TX; cultivated.

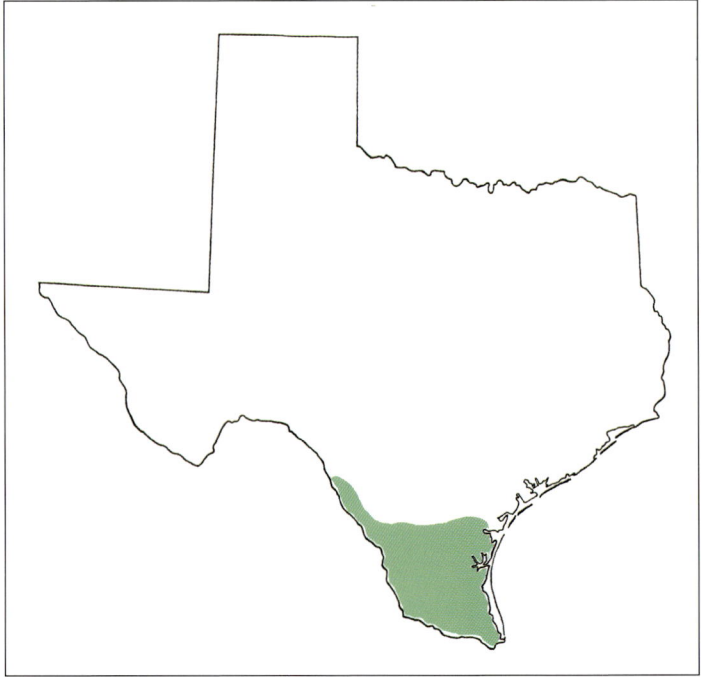

Map 96. *Mammillaria sphaerica* (pale mammillaria).

the yellow style; stigma lobes eight, yellow, 6–7.5 mm long.

Fruits. Fruit green, ovoid, ca. 1.3 cm long, persisting, ultimately turning reddish. Seeds black, pitted.

Full Name and Synonyms. *Mammillaria sphaerica* A. Dietr. *Cactus sphaericus* (A. Dietr.) Kuntze; *Dolichothele sphaerica* (A. Dietr.) Britton & Rose; *M. longimamma* DC. var. *sphaerica* (A. Dietr.) L. D. Benson.

CORYPHANTHA

Pincushion Cactus, Beehive Cactus, Cory Cactus

Coryphantha (Engelm.) Lem. is a genus of 60–70 species distributed from western Canada through the western United States and into Mexico, south to Oaxaca, and in Cuba. Most of the species occur in Mexico, with about 33 taxa in the United States. About 41 species occur in the CDR. About 17 species of *Coryphantha* are reported for Texas, and 14 of these, many of them endemic, occur in the Trans-Pecos. The size of the genus depends in part upon whether closely related taxa, such as the subgenus *Escobaria* and *Cochiseia* are combined with or segregated from *Coryphantha* (Taylor, 1986; Zimmerman, 1985). The genus name is derived from the Greek *coryph,* "summit," and *anthos,* "flower," in reference to the characteristic position of flowers on new growth at the stem apex. Generic synonyms reviewed in Powell and Weedin (2004).

It is widely recognized that *Coryphantha* and *Mammillaria* are closely related, and some workers prefer to combine them as a single genus. The most consistent distinction of *Coryphantha* from *Mammillaria* is the presence in *Coryphantha* of an areolar groove extending from the spine clusters down the tubercles to the tubercle axils. Another frequently cited difference between the two taxa is the production of flowers on new growth at the stem apex in *Coryphantha,* whereas in most *Mammillaria* species flowers are produced in a ring at least 1 cm below the stem apex. Intermediate flower positions are known in species of both genera. In one subgenus of *Coryphantha* (subgenus *Coryphantha*), areolar glands are present, but areolar glands are completely absent in subgenus *Escobaria* and the related genus *Mammillaria*. Areolar glands in this genus are domelike structures ca. 1 mm in diameter, and they may occur singly or in groups.

Recognition of *Escobaria* as a distinct genus is favored by Europeans and many American botanists as well. *Escobaria* is distinguished from *Coryphantha* by (1) pitted seeds; (2) fringed outer tepals; (3) areolar glands absent; (4) generally smaller vegetative and floral parts; (5) red fruits as in *Mammillaria*. In the present work we have treated *Escobaria* (species 1–9) as a subgenus of *Coryphantha* for the reasons reviewed by Zimmerman (1985). *Escobaria,* described as a genus by Britton and Rose (1919–23), commemorates the work of brothers Rómulo and Numa Escobar, of Mexico City and Juárez.

Key to the Trans-Pecos Species

1. Stems 1–2 cm long, 0.7–1.7 cm in diameter; spines appressed, relatively short and thick, the largest ones less than 6.5 mm long, with abruptly pointed or obtuse tips; flowers magenta; endemic in certain novaculite exposures, Marathon Basin, Brewster Co.

C. minima, p. 248

1. Stems usually considerably larger than 1–2 cm long and wide (individual stems sometimes this small); spines appressed or not, thin or thick, tapering gradually to a point; flowers yellow, magenta, or other colors; in various substrates including novaculite, but not restricted to the Marathon Basin (2).

2(1). Stems flat-topped or hemispheroidal, deep-seated, usually with a fleshy base below ground, when desiccated the stem top drawn to near ground level or below, 1.5–4 cm in diameter; plants often in flat-topped or depressed-hemispheric clumps 5–20 cm wide; flowers magenta; in or near the Marathon Basin, Brewster Co.

C. hesteri, p. 300

2. Stems typically cylindroid to short-cylindroid, with a fleshy base or not, not drawn below ground level when desiccated, usually 2 cm or more in diameter; flowers yellow, magenta, or other colors; Marathon Basin or elsewhere in the Trans-Pecos (3).

3(2). Central spines hooked (at least some of them)

C. scheeri var. *uncinata*, p. 338

3. Central spines straight or gently curved (4).

4(3). Radial spines 20–40 or more, often fewer than 20 in *C. vivipara* var. *vivipara* and *C. tuberculosa* var. *varicolor*, often poorly differentiated from outer centrals; spines overlapping and mostly to completely obscuring the stem (except in var. *vivipara* and var. *varicolor*; species of subgenus *Escobaria* (5).

4. Radial spines 9–15, rarely 15–20, except 16–27 in *C. echinus*; centrals (if present) usually clearly differentiated from radials, often a main, porrect, stout central; spines usually not obscuring the stems (except in some *C. echinus*); species of subgenus *Coryphantha* (11).

5(4). Flowers 2.1–6.5 cm in diameter (6).

5. Flowers 1.3–2 cm in diameter (7).

6(5). Flowers magenta; spine clusters persisting on tubercles near base of stems

C. vivipara, p. 302

6. Flowers pink, sometimes pinkish-white or white; spine clusters abscising from tubercles near base of stems, leaving corncoblike bases of older stems

C. tuberculosa, p. 322

7(5). Seeds black; stigmas green (8).

7. Seeds red-brown or orange-brown; stigmas white (except green in *C. pottsiana*) (10).

8(7). Plants with a thick taproot; floral remnant deciduous; spines snowy-white; usually in limestone crevices, S Brewster and Presidio counties, near the Rio Grande

C. duncanii, p. 296

8. Plants with diffuse roots; floral remnant persistent; spines snowy-white to brown; commonly other substrates and habitats (9).

9(8). Flowers 1.3–1.9 cm in diameter, the inner tepals 0.8–1.2 cm long; spines relatively few, white to brown; commonly in gravel or loam, *Larrea* and *Prosopis* desert, or among igneous rocks in Jeff Davis and Brewster counties

C. dasyacantha, p. 291

9. Flowers 0.6–1.6 cm in diameter, the inner tepals 4.5–9 mm long; spines relatively numerous, snowy-white; in the Trans-Pecos restricted to the upper Chisos Mts, igneous and limestone rocks

C. chaffeyi, p. 294

10(7). Stigma white; seeds larger (see text); pith and cortex containing lenticular druses 0.5–1 mm in diameter, evident to the naked eye (also present in C. *vivipara*); S Brewster and Presidio counties, rare in SW Pecos Co., Franklin Mts and Guadalupe Mts

C. sneedii, p. 308

10. Stigma green; seeds smaller (see text); pith and cortex containing only spheroidal druses 0.05–0.3 mm in diameter, not evident to the naked eye; Val Verde Co.

C. pottsiana, p. 320

11(4). Plants cespitose, forming mats or mounds 20–100 cm in diameter; pith and cortex mucilaginous; areolar grooves extending only one-half, sometimes three-fourths, the distance from spine clusters toward tubercle axils; central spines, at least the lowermost one, 3.5–7 cm long, usually striate or angular, gray to black, often flexible and slightly curved; flowers rose-pink or magenta

C. macromeris var. *macromeris*, p. 328

11. Plants (in the Trans-Pecos) of solitary or few stems (except for *C. echinus* at certain sites including S of the Chisos Mts); pith and cortex nonmucilaginous; areolar grooves extending from spine clusters to tubercle axils; central spines various but usually less than 4 cm long; flowers yellow, rose-pink to magenta (12).

12(11). Flowers rose-pink to magenta, the inner tepals sometimes whitish at bases; resembling *C. macromeris*, but stems solitary or in small clumps (in TX), central spines shorter, the porrect central terete, dark brown or black, and radial spines acicular (dorsiventrally flattened in *C. macromeris*); Brewster and Terrell counties, along and near the Rio Grande, and in adjacent Mexico

C. ramillosa, p. 332

12. Flowers yellow; not much resembling *C. macromeris;* wide distribution in the Trans-Pecos, or E Trans-Pecos (13).

13(12). Tubercles 1.5–3 cm long; radial spines 6–16; flowers golden, pale greenish to dull yellow

C. scheeri, p. 334

13. Tubercles 0.8–1.5 cm long; radial spines 8–27; flowers bright to golden-yellow (14).

14(13). Radial spines 16–27; plants mostly of solitary stems throughout most of the Trans-Pecos, multistemmed with clumps or mounds to 50–80 cm across in S Brewster and S Presidio counties

C. echinus, p. 339

14. Radial spines 8–15; plants usually multistemmed, in clumps 30–50 cm or more across; reported E Trans-Pecos, Pecos Co., mostly S-central TX

C. sulcata, p. 346

Coryphantha dasyacantha
Desert Pincushion Cactus

PLATES 253–56

Coryphantha dasyacantha is a Chihuahuan Desert endemic species essentially restricted to parts of Trans-Pecos Texas and probably adjacent Mexico. At least two morphotypes occur in the Trans-Pecos, one in the eastern Glass Mountains and southern Big Bend desert country, north along the Rio Grande to near El Paso, and the other in the Davis Mountains near Fort Davis to south of Alpine about 30 miles. Almost half of localities plotted by Benson (1982) are based on misidentified specimens of other species.

Our field investigations suggest that *C. dasyacantha* usually is rare throughout most of its range in the desert, although at a few sites near the Rio Grande in Presidio County it is locally common. At the Jeff Davis County and northern Brewster County sites, the taxon is not uncommon, but it is cryptic. *Coryphantha dasyacantha* is very closely related to *C. chaffeyi* and less so to *C. duncanii*; all share black, pitted seeds, red fruits, green stigma lobes, and other characters. Other Trans-Pecos coryphanthas, except *C. minima,* have brown or reddish-brown seeds. The specific epithet is after the Greek *dasy,* meaning "hairy" or "shaggy," and *akantha,* in reference to spines.

Distribution. *Larrea, Prosopis,* lechuguilla, and other desertscrub, gravel slopes and loamy flats, or oak-juniper woodland among rocks, igneous and limestone substrates, usually rare but locally common at some sites. El Paso, Hudspeth, Jeff Davis, Presidio, Brewster, and Pecos counties,

Plate 253. *Coryphantha dasyacantha* (desert pincushion cactus), S Presidio Co., TX; cultivated.

Plate 254. *Coryphantha dasyacantha* (desert pincushion cactus), S Presidio Co., TX; mature fruits; cultivated.

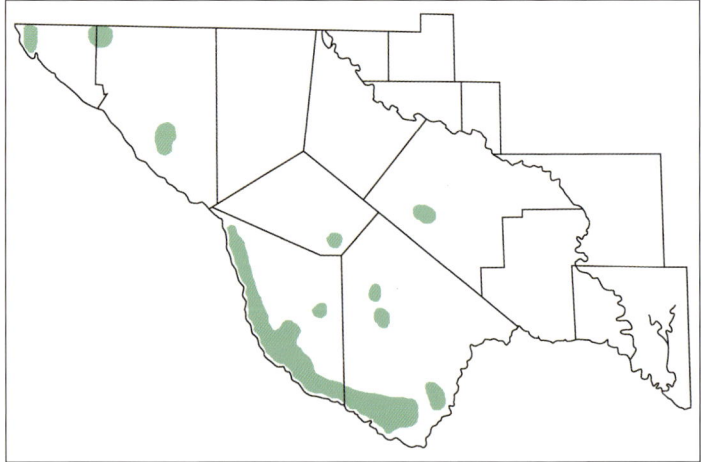

Map 97. *Coryphantha dasyacantha* (desert pincushion cactus).

mostly desert habitats, except in the Davis Mts and eastern Glass Mts. 2,500–5,500 ft. Mexico: to be expected in NE Chihuahua and N Coahuila. Map 97.

Vegetative Characters. Plants usually single-stemmed, although older plants may exhibit several stems. Young stems are globular, but become elongate to cylindroid with age. Stems are 4.5–10 cm long, 3–4.5 cm in diameter. Plants growing in the shade of small trees or shrubs in the desert may form the most elongated stems,

15–17 cm tall. Central spines number 4–9, the longest 1.2–1.7 cm, and radial spines typically 21–31 in number, 6–9 mm long. Usually 5–6 spreading centrals and several appressed subcentrals, the central spines slightly larger in diameter than the radial spines.

The morphotype of *C. dasyacantha* in Jeff Davis County and northern Brewster County also usually is single-stemmed, but clusters of 2–5 stems are occasional. In the field the Davis Mountain plants are depressed-globose or globose, usually 1–4 cm tall, 1.5–3 cm in diameter. The central spines usually are darker (often reddish-brown) at least on distal portions, although the radials are white, and the Davis Mountain plants of *C. dasyacantha* are not as white in aspect as the desert plants (Plates 255, 256).

Flowers. Flowering Mar–May. Flowers of *C. dasyacantha* and closely related species are among the smallest in subgenus *Escobaria*, noted for its smallish flowers, compared to subgenus *Coryphantha*. The flowers of *C. dasyacantha* are 1.5–2.5 cm

long and 0.9–1.5 cm in diameter. Flowers of the desert morphotype are slightly larger than those of the Davis Mountains morphotype, and flowers of the desert form basically are white or cream-colored. Inner tepals have white or cream-colored margins and sharply defined, usually brown, midregions. In size, shape, and color the flowers of the desert morphotype of *C. dasyacantha* are essentially indistinguishable from those of *C. duncanii*. Stamens show yellow anthers, ca. 0.5 mm long, and white or colorless filaments 5–8 mm long. Style greenish, 1–1.3 cm long, with 4–6 green stigma lobes, 1–2.5 mm long. Flowers of the Davis Mountains form are smaller, and the inner tepals are light pink with salmon-pink (darker) midregions.

Fruits. Fruits bright red, usually clavate, sometimes cylindroid to narrowly ellipsoid, 1.3–3.5 cm long, 4–6 mm in diameter. Floral remnant persistent. Seeds subspherical, 1–1.2 mm in longest diameter. Red fruits of the Davis Mountains morphotype are 1–1.7 cm long and broadly elliptic, much like those of *C. chaffeyi*.

Plate 255. *Coryphantha dasyacantha* (desert pincushion cactus), Jeff Davis Co., TX.

Plate 256. *Coryphantha dasyacantha* (desert pincushion cactus), Jeff Davis Co., TX; mature fruit; cultivated.

Full Name and Synonym. *Coryphantha dasyacantha* (Engelm.) Orcutt [= *Escobaria dasyacantha* (Engelm.) Britton & Rose]. *Mammillaria dasyacantha* Engelm.

Coryphantha chaffeyi
Chaffey's Pincushion Cactus
PLATES 257, 258

Coryphantha chaffeyi primarily is a Mexican species known in the United States only from the Chisos Mountains of Big Bend National Park. The specific epithet honors Dr. Elswood Chaffey, who collected the species near Cedros, Zacatecas, Mexico, in June 1910.

Distribution. Rock crevices, sometimes with *Selaginella*, open slopes, among rocks in grassy areas, oak-juniper-pinyon woodland, igneous rock and limestone substrates, upper Chisos Mts, Brewster Co.

4,700–7,300 ft. Mexico: Coahuila (Sierra de la Madera; Sierra de la Paila), N Zacatecas, N San Luis Potosí, NE Durango. Map 98.

Vegetative Characters. Stems are usually unbranched, although plants with clusters of 2–3 stems are not uncommon in the Chisos Mountains. Stems globular to short-cylindroid, 3–7 cm long, 3–4.5 cm in diameter, densely covered by white spines. In each areole are 30–50 spines, 8–10 centrals usually 0.7–1.5 cm long, and 22–40 radials, 5–9 mm long. In habit *C. chaffeyi* closely resembles *C. dasyacantha*.

Flowers. Flowering Mar–May. Flowers ca. 1.2 cm long and 0.7–1 cm in diameter, smaller than those of *C. dasyacantha*, and provide one of the most reliable means for distinguishing the taxa. Inner tepals pinkish to cream-white with prominent, broad,

pinkish-salmon to light brown midregions and narrow margins. Anthers yellow, 0.4–0.5 mm long. Filaments 4–6 mm long and pink. Style 7–8 mm long, greenish, supporting green stigma lobes only ca. 1 mm long.

Fruits. Brilliant red ripe fruits are oval, broadly ellipsoid, or broadly clavate, 0.8–1.1 cm long, 3–5 mm in diameter. Seeds 1–1.2 mm long, somewhat comma-shaped.

Full Name and Synonym. Coryphantha

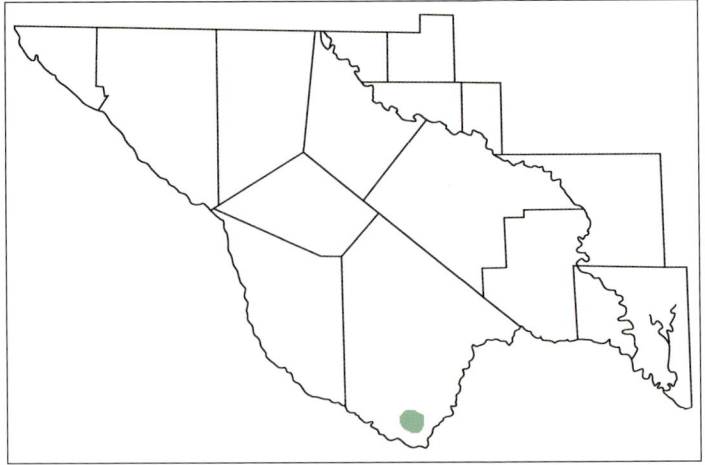

Map 98. *Coryphantha chaffeyi* (Chaffey's pincushion cactus).

chaffeyi (Britton & Rose) Fosberg [= *Escobaria chaffeyi* Britton & Rose]. *Escobaria dasyacantha* (Engelm.) Britton & Rose var. *chaffeyi* (Britton & Rose) N. P. Taylor.

Other Common Name. Bisquit cactus.

Coryphantha duncanii
Duncan's Pincushion Cactus
PLATES 259, 260

Coryphantha duncanii is a Chihuahuan Desert endemic species found primarily in crevices of layered limestone in the hottest parts of the Big Bend desert region. The small plants, densely covered with white spines, are difficult to notice against a background of white rocks. The species is named after Capt. Frank Duncan, a miner who owned a claim near the type locality.

Distribution. Locally common to rare, desert limestone hills. Near the Rio Grande (within 25 mi), Big Bend National Park NW to SE Presidio Co. 2,000–3,700 ft. Also Sierra Co., NM (rare, 4,500–5,000 ft). Not yet documented from Mexico (specimens from Coahuila are intermedi-

ate with a related species, not typical *C. duncanii*). Map 99.

Vegetative Characters. *Coryphantha duncanii* closely resembles the desert morphotype of *C. dasyacantha*, but the stems of *C. duncanii* are much smaller. Stems of *C. duncanii* usually are unbranched, although plants with 2–3 stems are not uncommon. Plants with 5–8 stems are rare, and plants with smaller, immature branches are seldom found. Plants are deep-seated in crevices, or less often in soil, the root system being dominated by a succulent taproot, sometimes carrotlike, up to 20–30 cm long. The usually inconspicuous stems are ovoid, spherical, to somewhat conical, usually 1.5–3.5 cm tall, 1–3 cm in diameter. The spines are snowy-white, or some of the larger spines, particularly near the apex, are tan to reddish-brown. The inner central spines, those most easily interpreted as central, are absent, or rarely there is one obvious central; 3–9 or more spines, 1–1.4 cm long, are interpreted as outer centrals. There are about 23–40 radial spines, ca. 9 mm long. Most likely to

Plate 259. *Coryphantha duncanii* (Duncan's pincushion cactus), Mariscal Mt, S Brewster Co., TX.

Plate 260. *Coryphantha duncanii* (Duncan's pincushion cactus), Mariscal Mt, S Brewster Co., TX; mature fruits; cultivated.

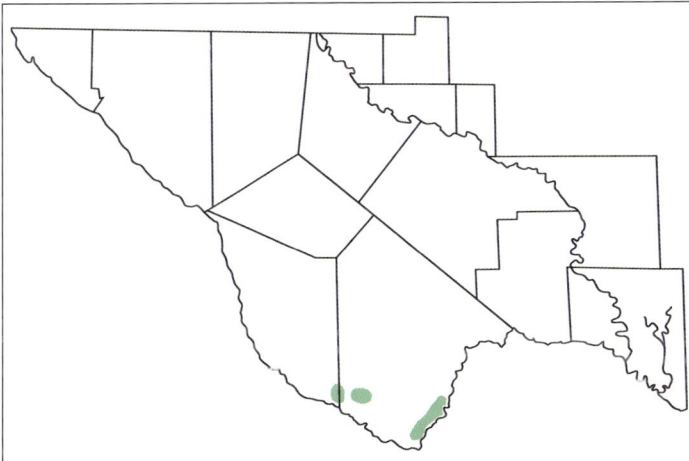

Map 99. *Coryphantha duncanii* (Duncan's pincushion cactus).

be confused with *C. duncanii* in the Trans-Pecos are smaller or immature plants of *C. tuberculosa* var. *tuberculosa,* particularly the more white-spined individuals of var. *tuberculosa,* which characteristically occur in rock crevices in habitats that appear suited for *C. duncanii. Coryphantha duncanii* is most easily distinguished from *C. dasyacantha* by its smaller stems, fleshy taproot, and 0(–1) central spine. Within the known range of *C. duncanii, C. dasyacantha* is most likely to be found in alluvial substrates, not the rock crevices favored by *C. duncanii.*

Flowers. Flowering Feb–Mar. Flowers of *C. duncanii* are similar to those of *C. dasyacantha* (larger than those of *C. chaffeyi*), 1.5–3 cm long, 1.3–1.9 cm wide. Inner tepal margins are white to cream-colored, rarely pinkish, and usually there are prominent pinkish, purplish, brown, or brownish-green midstripes. Anthers are yellow, and the filaments whitish or pinkish. The style is greenish, and the usually four stigma lobes are green.

Fruits. Bright red ripe fruits are cylindroid, ellipsoid, or clavate, 1.1–2 cm long, not juicy. The floral remnant falls early, leaving a shallow depression, a diagnostic trait for distinguishing the species and its two or more Mexican relatives from *C. dasyacantha* and all other similar species. The seeds are globose, ca. 1.2 mm in diameter.

Full Name and Synonyms. Coryphantha duncanii (Hester) L. D. Benson. [= *Escobaria duncanii* (Hester) Backeb.]. *Escobaria duncanii* (Hester) Backeb.; "*Mammillaria duncanii* (Hester)" Weniger, nom. nud.; *Escobaria dasyacantha* (Engelm.) Britton & Rose var. *duncanii* (Hester) N. P. Taylor.

Other Common Names. Duncan's cactus; Duncan snowball.

Coryphantha minima
Nellie's Pincushion Cactus
PLATES 261, 262

Coryphantha minima is remarkable for its marble-sized stems, unusual peglike spines, and substrate specificity. *Coryphantha minima* was federally listed as Endangered by the U.S. Fish and Wildlife Service in 1979 and by the Texas Parks and Wildlife Department in 1983. The specific epithet is derived from the Latin *minimus,* "smallest," in reference to the small stems.

Distribution. Restricted to outcrops on Caballos Novaculite and related chert-rich soils, associated with *Selaginella,* on rocky, grassy hills in the central Marathon Basin, Brewster Co. 3,700–4,300 ft. Endemic to the Trans-Pecos. Map 100.

Vegetative Characters. In the wild, the stems usually are unbranched, rarely with 2–3 stems, spherical to cylindroid, 1–2.5 cm tall, 0.7–1.7 cm in diameter. The spines are ca. 15–28 per areole, 3.5–5 mm long, all appressed against the stem. The most identifiable aspect of the spines is that they are relatively short, thick, and cylindroid or weakly clavate. Most cactus spines are needlelike and taper to a point, but the "peglike" spines of *C. minima* are acuminate or obtuse. In *C. minima* there are no projecting central spines.

Flowers. Flowering Apr–May. Magenta flowers when fully open are easily as large as or larger than most of the stems. Flowers are 1.3–1.6 cm long, 1.5–2.7 cm wide. Inner tepals characteristically are magenta, but sometimes rose-pink to light pink and sometimes even white basally, with weakly defined midstripes or distally with darker

Plate 261. *Coryphantha minima* (Nellie's pincushion cactus), Marathon Basin, Brewster Co., TX.

Plate 262. *Coryphantha minima* (Nellie's pincushion cactus), Marathon Basin, Brewster Co., TX; mature fruits, barely visible; cultivated.

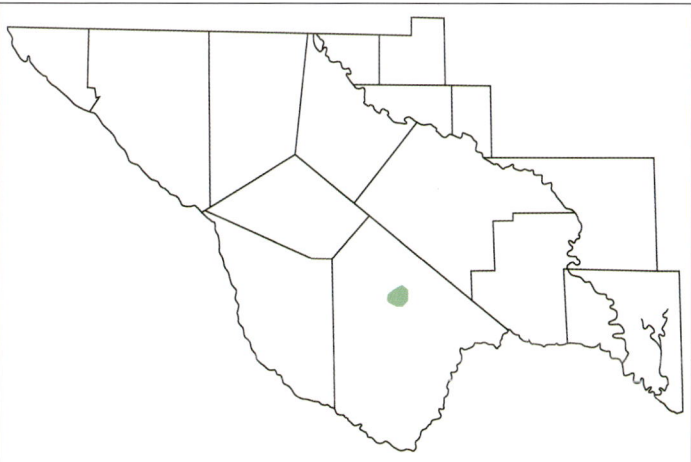

Map 100. *Coryphantha minima* (Nellie's pincushion cactus).

midlines. Filaments greenish, and anthers bright yellow or orange-yellow. Style greenish and rather short, with 4–8 green stigma lobes, 0.5–1.5 mm long.

Fruits. Fruits spherical, ovoid, or obconic, 1.5–6 mm long, 1.5–4 mm in diameter, either green at maturity or developing a yellowish tinge. Floral remnant persistent but fragile and easily lost by breakage, but not cleanly falling. Old fruits persistent among the spines, quickly drying and turning nearly white with thin, fragile pericarpels that ultimately disintegrate without ever deshiscing. Black, pitted seeds are visible through the thin, dry pericarpels before the fruit disintegrates. Seeds obovoid to weakly pyriform, 0.8–1 mm long.

Full Name and Synonyms. Coryphantha minima Baird [= *Escobaria minima* (Baird) D. R. Hunt]. *C. nellieae* Croizat; *Mammillaria nellieae* (Croizat) Croizat; *E. nellieae* (Croizat) Backeb. The synonymous epithet *nellieae* and source of the common name were after Mrs. Nellie Davis of Marathon, Texas.

Other Common Names. Dwarf cory cactus; Nellie's cory cactus.

Coryphantha hesteri
Hester's Pincushion Cactus
PLATES 263–65

Coryphantha hesteri was once thought to be a rare species restricted to novaculite substrates in the Marathon Basin of northern Brewster County and was listed as a class 3C species by the U.S. Fish and Wildlife Service. It has been found to be abundant and widespread in the Marathon Basin in both novaculite and limestone habitats. More recent studies have documented a wider distribution in novaculite, limestone, and igneous substrates and a greater population density than was previously known. *Coryphantha hesteri* is considered to be one of the several special diminutive cacti of the Trans-Pecos, even though its stems in some individuals may be larger than those of the other tiny cacti. The stems frequently occur in clumps. The type locality of *C. hesteri* is near Mount Ord ca. 10 miles southeast of Alpine,

Plate 263. *Coryphantha hesteri* (Hester's pincushion cactus), 9 mi SE of Alpine, Brewster Co., TX.

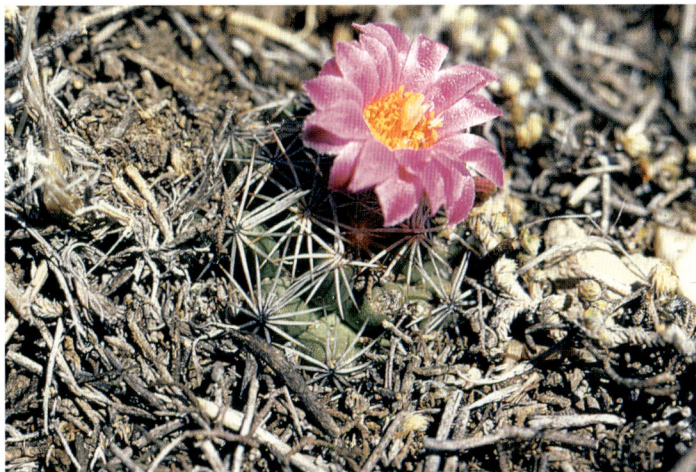

Plate 264. *Coryphantha hesteri* (Hester's pincushion cactus), Marathon Basin, Brewster Co., TX.

Plate 265. *Coryphantha hesteri* (Hester's pincushion cactus), N Brewster Co., TX; mature fruits, some spines cut away; cultivated.

Brewster County. The species is named after J. P. Hester, who collected the species in 1930 and made it available to Ysabel Wright.

Distribution. Crevices of rocks or rocky soil, grassland in Caballos Novaculite or limestone substrate, oak-juniper-pinyon woodland and grassland in igneous substrate. Mostly Brewster Co., Marathon Basin; localized in and near Del Norte Mts, S and SE of Alpine; Pecos Co., extreme S portion; Terrell Co., W of Sanderson. 3,600–5,300 ft. Endemic to the Trans-Pecos. Map 101.

Vegetative Characters. Stems are solitary or the plants cespitose, with few- or many-stemmed clumps common. Clumps 0.5–2 cm across are not uncommon, and large stem clusters to 30 cm or more across are present in the Marathon Basin. Stems are deep-seated, usually with a fleshy, enlarged taproot, and individually and collectively are flat-topped or depressed-hemispheric, or the stems may be hemi-

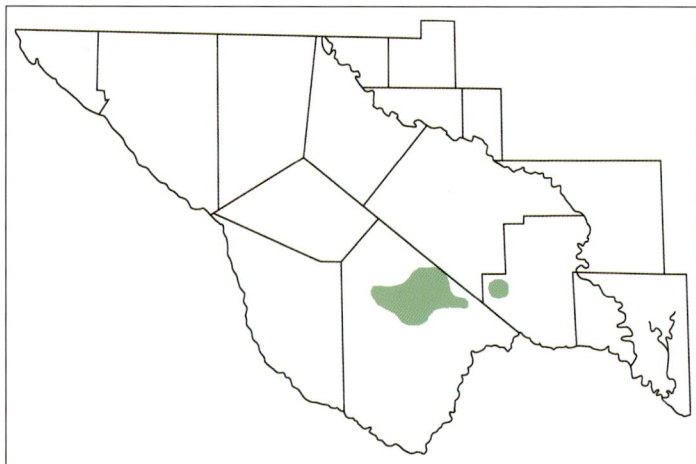

Map 101. *Coryphantha hesteri* (Hester's pincushion cactus).

spheroidal, spherical, to egg-shaped, 1–6.5 cm long aboveground, 1.5–4.7 cm in diameter. During dry periods, when stems are desiccated, they tend to withdraw below the ground or into crevices with little more than spine clusters being visible. Often the plants are obscured by vegetation. *Coryphantha hesteri* is one of the most cryptic cactus species in the Trans-Pecos, especially in grassland habitats.

The glabrous, straight spines typically number 13–18 per areole, with no manifest central spines. Spines white with red or brown tips, weathering to gray, or the centrals, when present, reddish to brown-gray for one-half or more of their length. Centrals, when present, 0.9–1.3 cm long, 0.2–0.3 mm thick. The longest radial spines, upper and upper lateral in position, are 0.7–1.3 cm long and 0.2–0.3 mm thick.

Flowers. Flowering Apr–Nov. Magenta to rose-pink flowers 1.8–2.5 cm long, 1.3–2 cm in diameter. Stamens 3–5 mm long with yellow anthers ca. 0.5 mm long and colorless to pinkish filaments. Style 6–10 mm long, greenish to greenish-yellow, often reddish above. Stigma lobes 5–6 in number, 1–3 mm long, white to cream or pale pinkish.

Fruits. Fruits green, sometimes turning greenish-red, globose to ovoid, 5–8 mm long, 3–6 mm in diameter, quickly drying. Floral remnant usually is persistent on the fruit. Seeds dark brown, pitted, globose, 0.9–1.1 mm in diameter.

Full Name and Synonym. *Coryphantha hesteri* Y. Wright [= *Escobaria hesteri* (Y. Wright) Buxb.]. *Mammillaria hesteri* (Y. Wright) Weniger, nom. nud.

Other Common Names. Hester's dwarf cactus; Hester cory cactus.

Coryphantha vivipara
Beehive Cactus

Coryphantha vivipara is the most widespread and abundant species of the genus. The plants are locally common and have been frequently collected. Seven varieties are recognized throughout its broad range, according to Zimmerman (1985), the treatment followed here. Three varieties of *C. vivipara* occur in Texas: var. *vivipara*, var. *neomexicana*, and var. *radiosa*. Two varieties are evident in the Trans-Pecos, as well as intermediates, with the third variety (var. *radiosa*) mostly east of the Trans-Pecos. The specific epithet includes the Latin *vivus*, "alive," and probably

pareo, "to bring forth," presumably from the branching stems of the type collection.

Full Name and Synonym. *Coryphantha vivipara* (Nutt.) Britton & Rose. [= *Escobaria vivipara* (Nutt.) Buxb.].

Key to the Varieties of *Coryphantha vivipara*

1. Spines 13–29 per areole, not obscuring the tubercles; central spines lustrous, reddish-brown or orange; stigmatic surfaces pale pink to bright magenta

C. vivipara var. *vivipara*, p. 303

1. Spines 24–55 per areole, obscuring the tubercles; central spines white, or if brown then opaque and drab; stigmatic surfaces usually white (but underlying parts variable, white to reddish)

C. vivipara var. *neomexicana*, p. 306

Coryphantha vivipara var. vivipara

Eastern Beehive Cactus

PLATES 266, 267

Coryphantha vivipara var. *vivipara* is the most widespread of the seven varieties of the species. It is common in north-central Texas but not often collected in the Trans-Pecos. It is to be expected in the eastern Trans-Pecos, where it may intergrade with *C. vivipara* var. *radiosa* (Plates 268, 269), a variety with its major distribution on the Texas Edwards Plateau.

Distribution. Various alluvial substrates, including gypsum, often with grasses, mesquite, and prickly pear. Mostly eastern Trans-Pecos, one collection in N Culberson Co., others in NE Brewster Co., Pecos Co., and Ward Co. just E of the Pecos River. 3,000–4,500 ft. From northern Edwards Plateau in TX, N through

the Great Plains into S Canada, W to NM, CO, WY, and MT. Map 102.

Vegetative Characters. Globose to ovoid stems unbranched or sometimes profusely branched. Stems 2.5–10 cm tall, 3–10.5 cm in diameter. Large lenticular druses, to 0.75–1 mm in diameter, are always present in older parts of the pith and cortex. A medullary vascular system (vascular bundles in the pith) is present in *C. vivipara*. The spines are straight and glabrous. Typically, there are four central spines, including usually one inner central and three outer centrals, often in a "bird's-foot" arrangement, one descending-porrect and three ascending, the longest 2–2.3 cm. Central spines usually bright reddish-brown, at least on the distal half, contrasting with gleaming white radial spines and bright green tubercles visible between the spines. Centrals may be horn-colored, or

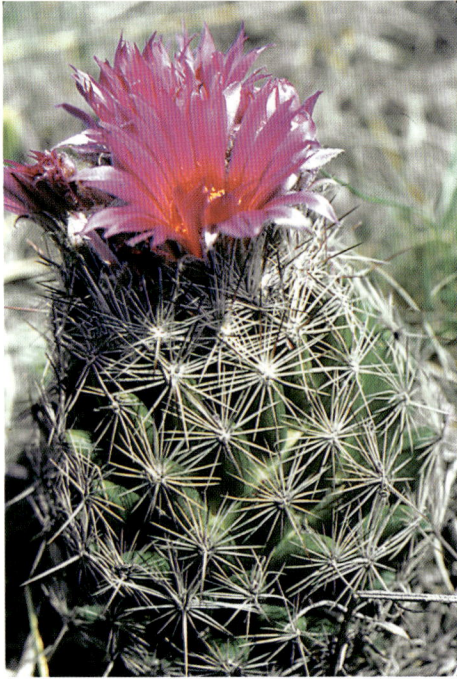

Plate 266. *Coryphantha vivipara* var. *vivipara* (eastern beehive cactus), Pecos Co., TX.

Plate 267. *Coryphantha vivipara* var. *vivipara* (eastern beehive cactus); Marathon Basin, Brewster Co., TX; cultivated.

Plate 268. *Coryphantha vivipara* var. *radiosa* (southeastern beehive cactus), Pecos Co., TX.

Plate 269. *Coryphantha vivipara* var. *radiosa* (southeastern beehive cactus), Brewster Co., TX.

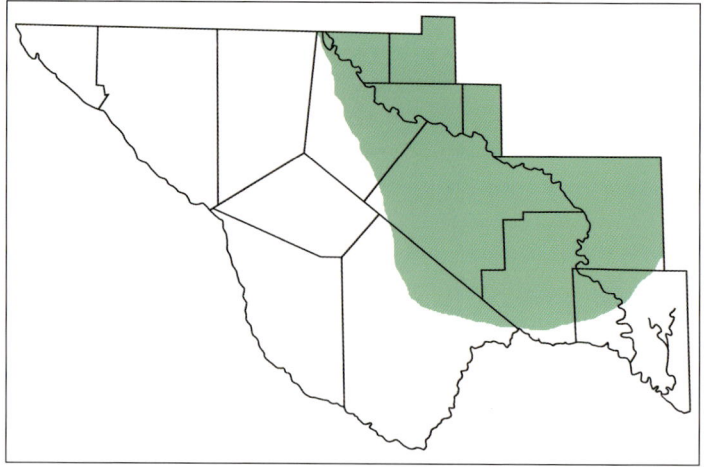

Map 102. *Coryphantha vivipara* var. *vivipara* (eastern beehive cactus).

rarely yellowish, whitish, or banded. In some plants a few of the radials may be colored like the centrals or have dark tips. Appressed radial spines are 10–26 in number, 0.7–2 cm long.

Flowers. Flowering May–Jun. Relatively large flowers usually are intense magenta but may be rose-pink or pale rose, 2.5–5 cm long, 3–6.7 cm in diameter. Stamen filaments are magenta to whitish or greenish-white. Anthers are ca. 1 mm long and yellow. The 5–13 stigma lobes are magenta to pale rose-pink (rarely white), and apiculate.

Fruits. Green, juicy fruits are ovoid to obovoid or ellipsoid, 1.2–2.8 cm long, 0.7–2 cm in diameter. Mature fruits may turn dull brownish-red or greenish-brown or even tan on some exposed portions. The floral remnant is persistent. The seeds are bright red-brown, pitted, 1–2.4 mm long, usually comma-shaped to subreniform, or the smallest seeds obovoid.

Full Name and Synonyms. Coryphantha vivipara var. *vivipara. Mammillaria vivipara* (Nutt.) Haw.; *Echinocactus viviparus* (Nutt.) Poselger.

Coryphantha vivipara var. neomexicana
New Mexico Beehive Cactus
PLATES 270, 271

Variety *neomexicana* is the most widely distributed taxon of *C. vivipara* in the Trans-Pecos, from El Paso County east to Pecos County. Intermediates between var. *neomexicana* and var. *radiosa* occur in Pecos and in northeastern Brewster and Terrell counties. See Zimmerman (1985) and Powell and Weedin (2004) for a taxonomic review of var. *neomexicana*. The varietal epithet *neomexicana* reflects the distributional center of the taxon in New Mexico.

Distribution. Desertscrub to mountain woodlands and basin grasslands, rocky exposures or alluvium, sedimentary, and igneous substrates. Every county of the Trans-Pecos except perhaps Val Verde Co., rare or nonexistent in S Presidio and Brewster counties. 2,300–5,200 ft. Across most of NM except extreme eastern portions, into SW CO, W-central AZ. Mexico: N Chihuahua, central Coahuila, gypsum. Map 103.

Plate 270. *Coryphantha vivipara* var. *neomexicana* (New Mexico beehive cactus), Jeff Davis Co., TX; cultivated.

Plate 271. *Coryphantha vivipara* var. *neomexicana* (New Mexico beehive cactus), Sutton Co., TX.

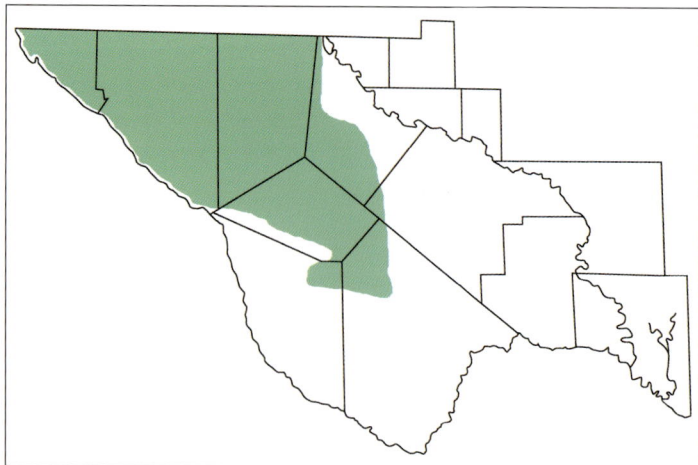

Map 103. *Coryphantha vivipara* var. *neomexicana* (New Mexico beehive cactus).

Vegetative Characters. Stems of var. *neomexicana* are globose, ovoid, to cylindroid, 6–20 cm long, 3–7.5 cm in diameter, with spines mostly or completely obscuring the tubercles. Spines characteristically are snowy-white, although centrals may be tan, reddish-brown, or dark reddish-brown, and radials may have dark tips. Areoles usually include 5–11 centrals, 25–35 radials, all straight and glabrous. Longest inner central 1.4–2 cm; radials 0.8–1.5 cm long.

Flowers. Flowering May–Jun, after summer rains. Magenta flowers are like those of var. *vivipara* except the stigma lobes are white.

Fruits. Fruits green, ovoid to obovoid, 1.2–2.8 cm long, and like those of var. *vivipara*. Bright red-brown, pitted seeds of var. *neomexicana* 1.3–3 mm long, significantly larger on average than those of var. *vivipara*.

Full Name and Synonyms. Coryphantha vivipara var. *neomexicana* (Engelm.) Backeb. [= *Escobaria vivipara* (Nutt.) Buxb. var. *neomexicana* (Engelm.) Buxb.].

Mammillaria vivipara (Nutt.) Haw. var. *radiosa* (Engelm.) Engelm. subvar. *neomexicana* Engelm.; *M. vivipara* (Nutt.) Haw. var. *neomexicana* (Engelm.) Engelm; *Coryphantha neomexicana* (Engelm.) Britton & Rose; *C. radiosa* (Engelm.) Rydb. var. *neomexicana* (Engelm.) Schelle.

Coryphantha sneedii
Sneed's Pincushion Cactus, Guadalupe Pincushion Cactus, Lee's Pincushion Cactus, Silverlace Cactus

Coryphantha sneedii is a variable species consisting of 9–10 geographic races, each of which is treated as a variety by Zimmerman (1985). Three of these are found within the political boundaries of Trans-Pecos Texas, and others are located on mountains visible across the boundaries in Chihuahua and New Mexico: the Organ Mountains, Doña Ana County; the San Andres Mountains, Doña Ana County; the Sacramento Mountains, Otero County, New Mexico; and just across the Rio Grande in Chihuahua, Mexico. The varieties of *C. sneedii* are seldom encountered, usually in the isolated rock outcrops high on mountainsides, except for the tall Big Bend variety, which is locally common at several frequently visited sites in and around Big Bend National Park. The specific epithet was taken after the original collector, J. R. Sneed, who collected the type specimen in the Franklin Mountains, El Paso County.

Coryphantha sneedii is sometimes sympatric with *C. vivipara* in the Trans-Pecos, most notably in the Guadalupe Mountains. To the experienced eye, plants of the two species usually can be distinguished by habit. More specific distinguishing traits have been outlined by Zimmerman (1985).

Full Name and Synonym. *Coryphantha sneedii* (Britton & Rose) A. Berger. [= *Escobaria sneedii* Britton & Rose].

Key to the Varieties of *Coryphantha sneedii*

1. Plants often profusely branched, dominated by immature stems, up to 200 or more, 1.3–3 cm in diameter; spines 62–95 per areole, all mostly appressed and conspicuously pubescent; inner central spines, the shortest ones, 1–8 mm long; NE end of Guadalupe Mts, in NM

   C. sneedii var. *leei*, p. 314

1. Plants less profusely branched, sometimes but not often dominated by immature stems, 1–100 or more stems, 1.4–6.5 cm in diameter; spines 31–68 per areole, not or only slightly appressed, glabrous or inconspicuously pubescent; inner central spines, the shortest ones, 2–22 mm long; SW end of Guadalupe Mts and elsewhere, mostly in TX (2).

2(1). Plants averaging 14–24 stems, up to 100 or more; spines shorter and more numerous (see text); Franklin Mts, El Paso Co., TX, and Franklin Mts and Bishop's Cap Mt, Doña Ana Co., NM

   C. sneedii var. *sneedii*, p. 310

2. Plants averaging 1–5 stems, up to 26; spines longer and fewer (see text); Guadalupe Mts, and eastern Pecos Co. to lower Big Bend region (3).

3(2). Stems mostly 3–6 cm long; Guadalupe Mts, high altitudes

   C. sneedii var. *guadalupensis*, p. 312

3. Stems mostly 7–25 cm long; lower Big Bend region, low altitudes

   C. sneedii var. *albicolumnaria*, p. 316

Coryphantha sneedii var. sneedii
Sneed's Pincushion Cactus

PLATES 272–74

Coryphantha sneedii var. *sneedii* is endemic to the CDR. As interpreted here, this rare taxon occurs only in the western tip of the Trans-Pecos and in adjacent New Mexico. Variety *sneedii* is listed as Endangered by the U.S. Fish and Wildlife Service (November 1979) and by the Texas Parks and Wildlife Department (April 1983).

Distribution. Limestone crevices, usually steep S-facing slopes, with desertscrub and *C. tuberculosa*. El Paso Co., Franklin Mts, TX (type locality), and Doña Ana Co., NM, Franklin Mts and Bishop's Cap Mt (at least in Doña Ana Co. always on or near Silurian-Ordovician-Cambrian limestone). 3,900–6,000 ft. Map 104.

Vegetative Characters. Plants of var. *sneedii* are cespitose, often forming mounded clumps of fewer than 25 densely white-spined stems, but in some clumps there are up to 100 or more stems. In most clumps only 3–5 stems obviously are full-sized, and the rest are immature stems of various sizes. Stems are globose when very young and become cylindroid or clavate with age, 3–13 cm long, 1.3–3 cm in diameter. In aspect the stems are bristly, although the radial spines are essentially appressed, snowy-white.

In vegetative condition one of the most diagnostic distinguishing features of *C. sneedii* (separating it from all related Texas species except *C. vivipara*) is the giant lenticular druses, 0.5–1 mm in diameter, present in the mature pith and cortical tissue. The stem tissue when exposed has a granular texture owing to the large druses, which are evident to the naked eye in fresh or dried tissue. All other species of *Coryphantha*, except *C. vivipara* and its allies, lack lenticular druses of that size. Other species of *Coryphantha* have relatively small spheroidal druses usually less than 0.4 mm in diameter, except for *C. hesteri*, which has lenticular druses to 0.4(–0.5) mm in diameter. There is no medullary vascular system in *C. sneedii*.

Of the 31–68 spines per areole, 0–6 (usually 2–3 in Texas) are inner central spines, and 5–18 (usually 13–14 in Texas)

Plate 272. *Coryphantha sneedii* var. *sneedii* (Sneed's pincushion cactus), Franklin Mts, El Paso Co., TX (photo by Richard D. Worthington).

Plate 273. *Coryphantha sneedii* var. *sneedii* (Sneed's pincushion cactus), Franklin Mts, El Paso Co., TX (photo by Richard D. Worthington).

Plate 274. *Coryphantha sneedii* var. *sneedii* (Sneed's pincushion cactus), Franklin Mts, El Paso Co., TX; mature fruits; cultivated.

are outer centrals. Shortest inner centrals are 2–13 mm long, averaging ca. 5 mm long, and the other inner centrals are 3–15.5 mm long. Outer central spines usually are 4–16.5 mm long. Inner centrals are porrect (usually one) or variously spreading or appressed. The largest central spines obviously are heavier than the radial spines and have bulbous bases. Fresh centrals are snowy-white, but the larger centrals usu-

Map 104. *Coryphantha sneedii* var. *sneedii* (Sneed's pincushion cactus).

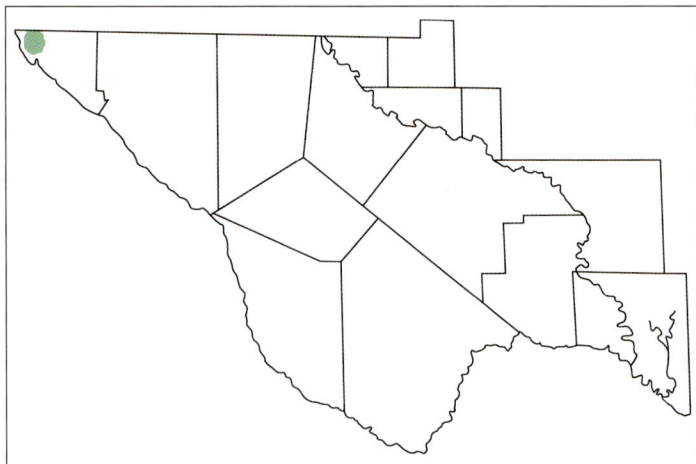

ally develop darker tips, tan to brownish-red or reddish-brown. The 24–46 (averaging 37 in Texas) radial spines are snowy-white, appressed, 3–12 mm long (averaging 5.5 mm long in Texas), slender. Outer radials closest to the stem are the most slender. By comparison, appressed radials in var. *sneedii* are not as appressed as those of var. *leei*. Additional details in spine morphology are available in Zimmerman (1985).

Flowers. Flowering Mar–May, after summer rains. Flowers basically whitish or pinkish, 1.2–2.5 cm long, and do not open widely. Inner tepals white with prominent midregions, broad or narrow, pink, magenta, dull lavender, brown, or greenish. Distal portions of inner tepals exhibit the deepest color. Outer tepals fimbriate in all varieties of *C. sneedii*. Filaments whitish to pale green, pink, or magenta. Stigma lobes 3–7, white, rarely pink or cream-colored, 1–3 mm long.

Fruits. Semisucculent green fruits obconic to cylindroid, 0.6–1.5 cm long, 0.25–0.6 cm in diameter. Fruits green when mature but with age may turn brownish-pink on exposed portions and become nearly dry. Floral remnant persistent. Pitted seeds red-brown or orange-brown, becoming darker when dry, 1.1–1.5 mm long.

Full Name and Synonym. *Coryphantha sneedii* var. *sneedii* [= *Escobaria sneedii* Britton & Rose]. *Mammillaria sneedii* (Britton & Rose) Cory.

Other Common Names. Sneed cory cactus; Sneed's carpet escobaria.

Coryphantha sneedii var. guadalupensis
Guadalupe Pincushion Cactus
PLATES 275, 276

This is the most recently described of the three varieties of *C. sneedii* in Texas, as "*Escobaria guadalupensis*," named after the mountain range in which it occurs. The type locality is from the highest peak in Texas in the Guadalupe Mountains. The Guadalupe Mountains lie mostly in southeastern New Mexico, extending just into Texas and harboring at least six of

Plate 275. *Coryphantha sneedii* var. *guadalupensis* (Guadalupe pincushion cactus); Guadalupe Mts, Culberson Co., TX (photo by Brent Wauer).

Plate 276. *Coryphantha sneedii* var. *guadalupensis* (Guadalupe pincushion cactus), Mesa Garden, NM; fruit; cultivated (photo by James F. Weedin).

the highest peaks in Texas in Guadalupe Mountains National Park.

Distribution. Mostly steep limestone slopes, frequently S-facing exposures, SW Guadalupe Mts, Culberson Co., mostly higher elevations in with oak-juniper-pinyon associations. 4,500–8,700 ft. Endemic to the Guadalupe Mts, extending a few miles into NM. Map 105.

Vegetative Characters. The plants high

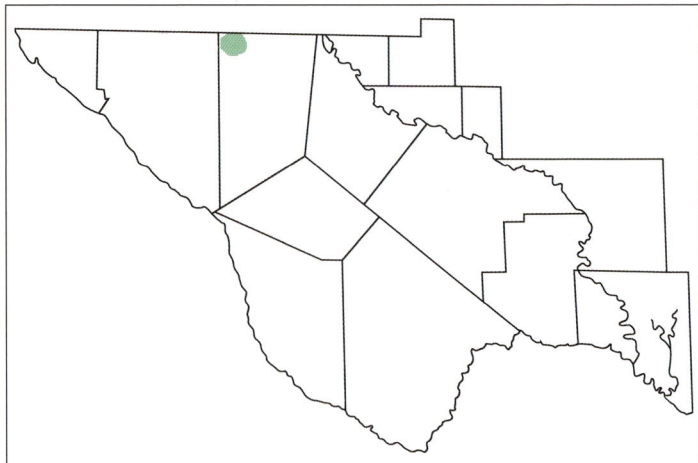

in the Guadalupe Mountains are similar in most characters to the largest adult stems of var. *sneedii*. They are mostly solitary or in small clumps of 2–3 or more robust stems. A few plants exhibiting tight, rounded clumps of stems have been observed in these populations.

Flowers and Fruits. Flowering Apr–May, after summer rains. Flower, fruit, and seed characters are within the range of variation described for var. *sneedii*. Flower color has been described as pale yellow or cream-colored to pink, with pink midregions, and as orangish-pink. Reportedly, the fruits are green when ripe.

Other Distinctions. Two other species of *Coryphantha* with stems densely white-spined occur in the Guadalupe Mountains: *C. tuberculosa* and *C. vivipara* var. *neomexicana*.

Full Name and Synonym. *Coryphantha sneedii* var. *guadalupensis* (S. Brack & K. D. Heil) A. D. Zimmerman, comb. nov. (forthcoming). [= *Escobaria guadalupensis* (S. Brack & K. D. Heil) var. *guadalupensis,* comb. nov., forthcoming]. *Escobaria guadalupensis* S. Brack & K. D. Heil.

Coryphantha sneedii var. leei
Lee's Pincushion Cactus
PLATES 277, 278

The main axis of the Guadalupe Mountains trends north and northwest in Otero and Eddy counties, but a smaller arm extends to the northeast from the Texas portion of the mountains into New Mexico, gradually terminating about 10 miles north of the New Mexico–Texas state line. The end of the northeast arm is the site of the famous Carlsbad Caverns and the general locality of *C. sneedii* var. *leei*. The varietal epithet honors W. T. Lee, who collected the taxon in Rattlesnake Canyon.

Distribution. Limestone outcrops (ledges on the N-facing slopes at the type locality), desertscrub, often dominated by lechuguilla, NE end of Guadalupe Mts, Carlsbad Caverns National Park, Eddy Co., NM. 3,750–5,500 ft. Map 106.

Vegetative Characters. Plants of var. *leei* have been described as profusely branching cushions of up to 200–250 stems tightly packed together and of various sizes. These are the types of plants

Plate 277. *Coryphantha sneedii* var. *leei* (Lee's pincushion cactus); S NM; cultivated.

Plate 278. *Coryphantha sneedii* var. *leei* (Lee's pincushion cactus), plants from S NM; fruits; cultivated.

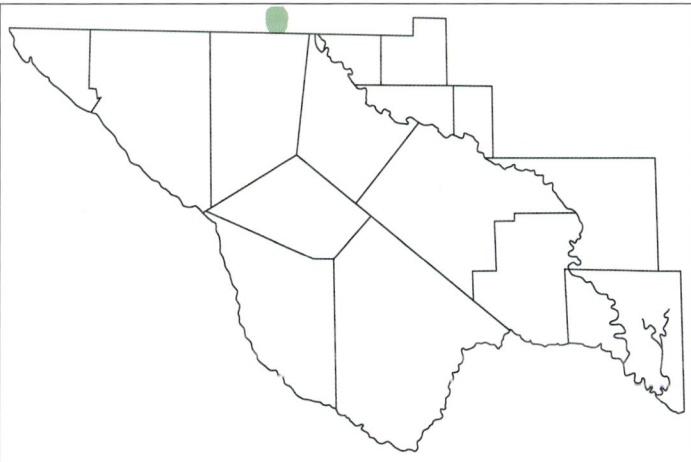

Map 106. *Coryphantha sneedii* var. *leei* (Lee's pincushion cactus).

that are familiar in horticulture and like the type specimen. The cushion plants are dominated by small, immature stems and represent an extreme form of the species. Use the relatively ordinary-looking old adult stems of var. *leei* for purposes of taxonomic comparison with other taxa (Powell and Weedin, 2004).

The distinctive vegetative features of var. *leei*, as compared to those of all other populations of the species, including var. *guadalupensis*, are the smaller and more numerous stems in greater numbers, shorter and more numerous spines that are more closely appressed, and more strongly pubescent spines on immature stems. The relatively few mature stems are 3.5–10 cm long, 1.3–3 cm in diameter, and most similar to the stems of *C. sneedii* var. *sneedii*. The stems are densely white-spined, and all the spines are tightly appressed (with the exception of some inner centrals), even reflexed toward the stem, resulting in a knobby surface of the stem. The stem surface is visible between the tubercles. On average, about five (ranging 1–12) inner central spines, usually one of them porrect. Outer centrals number 10–22, usually about 17, 3–10 mm long. Central spines snowy-white when fresh, rarely stramineous, with some of the largest spines dark-tipped. White radial spines 38–74 in number, commonly about 50, 3–8 mm long.

Flowers. Flowering Apr–May. Flowers 1.1–1.5 cm long and often described as pinkish-brown or pinkish-green, with deep pink midstripes. Inner tepal color variation may include pale rose-pink, reddish-purple to dull brownish-lavender, or pale tan with yellowish-pink midregions. Yellow anthers, ca. 0.5 mm long, on filaments 3–4 mm

long. Whitish style 9–10 mm long, supporting 3–6 white stigma lobes.

Fruits. Fruits ellipsoid to oblong or narrowly obovoid, 0.5–1.2 cm long, 3.5–5 mm in diameter, green or pale brownish-green at maturity, developing a pinkish or dull brownish-red tint on exposed surfaces. Seeds reddish-brown, drying darker, 1.2–1.5 mm long, more or less comma-shaped.

Full Name and Synonyms. *Coryphantha sneedii* var. *leei* (Rose ex Boed.) L. D. Benson. [= *Escobaria sneedii* Britton & Rose var. *leei* (Rose ex Boed.) D. R. Hunt]. *Escobaria leei* Rose ex Boed.; *E. sneedii* subsp. *leei* (Rose ex Boed.) D. R. Hunt.

Other Common Name. Lee's carpet escobaria.

Coryphantha sneedii var. albicolumnaria
Silverlace Cactus

PLATES 279–83

Coryphantha sneedii var. *albicolumnaria*, erroneously included with *C. tuberculosa* (as *C. strobiliformis* var. *durispina*) by Benson (1982), is by far the most widespread variety of *C. sneedii* in the Trans-Pecos. The varietal epithet includes the Latin *albi,* "white," *colum,* "pillar" (or "column"), and *aria,* a suffix forming an adjective ("connected with"), in reference to the cylindroid stems cloaked in white spines.

Distribution. Mostly if not entirely limestone, rocky-gravelly slopes, rocky outcrops, and alluvium; desertscrub, often associated with *Euphorbia antisyphilitica* Zucc., *Hechtia texensis*, and *Jatropha dioica* Cerv. Mostly lower Big Bend near the Rio Grande, rare in E Pecos Co. 1,850–4,450 ft. Mexico: Chihuahua; expected in Coahuila across from Big Bend National Park. Map 107.

Plate 279.
Coryphantha sneedii
var. *albicolumnaria*
(silverlace cactus) near
Terlingua, Brewster
Co., TX.

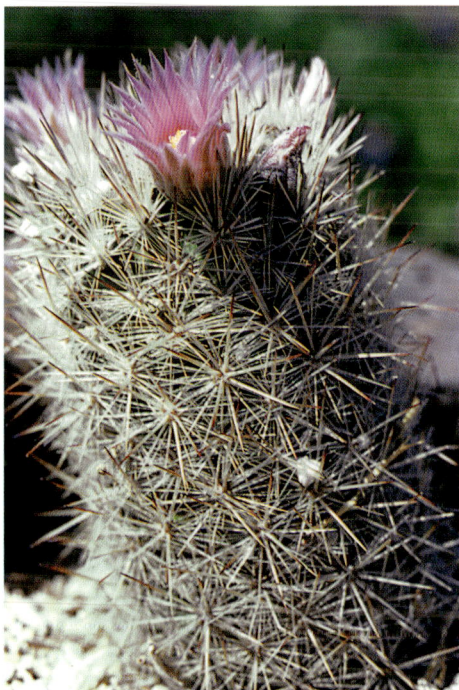

Plate 280.
Coryphantha sneedii
var. *albicolumnaria*
(silverlace cactus), from
S Brewster Co., TX;
cultivated.

Plate 281.
Coryphantha sneedii
var. albicolumnaria
(silverlace cactus), from
S. Brewster Co., TX;
white-flowered form;
cultivated.

Plate 282.
Coryphantha sneedii
var. albicolumnaria
(silverlace cactus),
S Brewster Co., TX;
mature fruits, red form;
cultivated.

Vegetative Characters. Plants of var. *albicolumnaria* are characterized by unbranched or few-branched, cylindroid stems densely covered by bristly white spines. Older plants may have 2–5 stems, these blunt apically, and presumably very old plants form loose clumps of up to 26 mature stems. Stems mostly erect, rarely decumbent, 7–20 cm long, 2.5–5.5 cm in diameter. Spines 35–55 per areole, includ-

Plate 283.
Coryphantha sneedii
var. *albicolumnaria*
(silverlace cactus), from
Solitario, Presidio Co.,
TX; mature fruits, green
form; cultivated.

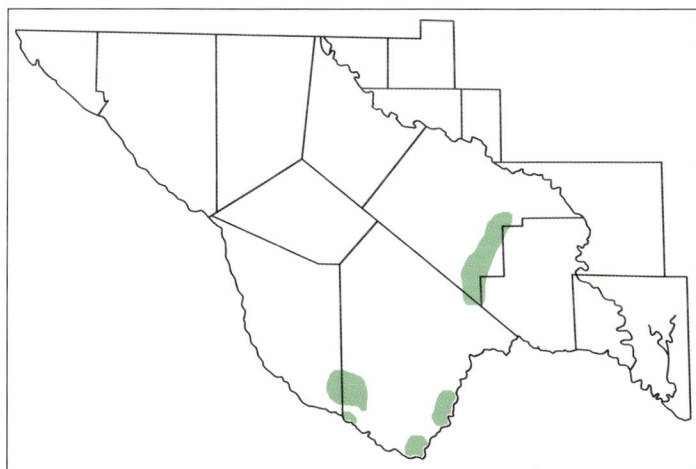

Map 107. *Coryphantha
sneedii* var.
albicolumnaria
(silverlace cactus).

ing 1–5 inner central spines, the longest 1.3–2.5 cm, and 9–16 outer centrals, the longest 0.9–2.6 cm, spreading in all directions. Radial spines 24–35, mostly appressed, 0.5–1.4 cm long. All spines snowy-white when fresh, or the central spines may be chalky-pink, and distally the larger centrals usually are chalky-pink, tan, reddish-brown, or blackish-purple.

Flowers. Flowering Mar–May. Usually

pink flowers 1.5–2 cm long and do not open widely. Inner tepals may be pale rose-pink, to deep pink, or even bright magenta or glossy milk-white. Sharply defined midstripes may or may not be present. Filaments usually white, supporting yellow anthers. Style relatively short and pinkish, with 3–7 short, white, or pale pink stigma lobes not much exserted above the anthers.

Fruits. Mature fruits either pale green or red, cylindroid or narrowly obovoid, 1–1.7 cm long, 0.5–0.7 cm in diameter. This is the only variety of *C. sneedii* in the Trans-Pecos that has brilliant red fruits in at least some of the populations. The green fruits usually are yellow or apricot-tinged at maturity, and the red fruits are either bright red or dull pinkish-red. Seeds pitted, red-brown, 1–1.5 mm long.

Other Distinctions. Within its range, var. *albicolumnaria* is most likely to be confused with *C. dasyacantha* and *C. tuberculosa*. See Powell and Weedin (2004) for a review of interspecific distinguishing characters, biosystematic discussion, and the involved taxonomic history of var. *albicolumnaria*.

Full Name and Synonyms. Coryphantha sneedii var. *albicolumnaria* (Hester) A. D. Zimmerman, comb. nov. (forthcoming). [= *Escobaria sneedii* var. *albicolumnaria* (Hester) comb. nov., forthcoming]. *Escobaria albicolumnaria* Hester; "*Mammillaria albicolumnaria* (Hester)" Weniger, comb. nud.; *Coryphantha albicolumnaria* (Hester) Zimmerman; *Escobaria tuberculosa* (Engelm.) Britton & Rose var. *durispina* auct. non Quehl.

Other Common Names. White column; white-spine cob cactus.

Coryphantha pottsiana
Runyon's Pincushion Cactus
PLATES 284, 285

To our knowledge *C. pottsiana* has not been documented in the Trans-Pecos (that is, west of the Pecos River), although reportedly it occurs near the mouth of the Pecos River. The taxon has been verified in similar habitats not more than 25 miles away on the west bank of the lower Devils River, now Amistad Reservoir. Both Benson (1982) and Weniger (1984) used the epithet *robertii* for this taxon before A. D. Zimmerman (forthcoming) determined that the oldest name is *pottsiana*. The name *pottsiana* presumably is after F. H. Potts or John Potts.

Distribution. Rocky limestone habitats, often in dense scrub, Val Verde Co., reported near the mouth of the Pecos River, relatively common near the W bank of the lower Devils River. 900–1,000 ft. Deep S TX, mostly in the Rio Grande valley, Webb and Duval counties S to Cameron Co. Mexico: adjacent NE Coahuila (N to Monclova), N Nuevo León, and Tamaulipas. Map 108.

Vegetative Characters. Plants in clumps of a few to several or numerous stems. Clumps may be 10–30 cm or more wide in Val Verde County and reportedly are up to 1 m wide in the lower Rio Grande valley, but usually they are small and inconspicuous. Individual stems ovoid to cylindroid, at maturity 3–6 cm long, 1.5–3 cm in diameter, densely covered with spines and bristly in aspect. In Val Verde County plants we have evaluated, there are 25–38 spines per areole, including 7–8 inner centrals, 1.3–1.5 cm long (longest one, 1.8–2.3 cm), and 20–30 outer centrals and radials to ca. 1 cm long. The central

spines spread in all directions, some of them porrect, either white with dark tips or dominated by reddish-brown or nearly black distal portions. Radial spines white or off-white, sometimes with brown tips.

Flowers. Flowering Feb–Mar. Pale purplish-brown flowers, 1.7–2.5 cm long, 1–1.5 cm in diameter. There appears to be considerable flower-color variation in *C. pottsiana.* In flowers we have examined

Map 108. *Coryphantha pottsiana* (Runyon's pincushion cactus).

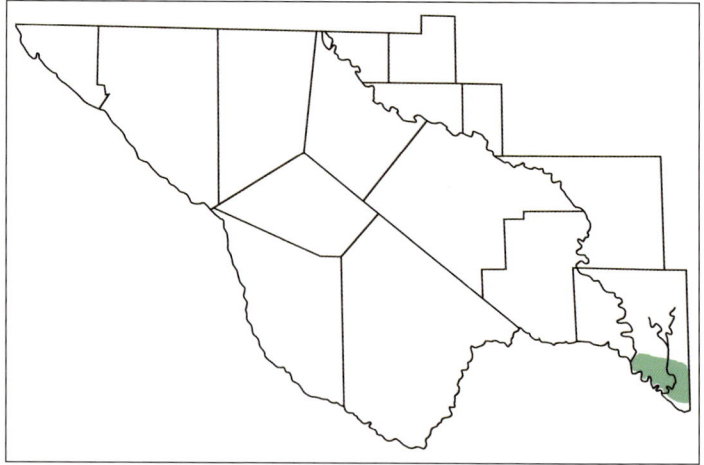

from Val Verde County populations, the inner tepals have broad midregions, purplish (maroon) or lavender to pale purple, or faintly so, with pale brownish margins. Yellow anthers supported by pink, purplish, magenta, or colorless filaments 5–6 mm long. Stamens sensitive. Stigma lobes 5–7, green, 1.5–2 mm long.

Fruits. Fruits red in Val Verde County populations, so far as known, and in plants of South Texas and adjacent Mexico. Fruits spherical to ovoid, 6–9 mm in diameter, juicy, the juice reddish and moderately sweet. Floral remnant persistent. Some populations in Mexico have green fruits. Seeds comma-shaped, 1–1.2 mm long, pitted, brown or reddish-brown, at least when fresh, perhaps drying darker. Hilum clearly basal.

Full Name and Synonyms. Coryphantha pottsiana (Poselger) A. D. Zimmerman [= *Escobaria pottsiana* (Poselger) comb. nov., forthcoming]. *Echinocactus pottsianus* Poselger; *Escobaria runyonii* Britton & Rose (non *Coryphantha runyonii* Brit-

ton & Rose); *C. emskoetteriana* (Quehl) A. Berger; *C. robertii* A. Berger; *C. muehlbaueriana* Boed.

Other Common Name. Runyon's escobaria.

Coryphantha tuberculosa
Cob Cactus, Varicolor Cob Cactus

Four varieties of *C. tuberculosa* are recognized, two of them restricted to Mexico in parts of Coahuila, Durango, and Chihuahua. Two varieties occur in the Trans-Pecos. Benson (1982) erroneously treated *C. tuberculosa* var. *varicolor* as a variety of the distantly related species *C. dasyacantha.* Benson treated *C. tuberculosa* as *C. strobiliformis* (Poselger) Moran while recognizing three varieties, two of which clearly belong with *C. sneedii.* See Powell and Weedin (2004) for a biosystematic review of *C. tuberculosa. Coryphantha tuberculosa* is the type species of *Escobaria,* regarded here as a subgenus of *Coryphantha.* The specific epithet, *tuberculosa,* is intended to be descriptive, after the Latin *tuberculum,* "knob," and *osa,* as a termination meaning "full of," apparently in reference to the bare tubercles collectively lending a corncob aspect at the base of old stems in many plants.

Full Name and Synomym. Coryphantha tuberculosa (Engelm.) A. Berger [= *Escobaria tuberculosa* (Engelm.) Britton & Rose].

Key to the Varieties of *Coryphantha tuberculosa*

1. Stems almost always branched, forming clumps of 3–50 or more stems; mature stems ovoid or cylindroid, often pointed at the apex, typically 3–4 cm in diameter; plants usually restricted to limestone habitats, or at least to near limestone
   C. tuberculosa var. *tuberculosa,* p. 323
1. Stems almost always solitary, rarely branched after injury, mature stems oblate or globose, in old age cylindroid, typically 3.6–6.5 cm in diameter; plants usually restricted to igneous or novaculite habitats
   C. tuberculosa var. *varicolor,* p. 326

Coryphantha tuberculosa var. tuberculosa
Cob Cactus

PLATES 286, 287

In the Trans-Pecos *C. tuberculosa* var. *tuberculosa* is a common species in desert limestone habitats, where potentially it might be confused in vegetative condition with several other densely spined species. When in flower, *C. tuberculosa* is readily distinguished from any other species in the Trans-Pecos; note the pale yellow anthers, white stigma lobes, and late flowering time.

Distribution. Limestone habitats, desert mountains, ridges, eroded outcrops, rarely extending into adjacent igneous habitats, often associated with lechuguilla. Throughout most of the Trans-Pecos, from El Paso Co. E to W Terrell Co., absent or uncommon in Jeff Davis and Reeves counties, in mostly S portions of Presidio and Pecos counties, reported but not reliably

documented in Val Verde Co. at the mouth of the Devils River. 1,700–5,500 ft. Southern NM, Doña Ana, Otero, and Eddy counties, possibly Luna and Sierra counties. Mexico: S to central Coahuila and NE Durango, E Chihuahua. Map 109.

Vegetative Characters. Plants frequently grow in crevices of limestone rock and almost always form clumps of branched stems. Probably less than 1% of adult plants are unbranched. Stems ovoid to cylindroid and sometimes slightly pointed at the apex, 5–16 cm long. Tubercles on lower stem above or below ground level are naked because the spine clusters and often the whole tubercle tips have abscised, revealing in many stems a corncob aspect. Spine color is a subtle but often useful trait in distinguishing *C. tuberculosa* from several other taxa; spines of Trans-Pecos varieties of *C. tuberculosa* are white, pale gray, or pale tan, not

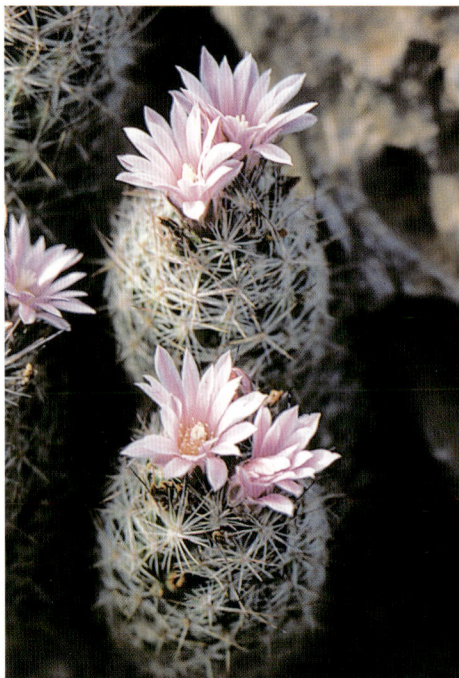

Plate 286. *Coryphantha tuberculosa* var. *tuberculosa* (cob cactus), Rosillos Mts, S Brewster Co., TX.

Plate 287. *Coryphantha tuberculosa* var. *tuberculosa* (cob cactus), from S Brewster Co., TX; mature fruits; cultivated.

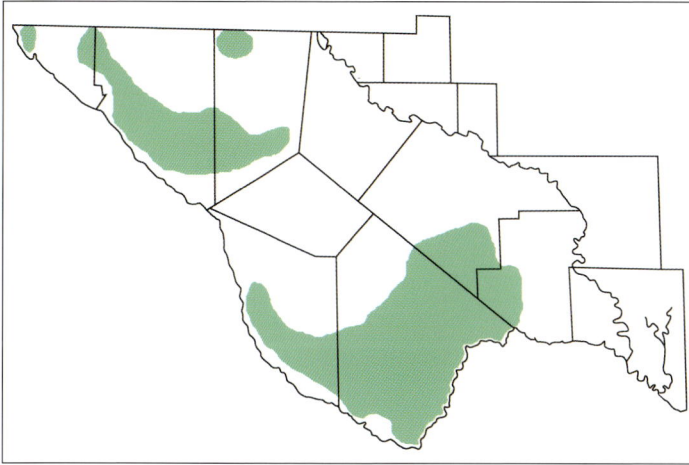

Map 109. *Coryphantha tuberculosa* var. *tuberculosa* (cob cactus).

the gleaming translucent white typical of certain other species, including *C. sneedii.* Tips of larger spines are pinkish-gray or tan, or darker gray-brown or reddish-brown. Typically in var. *tuberculosa* there are only 1–2 inner central spines, usually two in areoles of adult stems. If only one, it is in a lower position and 0.5–1.5 cm long. If two inner centrals, the one in the lower position is always the shortest and stoutest and either porrect or deflexed, and the longer upper one is 1.3–1.8 cm long and either ascending or porrect. In some plants, presumably with age, there appear to be four inner centrals (the upper ones may be interpreted as outer centrals), still with the lower one always the shortest and stoutest. In each areole usually there is only one porrect central. Outer centrals number 4–9, the upper three appressed with the radials and 1.3–1.8 cm long. Light gray radial spines 17–25, usually 0.7–1.3 cm long.

Flowers. Flowering Apr–Aug. Unique features of the flowers are best observed in fresh condition: (1) flowers open relatively late, about 3:00–4:00 p.m., remaining open until dusk or into early evening; (2) sterile distal portion of receptacular tube unusually long, almost tubular, between the stamens and the inner tepals; and (3) anthers and pollen very pale yellow, almost white, probably lighter in color than those of any other species of *Coryphantha.* Stigma lobes 4–6, white, 1.8–4 mm long. Pale pink to nearly white or pale rose-pink funnelform flowers 1.8–3.2 cm long, 2.1–4 cm in diameter. Typically, inner tepals are more intensely pigmented in their central regions, but a prominent midstripe seldom is evident.

Fruits. Dull to bright red fruits broadly ellipsoid to ovoid as exposed above the spines but seem to be broadly clavate as seen when separated from the plant. Fruits 1–2.5 cm long, 3.5–7 mm in diameter. A variety of *C. tuberculosa* in southeastern Chihuahua has green, maroon, or, at best, only rarely red fruits. Fruits of both Trans-Pecos varieties always red when fully ripe. In the Big Bend area, we have observed nearly full-sized green fruits positioned just inside (toward the apex from) bright red or dull red fruits on the same stem, as though nearly mature green fruits turn red abruptly at full maturity. Seeds reddish-brown, pitted, 0.95–1 mm long, always smaller than seeds of *C. sneedii* (0.9–1.6 mm long).

Full Name and Synonyms. *Coryphantha tuberculosa* var. *tuberculosa. Mammillaria tuberculosa* Engelm.; *Mammillaria strobiliformis* Scheer ex Salm-Dyck.

Other Common Names. Corncob escobaria; cob cory cactus.

Coryphantha tuberculosa var. varicolor
Varicolor Cob Cactus

PLATES 288, 289

Variety *varicolor* predominantly is a taxon of the igneous central mountains of the Trans-Pecos, but it extends south into the Chinati and Chisos mountains and into novaculite substrates. The varietal epithet is derived from the Latin prefix *vario*, "variable," and *color*, presumably in reference to the different colors in central and radial spines.

Distribution. Usually in volcanic mountains or other igneous rock habitats, oak-juniper woodlands and rocky sites in grasslands to lower elevations in desertscrub, and in novaculite exposures in the Marathon Basin and Solitario, associated with grassland or desertscrub, rarely extending onto limestone substrates. Mostly Presidio, Jeff Davis, and Brewster counties, rare in S Pecos Co. 2,300–5,500 ft. Known only from Trans-Pecos TX, but surely not endemic, considering its abundance on the peaks overlooking adjacent Coahuila and Chihuahua, Mexico. Map 110.

Plate 288. *Coryphantha tuberculosa* var. *varicolor* (varicolor cob cactus), from Jeff Davis Co., TX; cultivated.

Plate 289. *Coryphantha tuberculosa* var. *varicolor* (varicolor cob cactus), from Jeff Davis Co., TX; mature fruits; cultivated.

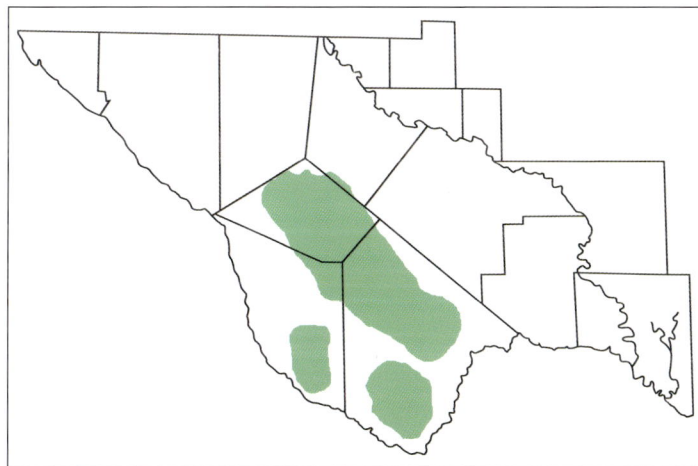

Map 110. *Coryphantha tuberculosa* var. *varicolor* (varicolor cob cactus).

Vegetative Characters. Plants single-stemmed, similar in aspect to the stems of var. *tuberculosa* except larger in diameter. Globose, oblate, or cylindroid stems 4–16 cm long. Plants growing in novaculite substrates tend to be smaller in size. In var. *varicolor* the green to gray-green stem surfaces are more visible through the spines than in var. *tuberculosa*, evidently because there are fewer, shorter, and more slender spines. Also, the tubercles are slightly farther apart on slightly broader stems than in var. *tuberculosa*. In var. *varicolor*, 1–2 inner central spines, usually two, with the shortest one porrect or descending and in a near-central position and an upper one (or three) longer, 1.1–1.7 cm, erect or ascending. The upper inner central spines also may be interpreted as outer centrals. Radial spines 15–24 in number, 7–12 mm long, appressed against the plane of the stem. In var. *varicolor* the spines are slightly fewer (radials), shorter, more slender, and the centrals with more color, than in var. *tuberculosa*, although spine clusters alone, in the absence of stems, do not provide for reliable identification of the varieties.

Flowers and Fruits. Flowering May–Aug. Pale pink to white flowers, fruits, and seeds are virtually identical to those of var. *tuberculosa*.

Other Distinctions. Plants that appear to be intermediate between var. *varicolor* and var. *tuberculosa* occur in some populations in the lower Big Bend region; most are in the form of solitary, relatively slender stems lacking the dense spination of var. *tuberculosa*, or solitary, relatively thick stems with the dense spination of var. *tuberculosa*. Plants of *C. tuberculosa* in novaculite substrates of the Marathon Basin are different in vegetative appearance from var. *varicolor* throughout most of its range. In novaculite the stems are usually flat-topped and short, and the spine covering is relatively smooth. See Powell and Weedin (2004) for further discussion of these morphotypes.

Full Name and Synonyms. *Coryphantha tuberculosa* var. *varicolor* (Tiegel) A. D. Zimmerman, comb. nov. (forthcoming) [= *Escobaria tuberculosa* (Engelm.) Britton & Rose var. *varicolor* (Tiegel) S. Brack & K. D. Heil]. *Coryphantha varicolor* Tiegel;

Escobaria varicolor Tiegel; *E. dasyacantha* (Engelm.) Britton & Rose var. *varicolor* (Tiegel) D. R. Hunt; *Coryphantha dasyacantha* (Engelm.) Orcutt var. *varicolor* (Tiegel) L. D. Benson; "*Mammillaria varicolor*" (Tiegel) Weniger, comb. nud.

Other Common Names. Varicolor cory cactus; mountain cob cactus.

Coryphantha macromeris var. macromeris
Big-Needle Pincushion Cactus
PLATES 290, 291

Coryphantha macromeris is one of the most widespread cactus species in the Trans-Pecos, occupying nearly all substrate types, including gypsum. In Texas two varieties of *C. macromeris*, var. *macromeris* and var. *runyonii* (Britton & Rose) L. D. Benson (Runyon's coryphantha) were recognized by Benson (1982), but Weniger (1984) treated var. *runyonii* as a distinct species (as *Mammillaria runyonii* Britton & Rose). Variety *runyonii* (Plates 292, 293) is restricted to the lower Rio Grande valley, on gravelly hillsides and in silt often under shrubs near the Rio Grande, from about Roma in Starr County to near Brownsville in Cameron County and in adjacent Tamaulipas, Mexico (Map 111). The South Texas variety forms particularly untidy clumps, with relatively short tubercles usually 1–2 cm long, the result of relatively rapid proliferation of axillary branches.

The epithet *macromeris* is after the Greek *makros*, "large," and *meros*, "a part," probably a reference to the elongated tubercles that characterize the taxon. The epithet *runyonii* is after Robert Runyon, an early (ca. 1920) botanist in the Rio Grande valley, who studied the taxon and

Plate 290. *Coryphantha macromeris* var. *macromeris* (big-needle pincushion cactus), **Study Butte, S Brewster Co., TX.**

Plate 291. *Coryphantha macromeris* var. *macromeris* (big-needle pincushion cactus), from **S Brewster Co., TX;** mature fruits; cultivated.

Plate 292. *Coryphantha macromeris* var. *runyonii* (Runyon's coryphantha), from **Starr Co., TX;** cultivated.

Plate 293. *Coryphantha macromeris* var. *runyonii* (Runyon's coryphantha), from Starr Co., TX; mature fruit; cultivated.

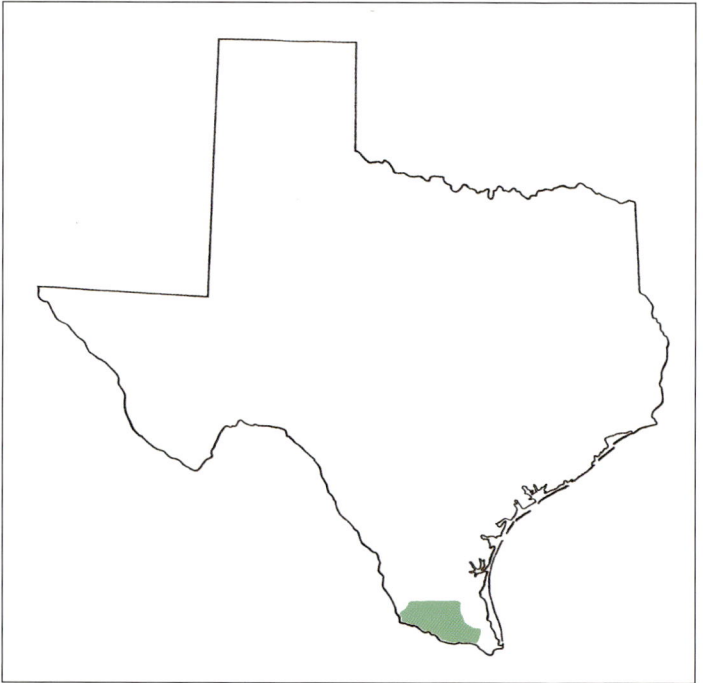

Map 111. *Coryphantha macromeris* var. *runyonii* (Runyon's coryphantha).

provided specimens for Britton and Rose, who described it as a distinct species. *Coryphantha macromeris* and the remaining coryphanthas in the current treatment (species 10–14) belong to subgenus *Coryphantha*, unless *C. macromeris* is taxonomically segregated as the monotypic *Lepidocoryphantha* Backeb.

Distribution. Commonly sandy alluvium, gravel benches, or clay, in the open

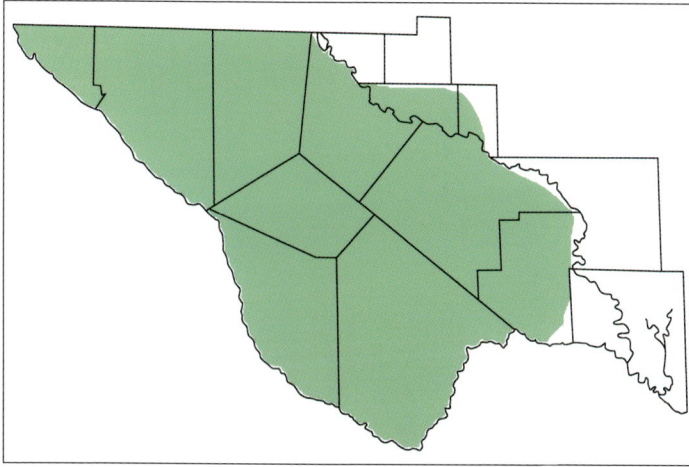

or under shrubs, various substrates but rarely on steep, rocky slopes or in crevices. Throughout the Trans-Pecos; most common in southern Presidio and Brewster counties. 1,500–4,000 ft. In TX E and NE to Glasscock and Ector counties. Southern NM, Chaves, Eddy, Lea, Doña Ana, Sierra, and Luna (near Columbus) counties. Mexico: NE Chihuahua, Coahuila, NE Durango, and N Zacatecas to Tamaulipas. Map 112.

Vegetative Characters. Mature plants of var. *macromeris* are clumped, forming many-stemmed low mats or hemispheric mounds 5–25 cm high, up to 1 m wide. Stems are hemispheroidal to short cylindroid, 5–23 cm long, 4–9 cm in diameter. Pith and cortex are mucilaginous. Tubercles are conspicuous, somewhat flaccid and soft-skinned, narrowly conoid-cylindroid, 1.5–3.5 cm long. *Coryphantha macromeris* is similar to *C. ramillosa* in the characters mentioned above except that *C. ramillosa* usually is single- or few-stemmed, the pith and cortex are not mucilaginous, and the tubercles are shorter. The

presence or absence of mucilage in a living plant usually can be determined by removing and examining a tubercle.

In *C. macromeris* there are 7–21 spines per areole, 3–7 central and usually 12–14 radial. In Trans-Pecos populations typically there are four central spines, usually 3–7 cm long. Lowermost central spine porrect or descending, and the other centrals diffuse or weakly appressed. Centrals commonly dark gray to black, but in some plants pale gray to tan or stramineous. Longest radial spines 1.5–3 cm, tan to stramineous or brown, sometimes gray to whitish in age. See Powell and Weedin (2004) for further distinctions and comparisons between *C. macromeris*, *C. ramillosa*, and *C. macromeris* var. *runyonii*.

Flowers. Flowering sporadically May–Sep. Bright rose-pink to magenta flowers 3–6 cm long, 4–7 cm in diameter. In the Trans-Pecos the inner tepals are more commonly rose-pink than magenta, often with a darker midline and pale margins. Filaments 1–1.7 cm long, greenish-white or purplish-pink distally, contrasting with yellow anthers 1–1.5 mm long. Style 2–2.7 cm long, supporting 7–13 white or cream-colored stigma lobes. Stigma lobes 4–6 mm long, often with slender, soft points at the tip.

Fruits. Fruits ovoid to obpyriform, 1.4–3 cm long, 1.2–1.8 cm in diameter, dark green at maturity. Fruit pulp whitish to pink. Ripe fruits have a pleasant, sweet aroma, sweeter in smell than a ripe banana, but the taste is not correspondingly sweet. Floral remnant persistent. Mature fruits have a few scales at the top (the basis for the generic name *Lepidocoryphantha*), near the base of the floral remnant. Reddish-brown seeds finely and weakly raised-reticulate, as seen under magnification, but otherwise appear to be nearly smooth, 1.2–1.5 mm long, more or less comma-shaped to globose.

Full Name and Synonyms. Coryphantha macromeris (Engelm.) Lem. var. *macromeris. Mammillaria macromeris* Engelm.; *Echinocactus macromeris* (Engelm.) Poselger; *Lepidocoryphantha macromeris* (Engelm.) Backeb.

Other Common Names. Long mamma; big nipple cory-cactus.

Coryphantha ramillosa
Whiskerbrush Pincushion Cactus
PLATES 294, 295

In Texas *C. ramillosa* is restricted to certain limestone habitats near the Rio Grande in Brewster and Terrell counties. The species was federally listed as Threatened in 1979 and listed in Texas as Threatened in 1983. Presumably the specific epithet is from the Latin *ramulosus*, "full of branches," but with reference to its

Plate 294. *Coryphantha ramillosa* (whiskerbrush pincushion cactus), from S Brewster Co., TX; cultivated.

Plate 295. *Coryphantha ramillosa* (whiskerbrush pincushion cactus), from S Brewster Co., TX; mature fruits; cultivated.

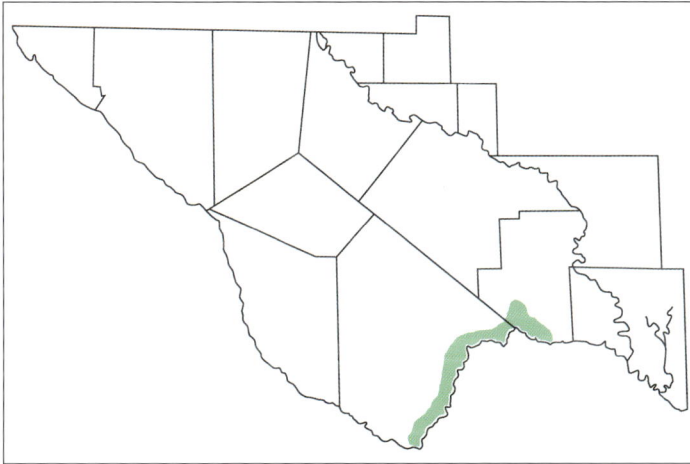

Map 113. *Coryphantha ramillosa* (whiskerbrush pincushion cactus).

superficial appearance in the field rather than its actual architecture.

Distribution. Desert limestone hills, ridges, benches, and mesa slopes, associated with lechuguilla. Brewster and Terrell counties, near the Rio Grande. 1,350–3,000 ft. Mexico: N Coahuila, adjacent to Big Bend National Park and southeastern Brewster Co., and probably adjacent to Terrell Co., S to near Cuatro Ciénegas. Map 113.

Vegetative Characters. One of the most conspicuous features reportedly distinguishing *C. ramillosa* from the regionally sympatric and superficially similar *C. macromeris* is the single or sparingly branched stems of *C. ramillosa* and the multibranched mats or mounds of *C. macromeris*. In Trans-Pecos populations, at sites where it is possible to observe several plants, the branching habit is a good distinguishing character for *C. ramillosa* and *C. macromeris*. Populations of *C. ramillosa* in Brewster County are predominated by plants with single stems, although some plants with 8–10 stems also are present in these populations. The same sparsely branching habit has been reported for populations in Terrell County, with about 60% of the plants unbranched, although plants with 22 or more stems are found south of Sanderson. In limestone hills across the Rio Grande from Big Bend National Park we have observed a preponderance of low-mounded, multibranched (20–25 stems) plants in one population, where single-stemmed plants were also present.

Stems of *C. ramillosa*, 4–9 cm long and 4–8 cm wide, are more compact and tidy in appearance than the stems of *C. macromeris*. Part of this stem aspect is the result of subtly distinctive spination. In *C. ramillosa* there are 14–27 spines per areole, with one main (inner), porrect central, 2.2–4 cm long, usually three upper (outer) centrals, and usually 13–16 radial spines, the longest 1–3 cm. The single porrect, terete inner central is different from the several (typically four) radiating, angular main centrals of *C. macromeris*. In *C. ramillosa* the pith contains a conspicuous medullary vascular system, absent in *C. macromeris*.

Flowers. Flowering Aug–Sep. Flowers of *C. ramillosa* are basically the same color as those of *C. macromeris* but a little smaller, 3.8–6.5 cm long, 3–5 cm in diameter. The glossy inner tepals usually are rose-pink but may be pale pink to deep rose-purple (nearly magenta). Outer tepals are entire, unlike the long-fimbriate outer tepals of *C. macromeris*. Filaments are whitish, 6–8 mm long, supporting bright yellow to pale orange anthers 1–1.3 mm long. Style 1.5–1.8 cm long. Stigma lobes number 6–7, white, 3.5–7 mm long.

Fruits. Fruits dark green to gray-green, obovoid, globose, or elliptic, 1.6–2.5 cm long, 1.2–1.6 cm in diameter. Fruits juicy, with colorless or greenish pulp. Ripe fruits have a sweet aroma, similar to that of *C. macromeris*, but the juice is not sweet. Floral remnant persistent. No scales on the fruit surface, even at the top, in contrast with the fruits of *C. macromeris*. Reddish-brown to golden-brown seeds globose to comma-shaped, 1.0–1.5 mm long. Seed coat shiny but microscopically finely raised-reticulate.

Full Name and Synonym. *Coryphantha ramillosa* Cutak. *Mammillaria ramillosa* (Cutak) Weniger.

Coryphantha scheeri
Long-Tubercled Coryphantha

Populations of *C. scheeri* in the Trans-Pecos always consist of widely spaced or scattered individuals. Rarely more than one or a few plants are found within several meters of each other. This widely spaced distributional pattern results from infestation by cactus-specialist beetle larvae (A. D. Zimmerman, unpub.). In Texas, *C. scheeri* is practically restricted to the Trans-Pecos, where it is widely distributed but poorly collected and is not well understood in terms of habit, morphological variability, and taxonomy.

Coryphantha robustispina is the older name for *C. scheeri* (Taylor, 1998) and must be taken up for the species. Taylor published the new combinations as subspecies of *C. robustispina* (that is, subsp. *scheeri*, subsp. *uncinata*). Because we recognize varieties of *C. robustispina*, instead of subspecies, and because the varietal combinations have not been published, temporarily we have retained use of the *C. scheeri* names. The basionym of *C.*

robustispina (Schott ex Engelm.) Britton & Rose is *Mammillaria robustispina* Schott ex Engelm. Variety *robustispina* occurs in southern Arizona.

The taxonomic history of *C. scheeri* is reviewed in Powell and Weedin (2004). The specific epithet honors Frederick Scheer (1792–1868), who evaluated and described the cacti for B. C. Seemann (1825–71) in *The Botany of the Voyage of H.M.S. Herald under the Command of Captain Henry Kellett . . . during the Years 1845–51 (1852–57).*

Full Name. *Coryphantha scheeri* (Muehlenpf.) L. D. Benson.

Key to the Varieties of *Coryphantha scheeri*

1. Central spines always straight; central Trans-Pecos and SE NM
$\qquad\qquad\qquad\qquad\qquad\qquad$ *C. scheeri* var. *scheeri*, p. 335
1. Central spines of immature and young adult plants always hooked downward (often straight in older plants); El Paso Co., SW NM, SE AZ
$\qquad\qquad\qquad\qquad\qquad\qquad$ *C. scheeri* var. *uncinata*, p. 338

Coryphantha scheeri var. scheeri
Long-Tubercled Coryphantha

PLATES 296, 297

Variety *scheeri* as treated here corresponds most closely with Benson's (1982) concept of var. *scheeri* and var. *valida*. By our definition of the taxon, only one variety of *C. scheeri* is present in the central mountain region of the Trans-Pecos and north along the Pecos River to southeastern New Mexico.

Distribution. Woodland and grassland in the central mountains, and degraded grassland-brushland, or less often in desertscrub, usually in alluvium derived from igneous or limestone. Pecos, Brewster, Presidio, Jeff Davis, Reeves, Culberson, and Hudspeth counties. 2,700–5,200 ft. In TX, E in Ward Co. near Royalty. Southeast NM in Eddy Co., near the Pecos River. Mexico: Chihuahua, near Ciudad Chihuahua (type locality of *scheeri*). Map 114.

Vegetative Characters. Stems of both Texas varieties of *C. scheeri* usually are single; rarely, there are a few basal branches. Stems globose to ovoid or cylindroid in older plants or flat-topped and not much protruding above the ground when young. Low, flat-topped stems prevail in mountain woodland habitats. Stems reach 5–10 cm or more tall, 5.5–8.5 cm in diameter. The tallest plant we have seen in the field, estimated at 25 cm, was growing in brushland near Tunis Spring in eastern Pecos County. Like those of *C. macromeris*, the stems are dominated by prominent tubercles, 1–2.5 cm long, 0.8–1.3 cm thick, ovoid to subcylindroid, dull gray-green, and also by stout spines that are few enough in number and spread far enough apart so that the stem surface and tubercles are not much obscured.

In the eastern var. *scheeri* the one main central spine is always straight, porrect or slightly ascending, 2–2.5 cm long, ca.

Plate 296. *Coryphantha scheeri* var. *scheeri* (long-tubercled coryphantha), from S Brewster Co., TX; cultivated.

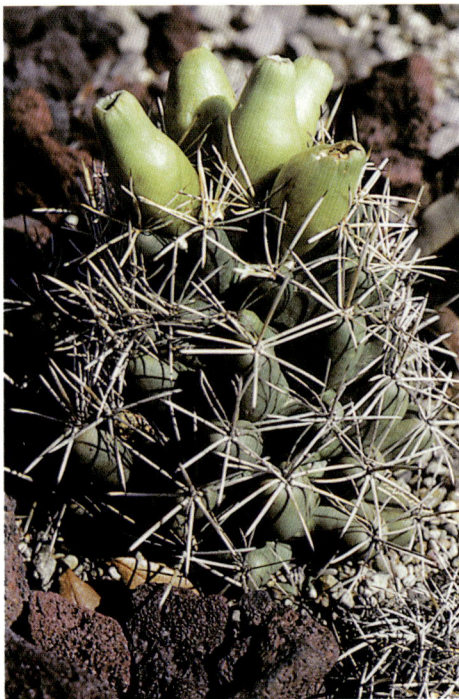

Plate 297. *Coryphantha scheeri* var. *scheeri* (long-tubercled coryphantha), from Brewster Co., TX; mature fruits; cultivated.

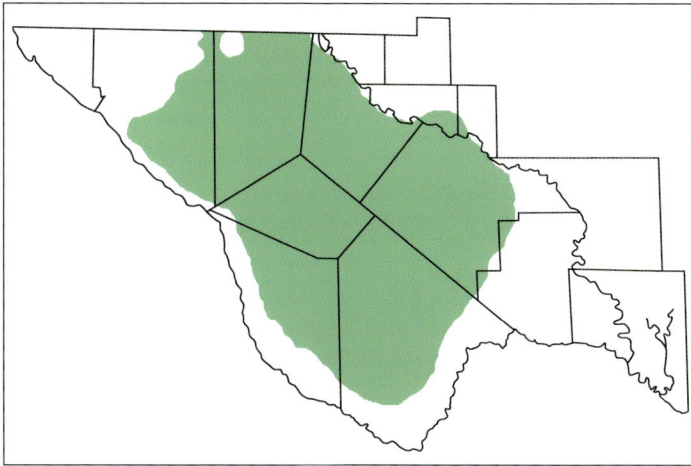

Map 114. *Coryphantha scheeri* var. *scheeri* (long-tubercled coryphantha).

1 mm wide, terete with a large-bulbous base, mostly pale stramineous, gray, or pinkish-gray, with dark red to red-brown at the tip. In some specimens there are 1–2 upper, smaller subcentrals. Radial spines of all varieties are stout and have large-bulbous bases, the lower radials being the largest. Radials terete or dorsiventrally flattened. In the eastern variety, they number 8–12, the largest 1.5–2 cm long. Radials appressed and spokelike, pale gray to nearly white to stramineous (in younger areoles), with dark tips.

Flowers. Flowering Apr–Jul. Yellow flowers with or without a reddish throat, 4–5.5 cm long, 3.5–5 cm in diameter. Specific shades of yellow or yellowish coloration of the flowers are different from the "clear" bright yellow of *C. echinus*, more nearly the color of *C. sulcata* flowers. Inner tepals dark golden-yellow, pale greenish-yellow, to translucent dull yellow, sometimes bronze-tinted basally or in a midstripe, sometimes turning pinkish-yellow or bronze before wilting. Outer tepals minutely fringed, unlike the entire outer tepals of *C. echinus*, *C. sulcata*, and *C. ramillosa*. Filaments reddish to reddish-purple or pale reddish, supporting yellow anthers. Stigma lobes yellow or sometimes cream-colored, cream-pink, or slightly orange-yellow, 6–11 in number, 3–7 mm long on a pale reddish style.

Fruits. Fruit light green at maturity, fusiform-cylindroid, 4–5 cm long, ca. 1 cm in diameter. Fruit surface microscopically papillate. Floral remnant deciduous, leaving bare the truncate-concave fruit apex. Ripe fruits are exceedingly aromatic, with the aroma of tropical fruits, something like the smell of ripe bananas, but richer. Seeds bright reddish-brown, 2.3–3.5 mm long, smooth.

Full Name and Synonyms. *Coryphantha scheeri* var. *scheeri*. *Mammillaria scheeri* Muehlenpf.; *M. robustispina* Schott ex Engelm.; *Coryphantha scheeri* Lem. var. *valida* (Engelm.) L. D. Benson, in part, as to lectotype.

Other Common Names. Needle "mulee"; Scheer cory cactus.

Coryphantha scheeri var. uncinata
El Paso Long-Tubercled Coryphantha

PLATES 298, 299

In Benson (1982) the specimens mapped as *C. scheeri* var. *valida* were from populations of all three currently recognized varieties of the species. Benson erroneously considered var. *uncinata* as a rare hooked-spined entity restricted to El Paso County. The varietal name *uncinata* is derived from the Latin *uncus*, "hook," in reference to the hooked tips of central spines in some plants.

Distribution. Grasslands and degraded grasslands in *Larrea*-dominated desertscrub. El Paso Co., parts of Hudspeth Co. 3,300–4,000 ft. Southwest NM, Doña Ana, Sierra, Luna, Grant, and Hidalgo counties; adjacent SE AZ, Cochise and Graham counties. Mexico: sight records in extreme northern Chihuahua; expected in extreme NE Sonora. Map 115.

Vegetative Characters. Variety *uncinata*, as recognized herein, always has strongly curved and/or hooked central spines on immature or young adult plants. Old adult plants usually have straight central spines. The direction of any curving or hooking in central spines is downward. This variety has 1–4 central spines, usually 14–16 radial spines, and all parts average slightly larger and more numerous than those of the eastern variety.

Flowers and Fruits. Flowering May–Sep. The flowers and fruits are similar in all varieties, except that they may be slightly larger in the relatively robust western varieties. Inner tepals golden-yellow to

Plate 298. *Coryphantha scheeri* var. *uncinata* (**El Paso long-tubercled coryphantha**), **El Paso Co., TX** (photo by Richard D. Worthington).

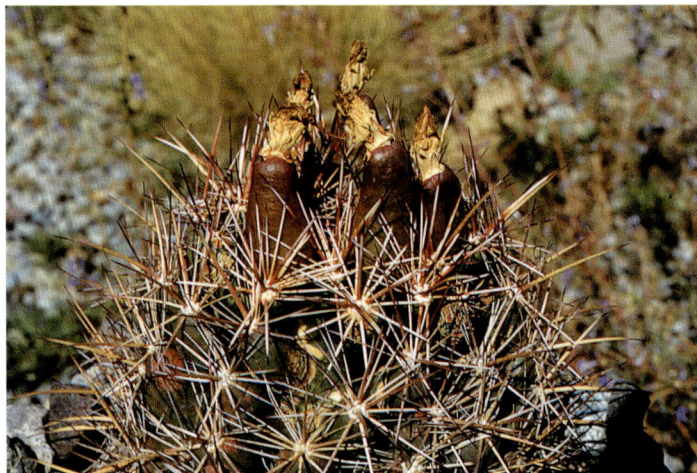

Plate 299. *Coryphantha scheeri* var. *uncinata* (El Paso long-tubercled coryphantha), Fabens, El Paso, TX (photo by Dale and Marian Zimmerman).

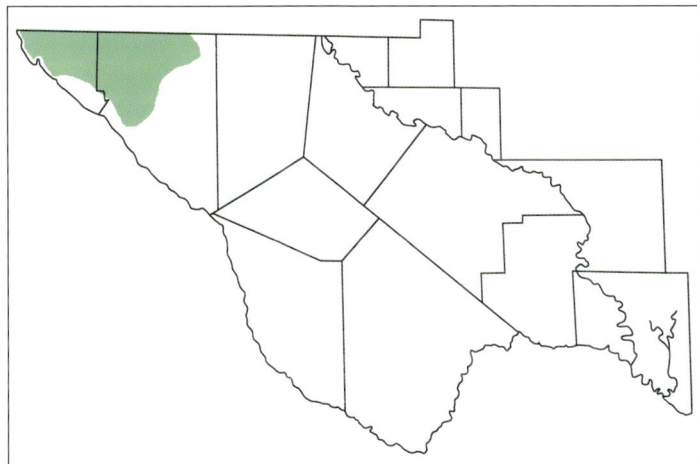

Map 115. *Coryphantha scheeri* var. *uncinata* (El Paso long-tubercled coryphantha).

dull yellow and sometimes bronze-tinged basally. With age, bases of inner tepals also may or may not be reddish, or with red streaks at the base, forming a reddish throat.

Full Name and Synonyms. Coryphantha scheeri var. *uncinata* L. D. Benson. *Mammillaria scheeri* Muehlenpf. var. *valida* Engelm. (in part, excluding lectotype); *Coryphantha scheeri* Lem. var. *valida*, sensu L. D. Benson (in part, excluding lectotype and straight-spined specimens of var. *robustispina*); *C. robustispina* subsp. *uncinata* (L. D. Benson) N. P. Taylor.

Coryphantha echinus
Sea-Urchin Cactus

All three varieties of *Coryphantha echinus* (Engelm.) Britton & Rose are endemic in the CDR, two of them in Texas mostly in the Trans-Pecos, but also elsewhere in Texas. The common name "sea-urchin cactus" is particularly appropriate for these plants with globose stems, appressed radial spines, and a single protruding central spine in each areole. The specific epithet is equally descriptive, as the Greek root *echinos* means "hedgehog" or "sea urchin."

Key to the Varieties of *Coryphantha echinus*

1. Plants usually remaining single-stemmed; older stems remaining globose to subglobose, 3–10 cm long, 3–5.5 cm wide; spines with a whitish aspect (central spines appearing later in life, sometimes absent); northern and/or higher altitudes, widespread in the Trans-Pecos

C. echinus var. *echinus*, p. 340

1. Plants usually becoming multistemmed; older stems ovoid to cylindroid, rarely remaining subglobose, 7–15 cm or more long, 5–7 cm wide; spines with a dark aspect (central spines appearing early in life); along and near the Rio Grande in southern Brewster and Presidio counties

C. echinus var. *robusta*, p. 342

Coryphantha echinus var. echinus
Sea-Urchin Cactus
PLATES 300, 301

Variety *echinus* is by far the most widely distributed of the two varieties of *C. echinus* in Texas. It is most common on and near limestone soils of the eroded Stockton Plateau in Pecos and Terrell counties and in eastern Brewster County. It also occurs in the limestone mountains of the southern Big Bend. The type locality is listed as "on the Pecos River."

Distribution. Throughout much of the southern and eastern Trans-Pecos, from the Chisos Mts and Terlingua area N and NE to Pecos Co., E to Terrell and Val Verde counties (rarely W to El Paso Co.?). 1,000–4,800 ft. Sporadically NE in TX to Howard and Coke counties. Mexico: formerly expected in N Coahuila, but now thought to be replaced by other varieties from the Rio Grande southward. Map 116.

Vegetative Characters. Plants of var. *echinus* usually recognizable at a glance, with whitish appressed radial spines and with a single porrect central spine in each areole. The plants can be viewed, with a little imagination, as having the aspect of a white sea urchin.

Spines 16–31 per areole, usually drab white with brown tips. Spine number includes 0–4 centrals and 16–27 radials. Usually by the age of sexual maturity there is one porrect, main (lower) central and 2–4 upper centrals (or subcentrals). The main, porrect central, 0.8–1.5 cm long, straight or slightly decurved but strongly decurved in some populations. Main central bulbous- or buttressed-based. Radials, to 0.8–1.2 cm long, appressed, pectinately arranged, often curved toward the stem surface.

Flowers. Flowering spring, summer. Flowers 3.5–5 cm long and wide, glossy-yellow inside but greenish or dull reddish externally. At times the whole throat of

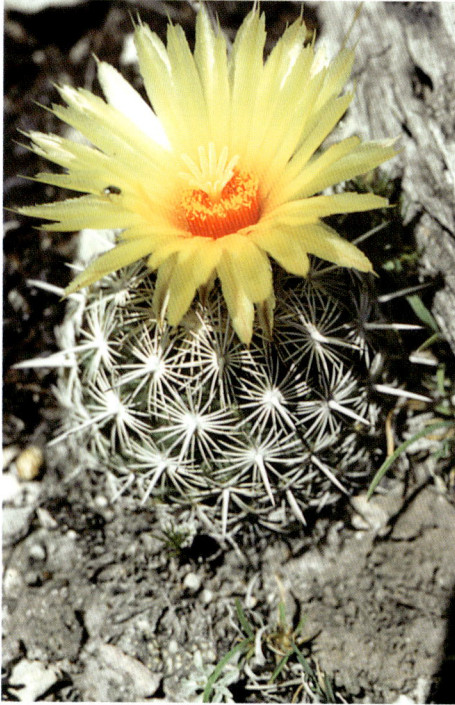

Plate 300. *Coryphantha echinus* var. *echinus* (sea-urchin cactus), Pecos Co., TX.

Plate 301. *Coryphantha echinus* var. *echinus* (sea-urchin cactus), from Pecos Co., TX; mature fruits; cultivated.

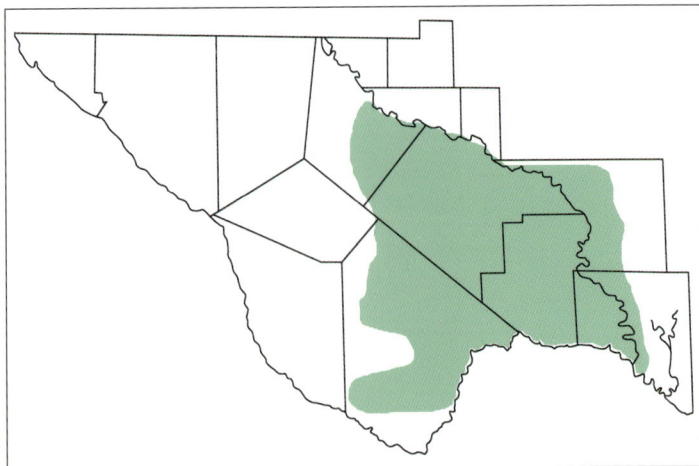

Map 116. *Coryphantha echinus* var. *echinus* (sea-urchin cactus).

the flower, including the base, may appear reddish, largely because the red color of stamen filaments, especially the distal portions, is reflected by the mirrorlike surface of the (actually yellow-pigmented) inner tepal bases. Anthers bright yellow. Style and stigma lobes pale yellow. Usually 10–12 stigma lobes, 3–4 mm long.

Fruits. Green to light green fruit 1.2–2.5 cm long, 1–1.9 cm in diameter, generally ovoid. Floral remnant persistent. Seeds reddish-brown, 1.7–1.9 mm long, with smooth, shiny surfaces.

Other Distinctions. The synonym *C. pectinata*, with its type locality "on the Pecos River," is based on a form of var. *echinus*, the "pectinate form," that lacks the lower (porrect) central spine in each areole. Essentially all immature plants of var. *echinus* lack porrect centrals, and occasional young adult and sometimes even sexually mature plants as well may lack the protruding centrals. Fully adult plants without porrect centrals, the "pectinate" form, seem to be otherwise identical to the common form of var. *echinus*. The pectinate form probably appears throughout

much of the population of var. *echinus*. We have seen specimens of the pectinate form from both 10 miles north and 20 miles south of Fort Stockton in Pecos County, and from 25 miles northwest of Comstock in Val Verde County, near the Pecos River.

Full Name and Synonyms. Coryphantha echinus var. *echinus. Mammillaria echinus* Engelm.; *Coryphantha pectinata* (Engelm.) Britton & Rose; *C. cornifera* (DC.) Lem. var. *echinus* (Engelm.) L. D. Benson; *M. scolymoides* Scheidw. and *C. scolymoides* (Scheidw.) A. Berger are based upon a Mexican species; reports from Texas are misidentifications.

Other Common Name. Hedgehog corycactus.

Coryphantha echinus var. robusta
Multistemmed Sea-Urchin Cactus
PLATES 302–4

Variety *robusta* is best exemplified by the relatively large cespitose plants, in mounds and mats, found in the gravel hills south of the Chisos Mountains and north of Lower Tornillo Creek. The varietal

Plate 302. *Coryphantha echinus* var. *robusta* (multistemmed sea-urchin cactus), S Tornillo Creek, Big Bend National Park, Brewster Co., TX; habit.

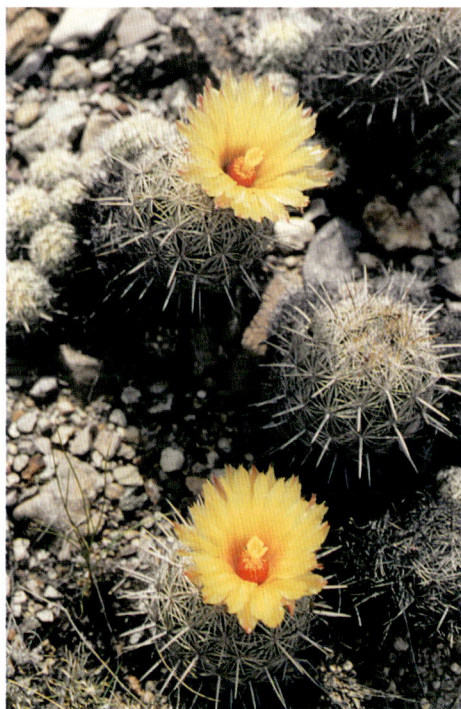

Plate 303. *Coryphantha echinus* var. *robusta* (multistemmed sea-urchin cactus), S Tornillo Creek, Big Bend National Park, Brewster Co., TX.

epithet, taken from the Latin *robustus*, "strong," alludes to the robust multistemmed plants.

Distribution. Gravel hills and benches, and other alluvial substrates of igneous and limestone origin, desertscrub and degraded desert grassland. Along and near the Rio Grande, southern Presidio and Brewster counties. 2,000–3,500 ft. Mexico: expected across the Rio Grande in Chihuahua (from Ojinaga) and Coahuila (from Boquillas). Map 117.

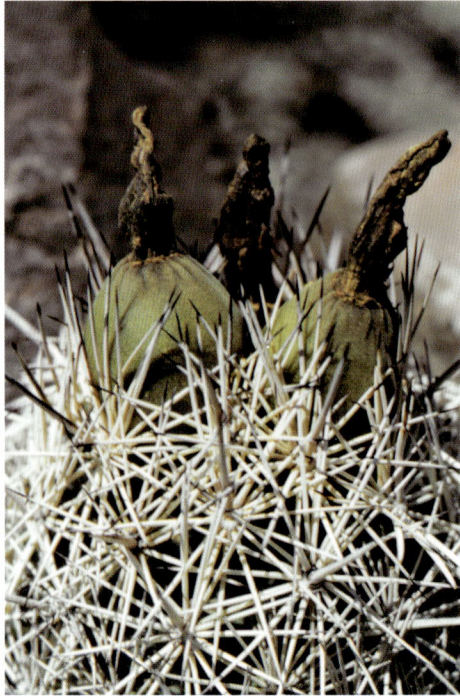

Plate 304. *Coryphantha echinus* var. *robusta* (multistemmed sea-urchin cactus), from S of Chisos Mts, Brewster Co., TX; mature fruits; cultivated.

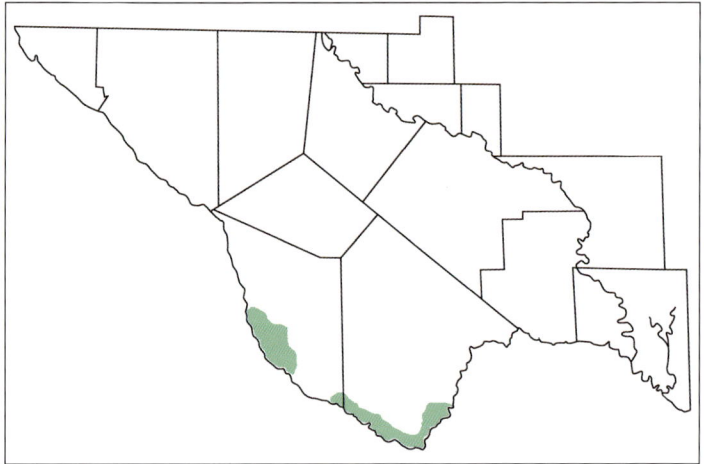

Map 117. *Coryphantha echinus* var. *robusta* (multistemmed sea-urchin cactus).

Vegetative Characters. Presumably older plants of var. *robusta* have 15–50 or more stems of various sizes, forming mounds or mats 15–30 cm or more high, 30–70 cm in diameter. Numerous small stems, 1–3 cm high and wide on these larger plants, originate near the bases of the larger stems in the mound or mat. Sexually mature stems are subglobose or ovoid to cylindroid, 7–15 cm or more long,

5–7 cm or more in diameter. In aspect, stems usually are darker than the stems of var. *echinus,* because of darker gray spine color. Older stem bases often are sheathed by nearly black spines. Porrect central spines gray to brownish-red or nearly black, or gray with nearly black tips. Radial spines usually gray.

In populations south of the Chisos Mountains, many older plants of var. *robusta* have senescent or dead stems in interior portions of the clumps. At the bases of senescent and dead stems, spine clusters tend to abscise en masse, with all of the spines turning very dark, remaining adherent, and forming a loose sheath around part of the stem bases. In some plants masses of decomposed centrally located stems contribute to a sprawling, matlike habit, with apparently healthy stems at the periphery of original tightly clumped mounds or mats.

In areoles of mature var. *robusta,* typically there are four central spines and 18–26 radial spines. The lower (main) central spine is porrect, straight or sometimes slightly decurved, 1.5–2.5 cm long, ca. 1 mm thick, and bulbous-based. The main central usually is gray, or gray with black speckles as seen under magnification, or on some stems is brownish-red to nearly black, particularly on the distal portion, or dark only at the tip. The three upper central spines, in a "bird's-foot" arrangement, are appressed in plane with the radial spines. The gray radial spines are appressed, usually with the distal portions curved toward the stem. The upper radial spines are 1.7–2.4 cm long, longer than the lateral (1.2–2.2 cm) and lower (1–1.6 cm) radials. The bases of radial spines are enlarged but not bulbous, and they are laterally compressed.

Flowers. Flowering Apr–May. Flowers are bright yellow like those of var. *echinus,* usually 5.5–6.5 cm long and wide. Stamens have reddish to reddish-orange filaments and yellow anthers ca. 1.5 mm long. Pale yellow style ca. 2.4 cm long, supporting 10–13 yellow to pale yellow stigma lobes. Filaments sensitive in response to mechanical stimulation.

Fruits. Fruits green to light green, 1.5–2.8 cm long, 1.3–1.9 cm in diameter, generally ovoid, closely resembling those of var. *echinus* but slightly larger. Floral remnant persistent. Reddish-brown seeds virtually identical to those of var. *echinus.*

Other Distinctions. In Big Bend National Park var. *robusta* is a typical associate of *Opuntia aggeria* and *Echinocereus chisoensis.* Variety *robusta* is more widely distributed than is *E. chisoensis* but more restricted than *O. aggeria.* The large cespitose habit of var. *robusta* may be the ecological manifestation of very old plants growing in relatively porous, deep substrates that are located in the hottest winter and summer habitats in the Chihuahuan Desert Region of Texas.

Plants in a population south of the Sierra Vieja in Presidio County (Plate 305), included here with var. *robusta,* are slightly different from the rest of *C. echinus.* The plants eventually branch (3–4 or more stems) with strongly decurved central spines. Single-stemmed plants also have been observed. This depauperate population presumably has been depleted over the years by commercial cactus harvesters.

Full Name. *Coryphantha echinus* var. *robusta* A. M. Powell.

Plate 305. *Coryphantha echinus* var. *robusta* (multistemmed sea-urchin cactus), from near Ruidosa, Presidio Co., TX; cultivated.

Coryphantha sulcata
Grooved Nipple Cactus
PLATES 306, 307

Weniger (1984) reported that *C. sulcata* extended west to the mouth of the Pecos River. Our own field investigations have not revealed the positive existence of *C. sulcata* in or near the eastern periphery of the Trans-Pecos. *Coryphantha sulcata* is traditionally endemic to Texas, but certain taxa from northeast Mexico are obviously related and nearly identical (e.g., *C. roederiana* Boed. and *C. obscura* Boed.). The specific epithet is after the Latin *sulcus*, meaning "a furrow," probably a reference to the grooved tubercles.

Distribution. Reported in limestone habitats near the mouth of the Pecos River, Val Verde Co., E Pecos Co., and N Brewster Co. 1,000–3,500 ft. Edwards Plateau E to Austin Co., and N to Somervell, Tarrant, and Denton counties, S to Duval Co. Map 118.

Vegetative Characters. Eastern plants of *C. sulcata* form clumps to 30 cm or more in diameter, potentially with dozens of stems (of different sizes). Larger stems are spherical or obovoid, often compressed at the apex, 4–10 cm long, 6–8 cm in diameter. Usually 9–16 spines per areole, with one, porrect, decurving, lower (main) central spine 1.1–1.7 cm long. Usually 1–2 erect upper centrals. Appressed radial spines number 8–15, all of similar length at 1–1.5 cm, slightly curved toward the stem. Rarely, up to 24 spines per areole, including 3–4 subcentrals. Spines yellowish in upper (younger) areoles and gray (with age) in lower areoles. Green tubercles and stem surface are not much obscured by the spines. In general aspect, individual stems of *C. sulcata* resemble expanded, heavily watered stems of *C. echinus*. The typical number of radial spines in *C. sulcata* is fewer than in *C. echinus* var. *echinus*.

Flowers. Flowering May, summer.

Plate 306. *Coryphantha sulcata* (grooved nipple cactus), from Somervell Co., TX; cultivated.

Plate 307. *Coryphantha sulcata* (grooved nipple cactus), Somervell Co., TX; mature fruit.

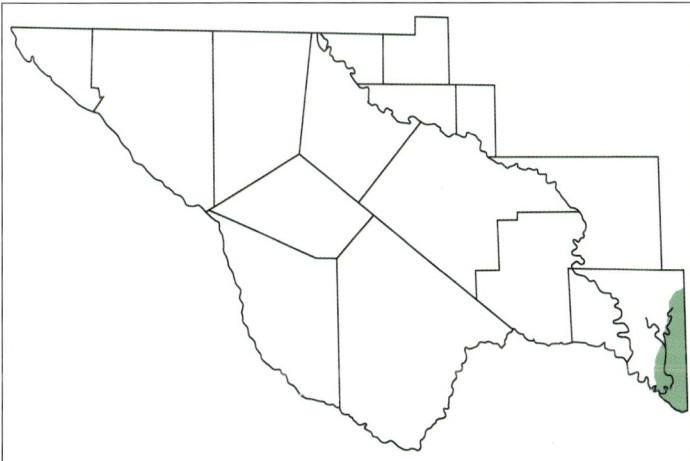

Map 118. *Coryphantha sulcata* (grooved nipple cactus).

Flowers usually golden-yellow or dark golden-yellow compared to the bright glossy yellow in flowers of *C. echinus*. Rarely, flowers of *C. sulcata* are greenish-yellow. These color differences are most easily appreciated when the two species are observed together in cultivation. Flowers of both species usually appear to have a red throat. Flowers 4–6 cm long, 3.5–8 cm in diameter. Exposed upper filaments 7–10 mm long, reddish distally, rarely greenish. Anthers ca. 1 mm long, yellow. Yellowish or greenish style ca. 1.7 cm long, usually longer than the stamens, supporting 7–11 cream-colored or greenish-yellow stigma lobes, ca. 3.5 mm long.

Fruits. Ellipsoid, fusiform, to broadly ovoid fruits usually green to dull red at maturity, 1.5–3.5 cm or more long, 1–1.5 cm in diameter. Floral remnant persistent. Seeds dark reddish-brown, ca. 2 mm long, smooth and shiny, somewhat comma-shaped.

Full Name and Synonym. Coryphantha sulcata (Engelm.) Britton & Rose. *Mammillaria sulcata* Engelm.

Another Coryphantha Species

Another Texas species of *Coryphantha* is summarized below.

Coryphantha missouriensis
Nipple Cactus
PLATE 308

In general *C. missouriensis* is cold-hardy but often short-lived in cultivation. The species tends to be very inconspicuous in cover of other vegetation, especially during the winter and in dry periods when desiccated stems are drawn to near ground level. The plants are most obvious during the late spring when the usually yellow flowers are produced in abundance. The specific epithet is after the Missouri River, not the state of Missouri (where *C. missouriensis* is rare or absent). The varietal epithet describes the cespitose or densely clumped low habit of var. *caespitosa*.

Distribution. South-central and N TX, mostly E of the Edwards Plateau, as far S as Bexar County, and E to Walker Co. (Map 119). North through OK, AR (near the Red River), KS, and W to E CO. Weni-

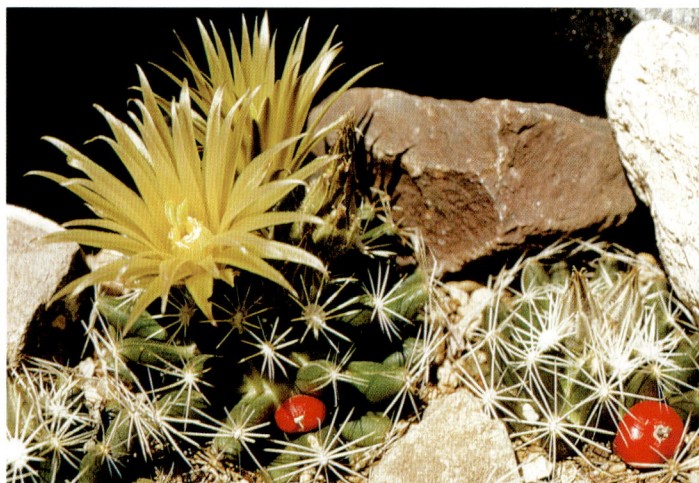

Plate 308. *Coryphantha missouriensis* var. *caespitosa* (nipple cactus), from Parker Co., TX, S of Weatherford; cultivated.

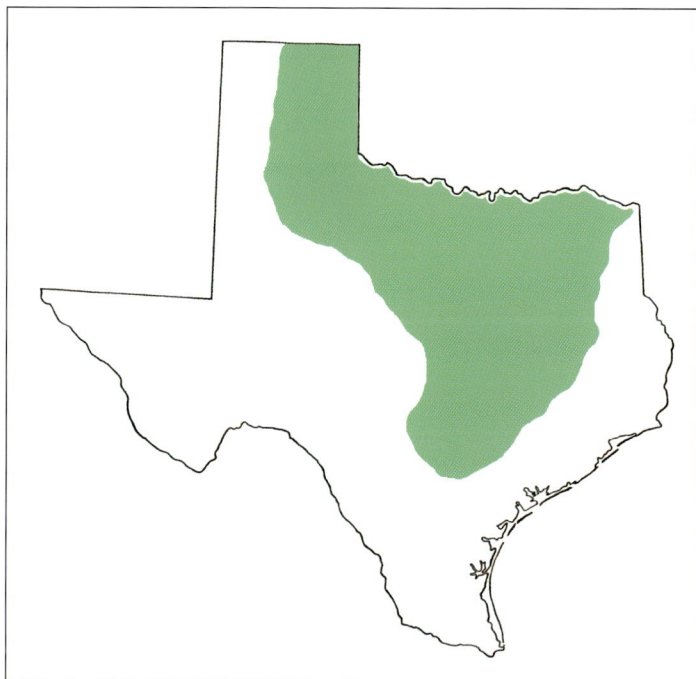

Map 119. *Coryphantha missouriensis* var. *caespitosa* (nipple cactus).

ger (1984) maintained that var. *caespitosa,* as *Mammillaria similis* Engelm., was "rather common near Austin, Texas," and "occasionally found in a band about 100 miles wide" from Waco to the Dallas–Fort Worth area.

Vegetative Characters. Plants usually branching, occasionally solitary, forming clumps 5–10 cm high, 15–30 cm or more in diameter. Roots fibrous, but stem bases usually deeply seated in soil. Stems dark green, the larger ones usually depressed-globose, 2.5–5 cm long, 3.5–5 cm in diameter. Tubercles 12–15 mm long. Areoles round or oval, 1.5–2 mm in diameter, white-woolly when young. Spines 10–17 per areole; central spines usually absent, rarely one, when present only slightly heavier and longer than the radials; radial spines 0.8–1.4 cm long, spreading, straight, slender, pubescent when young, gray to white, yellowish when young, often with brown tips.

Flowers. Flowers 2.5–5 cm in length and diameter when fully open, the inner tepals golden-yellow or greenish-bronze, or gold streaked with pink, or pale yellow; filaments whitish to yellowish or light green; anthers yellow; style much longer than stamens, green to yellowish; stigma lobes 4–6, 3–5 mm long, green or yellowish.

Fruits. Fruits globose to oval, 1–2 cm long, remaining green into the winter, then turning bright red, persisting into next flowering season at stem periphery. Seeds black, obovoid or subglobose, 1.5–2 mm in greatest diameter, pitted.

Full Name and Synonym. *Coryphantha missouriensis* (Sweet) Britton & Rose var. *caespitosa. Mammillaria similis* Engelm. var. *caespitosa* Engelm. (Engelm.) L. D. Benson.

Other genera of cacti found in Texas and not discussed elsewhere in the Trans-Pecos treatments are summarized below.

Acanthocereus tetragonus
Triangle Cactus
PLATES 309, 310

Acanthocereus (Engelm. ex A. Berger) Britton & Rose is a genus of 12 or fewer species in tropical and subtropical lowlands. The plants are remarkable for their sprawling, clambering, and branching habit (especially of prostrate stems) and fast growth. In South Texas stems may grow 5–6 feet under optimal conditions in a summer season. Plants of *A. tetragonus* are sensitive to frost, and aboveground parts may be killed at 0°C (32°F) or slightly below freezing. In the Trans-Pecos the plants are easily grown through the winter in heated greenhouses, where they occasionally produce flowers (one instance in September). The white flowers are large and showy, with minimal fragrance or a pineapple aroma. Texas plants of *A. tetragonus* usually have 3-angled stems (hence the common name, triangle cactus), although plants in Mexico and Central America may have 4–5–angled stems, the inspiration for the epithets *tetragonus* and *pentagonus*. Other common names used in Texas are barbwire cactus, night-blooming cereus, organo, and pitahaya.

Distribution. South TX in Kenedy, Willacy, Hidalgo, Cameron, and Webb counties (Map 120). Also S FL and doubtfully LA. Down the E coast of Mexico through Central America, and E along and near the coast in South America to Venezuela, and in both the Greater and Lesser Antilles.

Vegetative Characters. Plants with stems erect or sprawling, often clambering in other vegetation, ultimately arching over unless supported. Roots fibrous. Stems usually 3-angled, branching once or twice or more, individual growth increments 30–200 cm long, 2.5–5 cm in diameter, whole branched stems to 6–7 m long. Ribs sharply angled, 1.2–5 cm high, the stem axis slender. Areoles round, 1.5–3 mm in diameter, typically 2–5 cm apart, with short, whitish wool. Spines usually 7–8 per areole, light brown, aging gray, needlelike or slightly flattened, with bulbous bases; central spines 1–3, the porrect or slightly deflexed lower central 1.8–4 cm long, with 0–2 lateral centrals and 0–1 upper central; radial spines usually 5–7, light brown to gray, needlelike, mostly appressed.

Flowers. Flowers white, nocturnal, ca. 10 cm in diameter, 17–25 cm long; inner tepals white; filaments white, shorter than

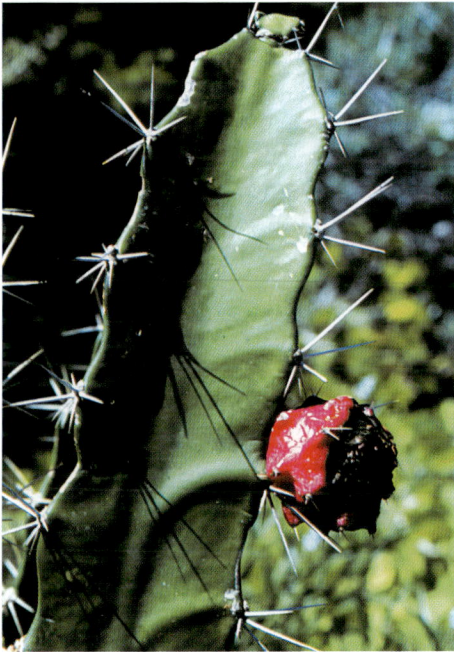

the tepals, anthers light yellow, ca. 3 mm long; style 1.7–2 cm long, stigma lobes 10–12, white.

Fruits. Fruit bright red, shiny, broadly elliptic to ovate, 3–7 cm long, 2.5–5 cm in diameter, with 1–4 spines per areole on low tubercles, the fruit pulp red, edible, moderately sweet. Seeds black, shiny, obovate, 2.5–3 mm long.

Full Name and Synonyms. *Acanthocereus tetragonus* (L.) Hummelinck. *Cereus tetragonus* L.; *Cactus pentagonus* L.; *Acanthocereus pentagonus* (L.) Britton & Rose; *Cereus pentagonus* (L.) Haw.

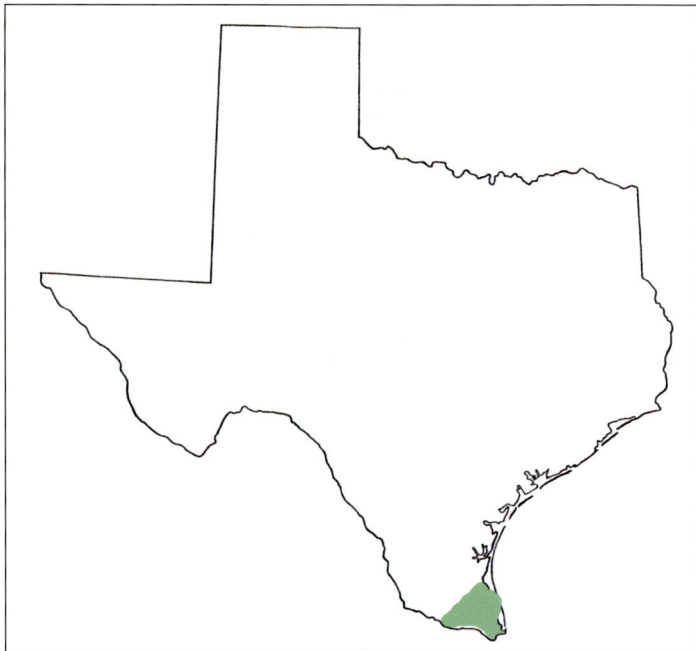

Astrophytum asterias
Sea-Urchin Cactus, Star Cactus
PLATE 311

The spineless plants of *A. asterias,* also known as star peyote, superficially resemble those of *Lophophora williamsii* (peyote) and reportedly possess some of the same alkaloids. *Astrophytum* Lem. is a genus of four species, all of them of Mexican distribution, two of them in the CDR; only the rare *A. asterias* barely extends into Texas. The genus name and specific epithet allude to either the myriad white dots on the stem surface (epidermal trichome tufts unique to this genus) or to the five-pointed "starlike" cross-sectional outline of the stem of *A. myriostigma* Lem., the type species, after the Greek *asteros,* "star," and *phyton,* "plant."

Distribution. Rare, reported to survive at sites in Starr County of the lower Rio Grande valley (Map 121), and in Tamaulipas and Nuevo León, Mexico, with an estimated total population of ca. 2,000 plants. Additional surveys are in progress.

Vegetative Characters. Plants unbranched. Roots diffuse. Stems green, depressed globose to globular, at maturity 5–15 cm in diameter, 2.5–6 cm high, retracting into the ground when desiccated. Ribs usually eight, very low (their crests nearly flat) separated by narrow but distinct vertical grooves. Stem surface dotted by ca. 1 mm clusters of short white hairs. Areoles circular, 3–5 mm in diameter, in straight rows, densely filled with yellow or gray wool. Spines absent in adults.

Flowers. Flowers yellow with orange throats, 4.5–5.4 cm long, 3.8–5.2 cm in diameter, opening widely; filaments yellowish, anthers yellow; style yellowish, with 10–12 yellow stigma lobes ca. 4 mm long.

Plate 311. *Astrophytum asterias* (sea-urchin cactus or star cactus), Starr Co., TX (photo by Jackie M. Poole).

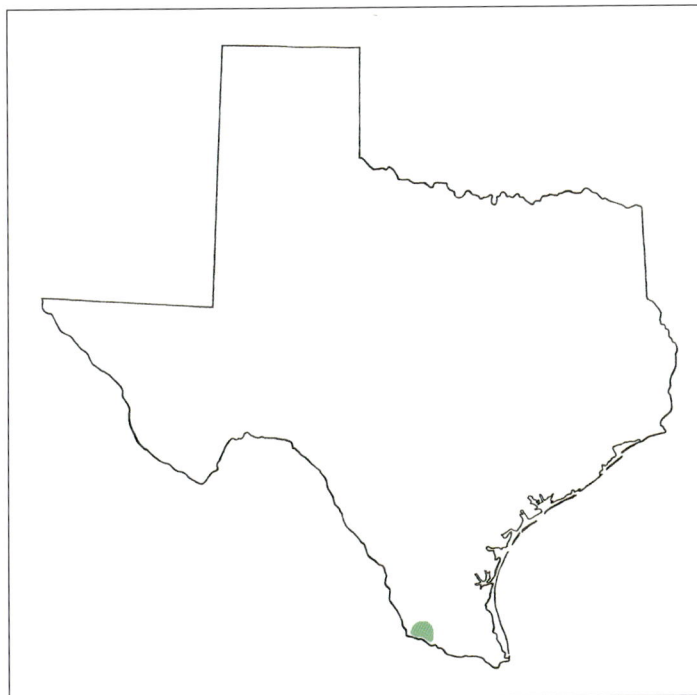

Map 121. *Astrophytum asterias* (sea-urchin cactus or star cactus).

Fruits. Fruit green or pinkish, fleshy, oval, 1.5–2 cm long, ca. 1.2 cm in diameter, densely covered with arcolar wool and spinescent bracts, drying after maturity, ultimately abscising. Seeds blackish or dark brown, shiny, 2–3 mm in broadest diameter.

Full Name and Synonym. *Astrophytum asterias* (Zucc.) Lem. *Echinocactus asterias* Zucc.

Pereskia aculeata
Lemonvine

PLATES 312, 313

The genus name commemorates the French scholar Nicolas Claude Fabri de Peiresc (1580–1637), although the nomenclaturally correct spelling of the genus name turned out to be *Pereskia*, probably reflecting the French pronunciation of Peiresc. The specific epithet is derived from the Latin *aculeus*, "needle," obviously a reference to the spines of this broad-leafed cactus.

Distribution. This tropical leafy cactus genus was unknown in Texas until it was reported in Willacy County (Map 122), lower Rio Grande valley, in 1996. The plants were located on a wooded tract of land surrounded by cotton fields, with stems climbing and spreading like vines in trees and hanging within 2 m of the ground. Previously the species, also known as Barbados gooseberry, was known to occur in Florida, southern Mexico, Central America, and the West Indies, where probably it is naturalized after cultivation (also escaped in South Africa), and in lowland tropical South America, its native territory. Presumably the plants are naturalized in Willacy County but probably not introduced directly at the site where they were discovered.

Vegetative Characters. Plants shrubs or lianas, forming vines 3–10 m long, the trunk 2–3 cm or more in diameter. Stems straggling to clambering; distal twigs to ca. 4 mm thick, green to reddish. Areoles on twigs ca. 2 mm in diameter, on the trunk

Plate 312. *Pereskia aculeata* (lemonvine); cultivated.

Plate 313. *Pereskia aculeata* (lemonvine); mature fruits; cultivated.

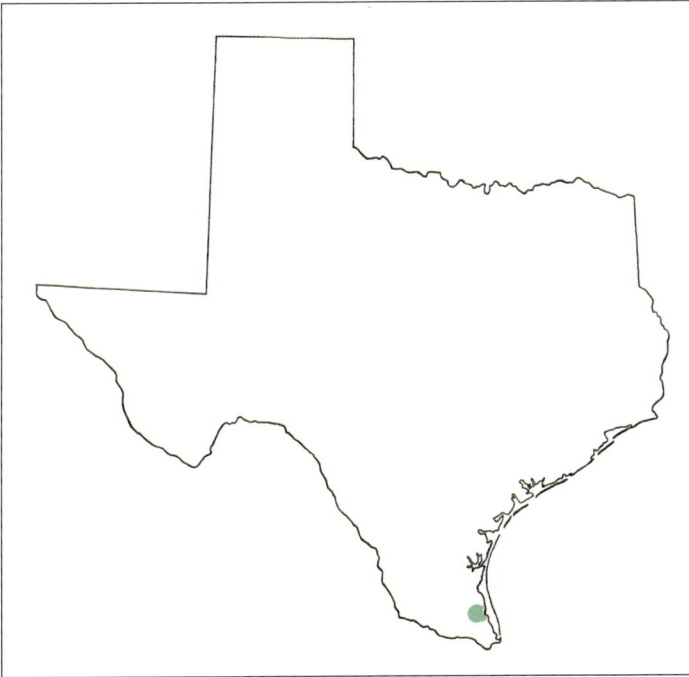

Map 122. *Pereskia aculeata* (lemonvine).

cushionlike and to 15 mm in diameter. Leaves flat, pinnately veined, weakly succulent, long-persistent, short-petiolate, broadly lanceolate to oblong or ovate, 4.5–7 cm long, 1.5–5 cm broad or smaller, green or purplish; petioles 3–7 mm long. Spines paired and clawlike on young shoots and twigs, later supplemented by long, straight spines (to 25 per areole) to 1–3.5 cm long, abruptly thickened at the base, brown to black, gray with age. Inflorescences of terminal clusters.

Flowers. Flowers whitish, heavily fragrant, nocturnal, 2.5–5 cm in diameter; receptacles cup-shaped or turbinate, 5–6 mm in diameter, on slender pedicels, bracteate; stamens numerous, 5–10 mm long, filaments whitish proximally, white or yellow to reddish distally, anthers yellow; ovary superior at anthesis but in fruit enclosed by the rim of the receptacular cup and filament bases; style 10–12 mm long, white, stigma lobes 4–7, white, 3–5 mm long, nearly erect.

Fruits. Fruit yellow to orange at maturity, globular to subglobular, fleshy, 1.5–2.5 cm in diameter, bearing 6–15 areoles with spreading spines 3–8 mm long. Seeds only 2–5 (solitary in each carpel, unusual for cacti), black, smooth, flat or weakly concave on one side, almost circular in general outline, 4.5–5 mm in diameter.

Full Name and Synonym. *Pereskia aculeata* Mill. *Cactus pereskia* L.

Selenicereus spinulosus
Queen of the Night

Distribution. Reported for South Texas (Map 123). Main distribution in Tamaulipas, Veracruz, Hidalgo, San Luis Potosí, Oaxaca, and Chiapas, Mexico.

Vegetative Characters. Plants clambering. Roots aerial. Stems many, usually 2–4 m long, light green, 1–2 cm in diameter. Ribs 4–6, acute. Areoles 1.5–2.5 cm apart. Spines brown, 1 mm long; central spine one; radial spines 6–7.

Flowers. Flowers nocturnal, white to pinkish, 10–12 cm long, to 8.5 cm or more in diameter.

Full Name and Synonyms. *Selenicereus spinulosus* (DC.) Britton & Rose. *Cereus spinulosus* DC.; *Selenicereus pseudospinulosus* Weing.

Map 123. *Selenicereus spinulosus* (queen of the night).

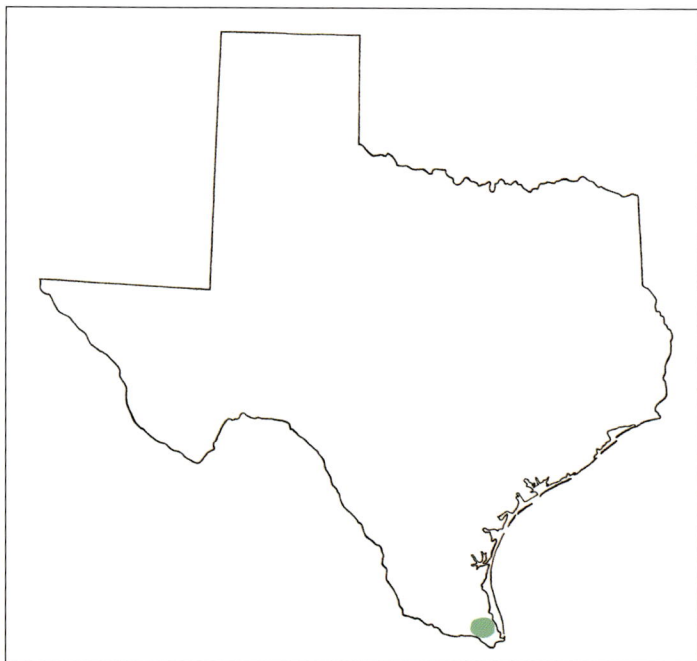

Glossary

Abaxial. The side of an organ away from the main axis; for example, the lower side or edge of a tubercle (opposite of adaxial).

Abscission. Separation of a plant part through disintegration of a special layer(s) of cells.

Acicular. Needlelike; slender, elongate, circular in cross section, and tapering to a pointed apex.

Acuminate. A shape abruptly narrowed into a long, pointed apex.

Acute. Tapering to an apex, at less than 45°.

Adaxial. Adjacent to the axis (facing the stem); said of the upper side or edge of a leaf or tubercle.

Adventitious. Formed in an unusual place; said especially of roots that sprout directly from stems or fruits instead of from preexisting roots.

Alluvium. A geologic substrate deposited by running water; clay, silt, sand, gravel, or loose rocks.

Annular. In the form of a ring, such as annular thickenings encircling stems or spines.

Annulate. Having ringlike bands; said of spines that show swollen growth increments, as in *Ferocactus wislizeni*.

Anther. The upper pollen-producing part of the stamen, consisting mostly of the pollen sacs.

Anthesis. The time of opening of a flower.

Apex. The uppermost point; the tip (pl., apexes or apices).

Apical. At the apex.

Apiculate. A shape ending abruptly in a short, protruding point (an apiculation), but not a hard prickly point (see Mucronate).

Appressed. Lying flat against the underlying surface; said of radial spines spreading horizontally instead of pointing outward.

Arachnoid. "Spidery"; said of certain plants having sticky or tangled cobwebby hairs; certain plant parts may be arachnid-like or arachnoid.

Arborescent. Treelike in habit.

Areolar Groove. On a cactus podarium (only in certain genera), a narrow continuation of the areole, extending adaxially from the spine cluster.

Areole. A modified nonphotosynthetic short shoot, borne at the upper end or protruding summit of each podarium; an axillary bud, although its subtending leaf is obsolete or rudimentary.

Aril. A structure attached to a seed, which develops as an outgrowth in the region of the hilum, from the funiculus, and partially or entirely envelops the seed; used with reference to the funicular covering (see Funiculus) of *Opuntia* seeds.

Arillate. Bearing an aril.

Aristate. Tipped with an awn or bristle; an awn is a terminal slender bristle on an organ.

Ascending. Angled upward.

Asexual. Reproducing without sex, as in vegetative reproduction or some other form of apomixis.

Attenuate. Gradually tapering toward the tip

into a point, involving a more or less flat-tened structure.

Auriculate. With earlike appendages.

Awl-Shaped. Tapering from the base to a sharp point.

Axil. The upper angle formed by the attachment of a leaf (or a similar structure) against its underlying surface.

Axillary. Developed in an axil; subtended by a leaf or similar structure.

Axis. The main stem or main line of develop-ment of a plant part.

Backcross. A cross between a hybrid and one or both of its parents.

Beaked. Having a firm, elongate, slender, pro-jecting structure.

Berry. A fleshy or pulpy fruit usually with several to many seeds embedded in the pulp; in most cases a cactus fruit is interpreted as a modified berry derived from an inferior ovary.

Betacyanin. One of the two color groups of betalains: red, magenta, pink, and so on.

Betalains. A class of water-soluble vacuolar pigments with almost the same colors as anthocyanins and anthoxanthins but a very different nitrogen-containing chemical structure; found only in Caryophyllales and certain fungi.

Betaxanthin. One of the two color groups of betalains: yellows and orange-reds.

Bisexual. When in reference to a flower, both male and female reproductive organs pres-ent; hermaphroditic.

Bract. A reduced, modified leaf, often subtend-ing a reproductive structure.

Bristle. A nearly rigid, slender, hair- or spine-like structure.

Bulbous. Swollen and bulblike.

Cambium. A layer of dividing cells in the stem, giving rise to xylem inside and phloem outside.

Canescent. Gray-pubescent with a dense cover-ing of hairs.

Capsule. A dry dehiscent fruit formed from more than one carpel.

Carpel. Specialized leaf or leaves forming a simple or compound pistil.

Caudex. Woody base of an otherwise herba-ceous plant.

Central. When in reference to spines, the individual(s) that originates anywhere inside the radials.

Cespitose, Caespitose. Having numerous stems that form a dense clump.

Cholla. Type of cactus plant characterized by erect or suberect habit and cylindroid stem joints.

Circumscissile. Opening along a circular horizontal line.

Cladode. A leaflike stem, such as the flattened green stem of a prickly pear.

Clavate. Club-shaped.

Cline. A character gradient, as between two populations or a series of contiguous popu-lations; clinal.

Clone. A group of individual plants propagated asexually from a single individual.

Column. A floral term, used in reference to the thin expanse of tissue extending from the top of the ovary up to the floor of the nectar chamber.

Columnar. Having the appearance of a column; said of cacti with large, columnlike stems, as in the saguaro of the Sonoran Desert.

Compressed. Flattened, usually laterally.

Concolorous. Of uniform color.

Conical. Cone-shaped, with the point of attach-ment at the broad base of the cone.

Cortex. Thick region of colorless, water-storing parenchyma cells in cactus stems inside the chlorenchyma and outside the vascular cylinder.

Cotyledons. In cacti the two first leaves of the embryo, apparent in most opuntioid seedlings, perhaps not evident in cactoid seedlings (not true leaves).

Crescentiform. Crescent-shaped.

Crested. A cactus stem or stems with abnormal fanlike apical growth.

Cultivar. A horticultural form maintained through cultivation by humans.

Cuticle. A waxy, noncellular layer on the outer surface of a plant organ.

Cyanic. Dark blue but also used here in reference to red, blue, purple, violet, and similar colors, as opposed to yellow, green, or white.

Cylindroid. In the form of a cylinder.

Deciduous. Falling away in senescence.

Declined. Turned downward.

Decumbent. Lying on the ground, with the apical end ascending.

Decurrent. Extending down and adnate to the stem, as the bases of some podaria.

Decurved. A curved axis with tip pointing downward.

Deep-Seated. With the stem base or most of the stem below ground level.

Deflexed. Abruptly bent downward from near the base.

Dehiscent. Opening by splitting along regular lines; seeds may be released after dehiscence of the fruit.

Deltoid, Deltate. Having a triangular shape.

Depressed. Pressed down endwise; more or less flattened.

Descending. Extending gradually downward.

Determinate Growth. Growing to a particular, predetermined size or form, then stopping.

Diffuse. Widely spread; scattered.

Dimorphic. Occurring normally in two dissimilar forms.

Dioecious. Having either male or female reproductive organs, or flowers, on different individuals; dioecy.

Diploid. Having two sets of chromosomes per nucleus; 2n.

Disarticulating. Separating joint from joint, usually with age.

Discoid. Circular and flat; disklike.

Distal. Away from the point of attachment; toward the apex.

Distinct. Separate.

Diurnal. Occurring during the daytime.

Divergent. Spreading away from each other.

Dog Cholla. Type of cactus plant characterized by ground-hugging, clumped or mounded habit and cylindroid stem joints.

Dorsiventral. Flattened and with definite dorsal and ventral surfaces.

Druses. Compound crystalline structures of calcium oxalate, oblong, spheroidal, or lens-shaped in most cacti.

Ellipsoid. Elliptic in three dimensions.

Elliptic. In the form of an ellipse; a flattened circle, about twice as long as wide, widest in the middle, narrowed to rounded or pointed ends.

Endemic. Occurring naturally only in a certain geographic area.

Entire. Surface with a smooth margin.

Epidermis. The cells forming the surface layer of a plant organ.

Epiphyte. A nonparasitic plant growing upon another plant above the ground.

Epithet. An adjective used as a noun; in nomenclature the second part of the binomial representing the specific epithet in a species or intraspecific name (e.g., *aggeria* in *Opuntia aggeria*).

Erect. Standing straight up or nearly so.

Erose. With an irregular margin.

Erumpent. Rupturing stem epidermis above areole, as in flower bud emergence in most *Echinocereus*.

Fasciation. Abnormal fanlike or undulate growth of the stem apex, as a result of growth from an altered apical meristem.

Felted. With intertwining, matted hairs.

Fibrous. Said of a root system in which there are several major roots of about equal caliber arising from the same area.

Filament. Of the stamen, the slender stalk that bears the anther.

Fimbriate. With a fringed margin.

Fissure. A natural or perhaps unnatural cleft, narrow opening, or crack in tissue or between organs or parts of an organism.

Flaccid. Limp, perhaps from loss of water in tissues; floppy.

Flexuous. Curving in and out; wavy.

Floral Remnant. The shriveled remains at the apex of a cactus fruit. These may be caducous or persistent.

Floral Tube. In Cactaceae, the funnelform section of the flower.

Flower. In Cactaceae, the specialized tip of the fertile long shoot, from the base of the ovary to the tepals, including stamens and pistil(s).

Foliaceous. Leaflike.

Fruit. The ripened ovary of a plant (whether juicy or dry), containing the mature seeds. In Cactaceae, the berrylike fruit, sensu stricto, is enclosed by surrounding "accessory tissues" (see Pericarpel), but the whole structure falls as one unit and is called the "fruit" for all practical purposes.

Funicular Envelope. The raised, hardened rim around the edge of seeds in many species of *Opuntia*; also known as the funicular girdle, pseudoaril, or aril-like envelope.

Funiculus. Attachment stalk of ovule to placenta.

Funnelform. Shaped like a funnel, with a tube gradually widening upward.

Fusiform. Spindle-shaped; narrowed to a point at both ends of a swollen middle.

Glabrous. Not hairy.

Gland. A tissue that secretes.

Glaucous. With a powdered bluish, white, or gray wax on the surface (easily rubbed off), as on the surfaces of unweathered cladodes or certain opuntias.

Globose, Globular. Essentially spherical; spheroidal.

Glochid. One of the small, retrorsely barbed, easily dislodged modified spines occurring in the areoles of *Opuntia*; glochids usually are manifestly smaller and otherwise different from spines.

Gynodioecious. Having pistillate and perfect flowers on separate plants.

Habit. The general aspect, shape, or appearance of a whole plant, in nature.

Habitat. The locality, site, and particular local environment occupied by an organism.

Hair. A slender projection of cells.

Hemispheric. Shaped like half a sphere; hemispheroidal; hemispherical.

Hexaploid. Having six sets of chromosomes per nucleus; $6n$.

Hilum. A depression or scar on the seed where the funiculus was attached; by definition the hilum is the basal point of the cactus seed, although it may appear to be lateral.

Holotype. The particular permanently preserved specimen upon which a taxon has been based and with which its scientific name is permanently associated; the type specimen, deposited in an herbarium.

Hooked. With a hook at the end, otherwise with a straight or slightly curved shaft.

Hypanthium. Floral cup or tube, extension of the receptacle rather than floral cup formed by fusion of perianth and stamens.

Imbricate. Overlapping like shingles.

Included. Not protruding beyond the surrounding organ, as in stamens not extending out of the corolla tube.

Indehiscent. Not regularly splitting open, as in some fruits.

Inferior Ovary. An ovary enclosed by and adnate with the floral tube, in most cacti including the receptacle and forming the pericarpel.

Inner Tepals. The innermost or petaloid tepals in cactus flowers (see Tepal).

Introduced. Native to another area and imported, often becoming established (naturalized) and spreading.

Introgression. Hybridization followed by the hybrids and their descendants crossing back to one or both parents.

Joint. One segment of a stem composed of several "joints" as used by some authors; "stem segment" should be used in place of "joint."

Keel. A longitudinal ridge, like the keel of a boat.

Lanceolate. Having the shape of a lance (i.e., 4–6 times as long as broad), broadest near attachment end, tapering to the apex.

Lateral. On the side; extending to the side.

Latex. Viscid, usually whitish liquid substance produced in specialized cells and canals in certain cactus species, such as in some *Mammillaria*.

Lectotype. A type specimen selected from one or more specimens cited in the original publication, if no holotype was indicated in the original published description, when the holotype is found to belong to more than one taxon, or if the holotype is missing (or otherwise selected according to the International Code of Botanical Nomenclature).

Lenticular. Lens-shaped, like a biconvex lens.

Linear. Narrow, considerably longer than wide, with parallel sides.

Lobe. A segment of an organ; a lobe may be almost any shape.

Long Shoot (fertile and sterile). Shoot or portion of a shoot with relatively long internodes, when on a plant with long-shoot–short-shoot organization.

Maturation. The process of becoming mature; ripening, as with a fruit exhibiting mature color and texture, and containing seeds.

Mature. Having completed natural growth and development; having undergone maturation.

Medullary Bundles. Vascular bundles that extend through the pith in certain cacti, or in most cacti in the cortex and usually outward to ribs, tubercles, and areoles.

Meristem. A region of tissue where the cells are actively dividing or have the potential to divide and produce new cells.

Meristematic. Having the potential or the function of cell division.

Micropyle. The opening in the integument through which the pollen tube enters the ovule.

Midregion. The middle area of a tepal or other structure, along and usually near the midrib.

Midrib. The major vein along the middle of a tepal or other structure.

Midvein. The central vein (vascular bundle) of a leaf, tepal, or other organ.

Morphotype. An individual plant or population distinguishable from another by one or more morphological features.

Mucilage. A viscid or slippery and viscous complex carbohydrate substance in cactus tissues, mostly stems, produced in specialized cells, reservoirs, or canals.

Mucilaginous. Secreting mucilage.

Mucronate. Bearing a mucro, which is a short, sharp, terminal projection of tissue similar in texture to the rest of the structure.

Mucronulate. Minutely mucronate.

Native. Naturally occurring in an area.

Nectar. A sugary fluid that attracts animals to plants.

Nectar Chamber. Floral chamber around the base of the style and below the lowermost stamen insertion, often containing nectar that is secreted from special surrounding tissues or the nectary.

Nectary. A gland or tissue producing nectar.

Nerve. A visible vein.

Nocturnal. Occurring during the hours of darkness; said of flowers that open at night and close during the day.

Nomen Nudum. An invalid name, published without sufficient information to satisfy criteria of valid publication, according to the International Code of Botanical Nomenclature; nom. nud.

Nothospecies. Any and all members of a hybrid species, including later-generation progeny and backcrosses; nothotaxon.

Ob-. A Latin prefix usually denoting inversion.

Obconic. Conic, but attached at the pointed end of a cone.

Oblanceolate. Lanceolate, but attached at the narrow end.

Oblate. Flattened from opposite ends.

Oblique. Diagonal.

Oblong. Rounded structure with more or less parallel sides and about two or three times as long as wide.

Obovate. Ovate, but attached at the narrow end.

Obovoid. Ovoid in three dimensions, but attached at the narrow end.

Obtuse. Blunt or rounded at the apex or base; greater than 45° angle at apex.

Octoploid. Having eight sets of chromosomes per nucleus; $8n$, as in *Opuntia ficus-indica*.

Orbicular. Approximately circular in outline; orbiculate.

Outer Tepals. The outer or sepaloid tepals in cactus flowers, transitional with inner (petaloid) tepals (see Tepal).

Oval. Having the shape of an egg; according to *Merriam-Webster's Collegiate Dictionary* (10th edition), broadly elliptical.

Ovary. The lower, expanded portion of the pistil, in cacti (except for some species of *Pereskia*) surrounded by pericarpel tissue and inferior in position; containing the ovules.

Ovate. Generally egg-shaped in outline: like elliptical, but wider at the proximal end.

Ovoid. Ovate, but in three dimensions.

Ovule. An immature potential seed, still in the ovary prior to fertilization and/or ripening;

anatomically, an integument surrounding a megasporangium and female gametophyte.

Pad. A term used in common reference to the strongly flattened stem segment of a prickly pear; cladode or phylloclade.

Papillae. Low, usually rounded projections.

Papillate. Having papillae.

Papillose. Minutely papillate.

Pectinate. Like the teeth of a comb.

Perianth. The sepals and petals collectively, or the tepals collectively.

Pericarpel. The stem tissue precisely surrounding the ovary, the external surface of which bears areoles in cacti that have areolate pericarpels.

Peripheral. On the margin.

Persistent. Remaining attached, perhaps beyond the usual period, before falling away, or tending not to fall away.

Petal. One of the separate parts of the corolla (the inner perianth organ), usually pigmented and showy; in Cactaceae the showy parts of flowers usually are inner tepals.

Petaloid. Petal-like in appearance or position, as in a petaloid perianth part, a tepal that is similar to a petal.

Phenology. Study of temporal aspects of recurrent natural phenomena, and their relation to climate, season, and weather (e.g., flowering phenology).

Phloem. Vascular tissue transporting sugars.

Phylloclad. The leaflike flattened stem segment (pad) of a prickly pear.

Pile. A reference to trichomes closely standing together like the pile of a carpet.

Pistil. The female reproductive structure of the flower, consisting of the stigma (in cacti with several stigma lobes), which is connected by a tubular style to the expanded ovary.

Pistillate. Flower with one (in Cactaceae) or more pistils, but no functional stamens.

Pith. The central tissue of a stem, usually composed of parenchyma cells and surrounded by a vascular cylinder.

Ploidy. General reference to number of sets of chromosomes in a nucleus.

Plumose. Featherlike, with fine hairs projecting from the sides of a shaft.

Podarium. Collective term for stem-surface protuberances in cacti (pl., podaria).

Pollen. The microscopic, often spheroidal microgametophytes produced, developed, and usually disseminated from the anther of a stamen.

Polyphyletic. Derived from two or more ancestral lineages, as in a species derived from hybridization between other species.

Polyploid. Having three or more sets of chromosomes per nucleus.

Poricidal. Opening by a pore.

Porrect. Perpendicular to the surface.

Prickly Pear. Type of cactus plant characterized by strongly flattened stem joints; nopal.

Project. Protrude outward, as in projecting spines.

Proliferous. In cacti the formation of buds from fruit areoles; reproducing by special buds or shoots.

Prostrate. Flat on the ground.

Proximal. Close to the point of attachment; toward the base.

Pseudoaril. Resembling an aril, e.g., the hardened funicular envelope or girdle surrounding the seeds of many species of *Opuntia*.

Pubescent. Hairy; most precisely meaning fine, soft hairs. The noun form, pubescence, is commonly used as a general term to indicate the presence of hair of any kind.

Pulp. The pulpy tissue in the fruit, or other organ.

Punctate. Covered with dots (glands) or pits.

Pyriform. Pear-shaped.

Radial. When in reference to spines, the individuals or series that radiate from the periphery of the spine cluster in a plane that is more or less parallel with the stem surface.

Raphe. Marginal ridge on an ovule or seed.

Receptacle. The stem tip that supports the floral organs.

Receptacular Tube. Tubular, funnelform, or cuplike part of the cactus flower that extends from just above the pericarpel and column to the outer tepals; essentially the "floral tube."

Recurved. Curving backward and downward.

Reflexed. Bent or extending abruptly downward or backward.

Relict. Persisting in only a part of its previous range; relic.

Reniform. Kidney-shaped.

Reticulate. Netlike; reticulum.

Rhizome. A horizontal underground stem.

Ribs. Confluent podaria forming longitudinal files on the stem surface in cacti, arising from stem tissue.

Rind. The wall of the cactus fruit; outer "skin" of the stem.

Root. Underground anchoring and absorptive organ of a plant, anatomically different from the stem, with no nodes or internodes; a storage root is enlarged with reserve food.

Rosette. Crowded basal cluster, usually said of leaves.

Rotate. Saucerlike.

Rudiment. A vestige of an organ or structure.

Rufous. Reddish-brown.

Rugose. Wrinkled.

Scabrous. Rough to the touch; bearing minute projections such as short, stiff hairs.

Scale. A thin scarious bract, usually a vestigial leaf.

Scarious. Membranous (thickish) and translucent.

Seed. A mature ovule, with a hardened seed coat and an embryo inside.

Seed Coat. The hardened outer wall (testa) of a seed, derived from the integument.

Sensu Lato. In a broad sense.

Sensu Stricto. In a narrow sense.

Sepal. One of the separate parts of the corolla (the outer perianth organ), usually not as pigmented and as showy as petals; in Cactaceae the showy parts of flowers are inner tepals, and nonshowy parts are outer tepals.

Sepaloid. Sepal-like in appearance or position as in a sepaloid perianth part; a tepal that is similar to a sepal.

Serrate. With marginal teeth, forward projecting and acute, resembling those of a saw; saw-toothed.

Sheath. On spines the paper-thin epidermal layer that separates from the lignified spine core.

Shoot. The aboveground vegetative part of a plant; including apical meristems, stems, and leaves.

Short Shoot. The highly contracted lateral branch of a shoot, when on a plant with long-shoot and short-shoot organization.

Shrub. A woody plant with several stems arising from ground level; usually shrubs are smaller than trees.

Skin. Thin outer layer of the cactus stem, consisting of the cuticle, epidermis, and hypodermis.

Species. A living population or population system of genetically closely related individuals; the most commonly used intraspecific categories are subspecies and varieties.

Spheroidal. Sphere-shaped.

Spine. A hard, sharp-pointed structure derived from leaf tissue; the spines of cacti are specialized leaves that develop from the bud in the areole or from secondary buds derived from the areole.

Spine Cluster. The characteristic spine configuration in a single areole, this remaining intact upon abscission.

Spiniferous. Bearing spines; spinescent.

Spinose. Spinelike or ending in a spine; used by some authors to mean bearing spines.

Spreading. Extending in several directions.

Stamen. The male reproductive structure of a flower, consisting of a pollen-producing anther with pollen sacs and a slender supporting filament.

Staminate. Flower with stamens but no functional pistil.

Stem. The plant axis aboveground, or usually so, with nodes and internodes; anatomically different from the root; the stems of opuntioid cacti are jointed.

Stigma. The terminal part of the pistil of a flower, usually supported by the style; functioning in pollen reception; in cacti comprising 3–20 cylindroid lobes or branches.

Stramineous. Straw-colored.

Striate. With longitudinal lines.

Strigose. Clothed with sharp, stiff, straight, appressed hairs.

Strophiole. An outgrowth at or near the hilum of some seeds.

Style. The tubular organ of the pistil, connecting the stigma and ovary.

Sub-. As a prefix, meaning somewhat or slightly.

Subgenus. A subgeneric taxonomic category; a group of related species, series, or sections; the category is not used formally unless there are at least two subgenera recognized.

Subspecies. An intraspecific taxonomic category with rank between that of species and variety.

Subtending. Positioned below or to the outside.

Subulate. Awl-shaped; flattened in cross section and tapering to an apical point.

Succulence. The condition in plants where relatively abundant water-storing tissues allow drought tolerance; succulent.

Sulcate. Grooved or furrowed lengthwise.

Superior Ovary. An ovary above the origin of the floral tube or perianth.

Sympatric. Occurring together in the same geographic area; in reference to populations, species, or subspecific taxa.

Synonym. A name or a name combination that has been applied to a taxon but is not the correct name for the taxon in a given taxonomic treatment; a discarded name is said to be in synonymy.

Taproot. The primary root usually of larger size than its branch roots.

Taxon. A taxonomic unit of any rank.

Tepal. Perianth segment not clearly differentiated as calyx (sepals) or corolla (petals).

Terete. Circular in transverse section.

Testa. Seed coat.

Tetraploid. Having four sets of chromosomes per nucleus; 4n.

Tomentose. Woolly-pubescent; usually covered with matted, soft, wool-like hairs that are not straight.

Transverse. Across, at right angles to the axis.

Tree. A relatively large, usually branched and woody plant with a single main trunk.

Trichome. A usually multicellular plant hair; a trichome arises from the epidermis of a plant organ.

Truncate. Having an abrupt ending, as in an apex at right angles to the axis.

Tuber. A thickened underground stem, functioning as a storage organ.

Tuberculate. Beset with tubercles.

Tubercle. Individually protruding podarium on stem surface in cacti, arising from both leaf and stem tissue, bearing an areole at the apex; podarium.

Tubercle Groove. A narrow sulcus or indentation on the adaxial surface of a tubercle.

Tuberous. Having the aspect of a tuber but not necessarily derived from a stem.

Turbinate. Top-shaped; inversely conical.

Turgid. Swollen; inflated with water.

Type. See Holotype.

Type Locality. The locality at which the type specimen of a taxon was collected.

Type Species. The species of a genus with which the generic name is permanently associated.

Type Specimen. The holotype, or designated type according to the International Code of Botanical Nomenclature.

Umbilicus. A cuplike depression at the apical end of fruits in *Opuntia*.

Uncinate. Hooked.

Unisexual. An individual with either staminate or pistillate reproductive structures but not both.

Upright. Standing vertical or nearly so.

Urceolate. Urn-shaped.

Variety. A subspecific taxonomic category, the lowest-ranking taxon commonly recognized.

Vascular Bundle. A strand composed of xylem and phloem and often a vascular cambium; a vascular trace is a vascular bundle that extends into a leaf, branch, or flower; a vein.

Vascular Cylinder. The cylinder of vascular bundles formed in a stem or root, best observed in cross section.

Vascular Tissue. Tissues of a plant that transport water in xylem and sugars in phloem.

Vegetative. Pertaining to any part of a plant except the flower and fruit.

Vegetative Reproduction. Asexual reproduction from any plant organ but not from seed after sexual reproduction.

Vein. Threadlike conducting tissue in a leaf or leaflike structure; vascular bundle.

Velutinous. Velvety vestiture; clothed with dense, straight, erect hairs.

Venation. The arrangements of veins as seen in a leaf or leaflike structure.

Vestige. A rudiment.

Vestigial. Rudimentary; poorly developed.

Wood Skeleton. A dried cylinder of wood, as in some chollas, exhibiting spaces where soft tissues were present in the living stem.

Woolly. With long, strongly interlaced hairs, collectively matted.

Xeric. Arid; dry.

Xerophyte. A plant that is adapted to dry or xerophytic conditions.

Xylem. Water- and mineral-conducting elements of the vascular system.

Bibliography

Anderson, E. F. 1960. A revision of *Ariocarpus* (Cactaceae). I. The status of the proposed genus *Roseocactus*. *Amer. J. Bot.* 47:582–89.

———. 1962. A revision of *Ariocarpus* (Cactaceae). II. The status of the proposed genus *Neogomesia*. *Amer. J. Bot.* 49:615–22.

———. 1963. A revision of *Ariocarpus* (Cactaceae). III. Formal taxonomy of the subgenus *Roseocactus*. *Amer. J. Bot.* 50:724–32.

———. 1964. A revision of *Ariocarpus* (Cactaceae). IV. Formal taxonomy of the subgenus *Ariocarpus*. *Amer. J. Bot.* 51:144–51.

———. 1965. A taxonomic revision of *Ariocarpus* (Cactaceae). *Cact. Succ. J. (U.S.)* 37:39–49.

———. 1980. *Peyote the divine cactus*. Tucson: University of Arizona Press.

———. 1986. A revision of the genus *Neolloydia* B. & R. (Cactaceae). *Bradleya* 4:1–28.

———. 1987. A revision of the genus *Thelocactus* B. & R. (Cactaceae). *Bradleya* 5:49–76.

———. 1996. *Peyote the divine cactus*. 2nd ed. Tucson: University of Arizona Press.

———. 1999a. *Ariocarpus*: Some reminiscences. *Cact. Succ. J. (U.S.)* 71:180–90.

———. 1999b. Some nomenclatural changes in the Cactaceae, subfamily Opuntioideae. *Cact. Succ. J. (U.S.)* 71:324–25.

———. 2001. *The cactus family*. Portland, OR: Timber Press.

———, and W. A. Fitz Maurice. 1997. *Ariocarpus* revisited. *Haseltonia* 5:1–20.

———, and M. E. Ralston. 1978. A study of *Thelocactus* (Cactaceae). I. The status of the proposed genus *Gymnocactus*. *Cact. Succ. J. (U.S.)* 50:216–24.

Anthony, M. 1954. Ecology of the *Opuntiae* in the Big Bend region of Texas. *Ecology* 35:334–47.

———. 1956. The Opuntiae of the Big Bend region of Texas. *Amer. Mid. Nat.* 55:225–56.

Backeberg, C. 1951. Some results of twenty years of cactus research. *Cact. Succ. J. (U.S.)* 23:150–51.

———. 1961. *Die Cactaceae*. Vol. 5. Jena, Germany: Gustav Fischer.

Baker, M. A., and R. A. Johnson. 2000. Morphometric analysis of *Escobaria sneedii* var. *sneedii*, *E. sneedii* var. *leei*, and *E. guadalupensis* (Cactaceae). *Syst. Bot.* 25:577–87.

Benson, L. 1969a. The cacti of the United States and Canada—new names and nomenclatural combinations. *Cact. Succ. J. (U.S.)* 41:124–28, 185–90, 233–34.

———. 1969b. Cactaceae. Pp. 221–317 in *Flora of Texas*, vol. 2., pt. 2, edited by C. L. Lundell et al. Renner: Texas Research Foundation.

———. 1970. Cactaceae. Pp. 1087–1113 in *Manual of the vascular plants of Texas*, edited by D. S. Correll and M. C. Johnston. Renner: Texas Research Foundation.

———. 1982. *The cacti of the United States and Canada*. Stanford, CA: Stanford University Press.

Blum, W., D. Felix, T. Oldach, and J. Oldach. 2004. The *Echinocereus reichenbachii-fitchii*-complex. *Echinocereus* Study Group of the German Cactus Society, D-95615 Marktredwitz. Printed in Germany, Dinges & Frick, Wiesbaden.

Blum, W., M. Lange, W. Rischer, and J. Rutow. 1998. *Echinocereus*. Monograph. Belgium: n.p.

Boke, N. H., and R. G. Ross. 1978. Fasciation and dichotomous branching in *Echinocereus* (Cactaceae). *Amer. J. Bot.* 65:522–30.

Bravo-Hollis, H. 1978. *Las Cactáceas de México*. Vol. 1. México, D.F.: Universidad Nacional Autónoma de México.

———, and R. Sánchez-Mejorada. 1991a. *Las Cactáceas de México*. Vol. 2. México, D.F.: Universidad Nacional Autónoma de México.

———. 1991b. *Las Cactáceas de México*. Vol. 3. México, D.F.: Universidad Nacional Autónoma de México.

Breckenridge III, F. G. 1981. A systematic study of the *Echinocereus enneacanthus* complex (Cactaceae). Master's thesis, Sul Ross State University, Alpine, Texas.

———, and J. M. Miller. 1982. Pollination biology, distribution, and chemotaxonomy of the *Echinocereus enneacanthus* complex (Cactaceae). *System. Bot.* 7:365–78.

Britton, N. L., and J. N. Rose. 1919–23. *The Cactaceae*. 4 vols. Carnegie Institute Washington Publication 248. Washington, DC: Carnegie Institute.

———. [1937] 1963. *The Cactaceae*. Vol. 3. Reprint, New York: Dover.

Brummitt, R. K., and C. E. Powell. 1992. *Authors of plant names*. Kew, England: Royal Botanic Gardens.

Champie, C. 2003. Discovery of a new variant of *Cylindropuntia leptocaulis* (DC.) F. M. Knuth. *Cact. Succ. J. (U.S.)*. 75:271–73.

Correll, D. S., and M. C. Johnston. 1970. *Manual of the vascular plants of Texas*. Renner: Texas Research Foundation.

Crook, R., and R. Mottram. 1995. *Opuntia* in-dex: Part 1: Introduction and A–B. *Bradleya* 13:89–118.

———. 1996. *Opuntia* index: Part 2: Nomenclatural note and C–E. *Bradleya* 14:99–144.

———. 1997. *Opuntia* index: Part 3: Nomenclatural note and F. *Bradleya* 15:98–112.

———. 1998. *Opuntia* index: Part 4: G–H. *Bradleya* 16:119–36.

———. 1999. *Opuntia* index: Part 5: Nomenclatural note and I–L. *Bradleya* 17:109–31.

———. 2000. *Opuntia* index: Part 6: M–O. *Bradleya* 18:113–40.

Crozier, B. S. Subfamilies of Cactaceae Juss., including Blossfeldioideae subfam. nov. *Phytologia* 86:52–64.

Diamond, D. D., D. H. Riskind, and S. L. Orsell. 1988. A framework for plant community classification and conservation in Texas. *Texas J. Sci.* 39:203–21.

Diggs Jr., G. M., B. L. Lipscomb, and R. J. O'Kennon. 1999. Shinners & Mahler's illustrated flora of north central Texas. *Sida*, Botanical Miscellany, no. 16. Fort Worth: Botanical Research Institute of Texas; Sherman, TX: Austin College.

Earle, W. H. 1963. *Cacti of the southwest*. Tempe, AZ: Daily News.

———. 1980. *Cacti of the southwest*. 2nd ed. Phoenix, AZ: Desert Botanical Garden.

Emory, W. H. 1859. Report of the United States and Mexican boundary survey: Cactaceae by George Engelmann. Vol. 2, 1–78. Washington, DC.

Engelmann, G. 1856 [1857]. Synopsis of the Cactaceae of the territory of the United States and adjacent regions. *Proc. Amer. Acad. Arts* 3:259–346.

———. 1859. Cactaceae of the boundary, from the report of the U.S. and Mexican Boundary Survey and the order of Lt. Col. W. H. Emory, Maj. 1st Cav. and U.S. Commissioner. Vol. 2, pt. 1, 1–78. Washington, DC.

Everitt, J. H., and D. L. Drawe. 1993. *Trees, shrubs, and cacti of South Texas*. Lubbock: Texas Tech University Press.

Ferguson, D. J. 1986. *Opuntia chisosensis* (Anthony) comb. nov. *Cact. Succ. J. (U.S.)* 58:124–27.

———. 1987. *Opuntia cymochila* Eng. & Big: A species lost in the shuffle. *Cact. Succ. J. (U.S.)* 59:256–60.

———. 1988. *Opuntia macrocentra* Eng. and *Opuntia chlorotica* Eng. & Big. *Cact. Succ. J. (U.S.)* 60:155–60.

———. 1989. Revision of the U.S. members of the *Echinocereus triglochidiatus* group. *Cact. Succ. J. (U.S.)* 61:217–24.

———. 1991. In defense of the genus *Glandulicactus* Backeb. *Cact. Succ. J. (U.S.)* 63:87–91.

———. 1992. The genus *Echinocactus* Link & Otto, subgenus *Homalocephala* (Britton & Rose) stat. nov. *Cact. Succ. J. (U.S.)* 64:169–72.

Fischer, P. C. 1962. Taxonomic relationship of *Opuntia kleiniae* de Candolle and *Opuntia tetracantha* Toumey. Master's thesis, University of Arizona, Tucson.

———. 1971. Taxonomical and ecological relationship of the *Coryphantha vivipara* complex in the Cactaceae. Ph.D. diss., University of California, Berkeley.

———. 1980. The varieties of *Coryphantha vivipara*. *Cact. Succ. J. (U.S.)* 52:186–91.

Flora of North America. 2003. Vol. 4. New York: Oxford University Press.

Glass, C., and R. Foster. 1975. The genus *Echinomastus* in the Chihuahuan Desert. *Cact. Succ. J. (U.S.)* 47:218–23.

———. 1977. The genus *Thelocactus* in the Chihuahuan Desert. *Cact. Succ. J. (U.S.)* 49:213–20, 244–51.

———. 1978. A revision of the genus *Epithelantha*. *Cact. Succ. J. (U.S.)* 50:184–87.

Grant, V., and K. A. Grant. 1979. Systematics of the *Opuntia phaeacantha* group in Texas. *Bot. Gaz.* 140:199–207.

Griffith, M. P. 2000. Breeding systems and natural interspecific hybridization in *Opuntia* of the northern Chihuahuan Desert Region. Master's thesis, Sul Ross State University, Alpine, Texas.

———. 2001. A new Chihuahuan Desert hybrid prickly pear, *Opuntia* X *rooneyi* (Cactaceae). *Cact. Succ. J. (U.S.)* 73:307–10.

———. 2002. *Grusonia pulchella* classification and its impacts on the genus *Grusonia*: Morphological and molecular evidence. *Haseltonia* 9:86–93.

Griffiths, D. 1908–11. Illustrated studies in the genus *Opuntia*. Vols. 1–4. Missouri Botanical Garden annual reports 19–22: vol. 1 (1908) 19:259–72, plus plates 21–28; vol. 2 (1909) 20:81–97, plus plates 2–13; vol. 3 (1910) 21:165–74, plus plates 19–28; vol. 4 (1911) 22:25–36, plus plates 1–17.

Hardy, J. E. 1997. Flora and vegetation of the Solitario Dome, Brewster and Presidio counties, Texas. Master's thesis, Sul Ross State University, Alpine, Texas.

Heil, K. D., and S. Brack. 1986. The cacti of Guadalupe Mountains National Park. *Cact. Succ. J. (U.S.)* 58:165–77.

———. 1988. The cacti of Big Bend National Park. *Cact. Succ. J. (U.S.)* 60:17–34.

———, and J. M. Porter. 1994. *Sclerocactus* (Cactaceae): A revision. *Haseltonia* 2:20–46.

Hernández, H. M., and R. T. Bárcenas. 1995. Endangered cacti in the Chihuahuan Desert. I. Distribution patterns. *Conservation Biology* 9:1176–88.

Hershkovitz, M. A., and E. A. Zimmer. 1997. On the evolutionary origins of cacti. *Taxon* 46:217–32.

Hester, J. P. 1945. *Echinomastus mariposensis* sp. nov. *Desert Plant Life* 17:59.

Hoffman, M. T. 1992. Functional dioecy in *Echinocereus coccineus* (Cactaceae): Breeding system, sex ratios, and geographic range of floral dimorphism. *Amer. J. Bot.* 79:1382–88.

Hunt, D. R. 1978. Amplification of the genus *Escobaria*. *Cact. Succ. J. (Gr. Brit.)* 40:13, 30.

————, and N. Taylor. 1986. The genera of Cactaceae: Towards a new consensus. *Bradleya* 4:65–78.

Ideker, J. 1996. *Pereskia aculeata* (Cactaceae), in the lower Rio Grande valley of Texas. *Sida* 17:527.

IOS Working Party. 1986. The genera of the Cactaceae: Towards a new consensus. *Bradleya* 4:65–78.

Kolle, D. O. 1978. A populational study of *Echinocereus chloranthus* (Cactaceae) in Trans-Pecos Texas. Master's thesis, Sul Ross State University, Alpine, Texas.

Leuck II, E. E. 1980. Biosystematic studies in the *Echinocereus viridiflorus* complex. Ph.D. diss., University of Oklahoma, Norman.

————, and J. H. Miller, 1982. Pollination biology and chemotaxonomy of the *Echinocereus viridiflorus* complex (Cactaceae). *Amer. J. Bot.* 69:1669–72.

Leuenberger, B. E. 1986. *Pereskia* (Cactaceae). *Mem. N.Y. Bot. Gard.* 41:1–144.

Lindsay, G. S. 1955. Taxonomy and ecology of the genus *Ferocactus*. Ph.D. diss., Stanford University, Stanford, California.

Lockwood, M. W. 1995. Notes on life history of *Ancistrocactus tobuschii* (Cactaceae) in Kinney County, Texas. *Southwest. Nat.* 40:428–30.

Mellen, G. 1991. The *Echinocereus fendleri* controversy. *Cact. Succ. J. (U.S.)* 63:208–12.

Meyer, B. N., and J. L. McLaughlin. 1981. Economic uses of *Opuntia Cact. Succ. J. (U.S.)* 53:107–12.

Moore, W. O. 1967. The *Echinocereus enneacanthus–dubius–stramineus* complex (Cactaceae). *Brittonia* 19:77–94.

Nobel, P. S. 1988. *Environmental biology of agaves and cacti.* New York: Cambridge University Press.

————. 1994. *Remarkable agaves and cacti.* New York: Oxford University Press.

Nyffler, R. 2002. Phylogenetic relationships in the cactus family (Cactaceae) based on evidence from *trn*K/*mat*K and *trn*L–*trn*F sequences. *Amer. J. Bot.* 89:312–26.

Parfitt, B. D. 1985. Dioecy in North American Cactaceae: A review. *Sida* 11:200–206.

————. 1991. Biosystematics of the *Opuntia polyacantha* complex (Cactaceae) of western North America. Ph.D. diss., Arizona State University, Tempe.

————. 1998. New nomenclatural combinations in the *Opuntia polyacantha* complex. *Cact. Succ. J. (U.S.)* 70:188.

————, and D. J. Pinkava. 1988. Nomenclatural and systematic reassessment of *Opuntia engelmannii* and *O. lindheimeri* (Cactaceae). *Madroño* 35:342–49.

Pendley, G. K. 2001. Seed germination experiments in *Opuntia* (Cactaceae) of the northern Chihuahuan Desert. *Haseltonia* 8:42–50.

Pickett, C. H., and W. D. Clark. 1979. The function of extrafloral nectaries in *Opuntia acanthocarpa* (Cactaceae). *Amer. J. Bot.* 66:618–25.

Pinkava, D. J. 2003a. Cactaceae, subfam. Opuntioideae. Pp. 102–3 in *Flora of North America,* vol. 4. New York: Oxford University Press.

————. 2003b. *Cylindrountia.* Pp. 103–18 in *Flora of North America,* vol. 4. New York: Oxford University Press.

————. 2003c. *Grusonia.* Pp. 118–23 in *Flora of North America,* vol. 4. New York: Oxford University Press.

————. 2003d. *Opuntia.* Pp. 123–48 in *Flora of North America,* vol. 4. New York: Oxford University Press.

————, and B. D. Parfitt. 1988. Nomenclatural changes in Chihuahuan Desert *Opuntia* (Cactaceae). *Sida* 13:125–30.

————, J. P. Rebman, and M. A. Baker. 2001. Nomenclatural changes in *Cylindropuntia* and *Opuntia* (Cactaceae) and notes on interspecific hybridization. *J. Arizona-Nevada Acad. Sci.* 33:162–63.

Poole, J. M., and D. H. Riskind. 1987. *Endangered, threatened or protected native plants of Texas*. Austin: Texas Parks and Wildlife Department.

Porter, J. M., M. S. Kinney, and K. D. Heil. 2000. Relationships between *Sclerocactus* and *Toumeya* (Cactaceae) based on chloroplast *trn*L–*trn*F sequences. *Haseltonia* 7:8–17.

Powell, A. M. 1995. Second generation experimental hybridizations in the *Echinocereus* X *lloydii* complex (Cactaceae), and further documentation of dioecy in *E. coccineus*. *Pl. Syst. Evol.* 196:63–74.

———. 1998a. *Trees and shrubs of the Trans-Pecos and adjacent areas*. Austin: University of Texas Press.

———. 1998b. Plant communities of the Chihuahuan Desert Region. *Wildflower* 14:38–42.

———. 1998c. Third generation experimental hybrids in the *Echinocereus* X *lloydii* complex (Cactaceae). *Haseltonia* 6:91–95.

———. 2000. *Grasses of the Trans-Pecos and adjacent areas*. Marathon, TX: Iron Mountain Press.

———. 2002. Experimental hybridization between *Echinomastus intertextus* and *E. warnockii* (Cactaceae). *Haseltonia* 9:80–85.

———, and J. F. Weedin. 2001. Chromosome numbers in Chihuahuan Desert Cactaceae. III. Trans-Pecos Texas. *Amer. J. Bot.* 88:481–85.

———, and J. F. Weedin. 2004. *Cacti of the Trans-Pecos and adjacent areas*. Lubbock: Texas Tech University Press.

———, A. D. Zimmerman, and R. A. Hilsenbeck. 1991. Experimental documentation of natural hybridization in Cactaceae: Origin of Lloyd's hedgehog cactus, *Echinocereus* X *lloydii*. *Pl. Syst. Evol.* 178:107–22.

Ralston, B. E. 1987. A biosystematic study of the *Opuntia schottii* complex (Cactaceae) in Texas. Master's thesis, Sul Ross State University, Alpine, Texas.

———, and R. A. Hilsenbeck. 1989. Taxonomy of the *Opuntia schottii* complex (Cactaceae) in Texas. *Madroño* 36:221–31.

———. 1992. *Opuntia densispina* (Cactaceae): A new club cholla from the Big Bend region of Texas. *Madroño* 39:281–84.

Raun, G. G. 1996. Distribution and population density of *Coryphantha hesteri* in the Big Bend region of Texas. *Cact. Succ. J. (U.S.)* 68:115–18.

———. 1997. *Echinomastus* in the Trans-Pecos region of Texas. *Cact. Succ. J. (U.S.)* 69:122–26.

Reeves, B. 1994. *Toumeya papyracantha*—what and where next. *Cact. Succ. J. (U.S.)* 66:184–88.

Robinson, H. 1973. New combinations in the Cactaceae subfamily Opuntioideae. *Phytologia* 26:175–76.

Schmidly, D. J. 1977. *The mammals of Trans-Pecos Texas*. College Station: Texas A&M University Press.

Schulz, E. D., and R. Runyon. 1930. Texas cacti. *Proceedings Texas Acad. Sci.* 14:1–181.

Scobell, S. A. 1999. Pollination ecology of claret cup cactus along an elevation gradient. Master's thesis, Indiana State University, Terre Haute.

Scogin, R. 1985. Nectar constituents of the Cactaceae. *Southwest. Nat.* 30:77–82.

Smith, D. W. 1973. A taxonomic and distributional study of the Cactaceae in Brewster County, Texas. Master's thesis, Sul Ross State University, Alpine, Texas.

Stuppy, W. 2002. Seed characters and generic classification of the Opuntioideae (Cactaceae). *Succulent Plant Research* 6:25–58. Sherborne, England: David Hunt.

Sutton, K., J. T. Baccus, and M. S. Traweek Jr. 1997. Habitat of *Ancistrocactus tobuschii* (Tobusch fishhook cactus, Cactaceae) on the Edwards Plateau of Central Texas. *Southwest. Nat.* 42:441–45.

Suzán, H., G. P. Nabhan, and D. T. Patten. 1994. Nurse plant and floral biology of a rare night-blooming cereus, *Peniocereus striatus* (Brandegee) F. Buxbaum. *Conservation Biology* 8:461–70.

Taylor, N. P. 1978. Review of the genus *Escobaria*. *Cact. Succ. J. (Gr. Brit.)* 40:30–37.

———. 1979. Further notes on *Escobaria*. *Cact. Succ. J. (Gr. Brit.)* 41:17–20.

———. 1984. A review of *Ferocactus* Britton & Rose. *Bradleya* 2:19–38.

———. 1985. *The genus* Echinocereus. Portland, OR: Timber Press.

———. 1986. The identification of *Escobarias* (Cactaceae). Botley, Oxford, England: British Cactus and Succulent Society.

———. 1987. Additional notes on some *Ferocactus species*. *Bradleya* 5:95–96.

———. 1998. *Coryphantha robustispina* (Engelm.) Britton & Rose: The correct name for the taxon variously known as *C. scheeri* Lemaire and *C. muehlenpfordtii* Britton & Rose (nom. illeg.). *Cactaceae Consensus Initiatives* 6:17–20.

———, and J. Y. Clark. 1983. Seed-morphology and classification in *Ferocactus* subg. *Ferocactus*. *Bradleya* 1:3–16.

Wallace, R. S. 1995. Molecular systematic study of the Cactaceae: Using chloroplast DNA variation to elucidate cactus phylogeny. *Bradleya* 13:1–12.

———, and S. L. Dickie. 2002. Systematic implications of chloroplast DNA sequence variation in subfam. Opuntioideae (Cactaceae). *Succulent Plant Research* 6:9–24. Sherborne, England: David Hunt.

Warnock, B. H. 1970. *Wildflowers of the Big Bend Country Texas*. Alpine, TX: Sul Ross State University.

———. 1977. *Wildflowers of the Davis Mountains and Marathon Basin Texas*. Alpine, TX: Sul Ross State University.

Wauer, R. H. 1980. *Naturalist's Big Bend*. Rev. ed. College Station: Texas A&M University Press.

———. 1997. *For all seasons: A Big Bend journal*. Austin: University of Texas Press.

Weedin, J. F., and A. M. Powell. 1978. Chromosome numbers in Chihuahuan Desert Cactaceae: Trans-Pecos Texas. *Amer. J. Bot.* 65:531–37.

———. 1980. IOPB chromosome number reports LXIX. Edited by A. Löve. *Taxon* 29:703–30.

———, A. M. Powell, and D. O. Kolle. 1989. Chromosome numbers in Chihuahuan Desert Cactaceae. II. Trans-Pecos Texas. *Southwest. Nat.* 34:160–64.

Weniger, D. 1969. The small-flowered echinocerei of Texas and New Mexico. *Cact. Succ. J. (U.S.)* 41:34–43.

———. 1970. *Cacti of the Southwest*. Austin: University of Texas Press.

———. 1984. *Cacti of Texas and neighboring states*. Austin: University of Texas Press.

Worthington, R. D. 1986. Observations on the flowering cacti from the vicinity of El Paso, Texas. *Cact. Succ. J. (U.S.)* 58:213–17.

———. 1989. *An annotated checklist of the native and naturalized flora of El Paso County, Texas*. El Paso Southwest Botanical Miscellany no. 1. El Paso, TX: Richard D. Worthington, Floristic Inventories of the Southwest Program.

———. 1995. *Biota of the Franklin Mountains*. Pt. 2. El Paso, TX: Floristic Inventories of the Southwest Program.

Zimmerman, A. D. 1985. Systematics of the genus *Coryphantha* (Cactaceae). Ph.D. diss., University of Texas at Austin.

———. 1993. Systematics of *Echinocereus* X *roetteri* (Cactaceae), including Lloyd's hedgehog-cactus. Pp. 270–88 in *Proceedings of the Southwestern Rare and Endangered Plant Conference*, edited by R. Sivinski and K. Lightfoot. Santa Fe: New Mexico Forestry and Resources Conservation Division.

———, and B. D. Parfitt. 2003a. *Echinocereus*. Pp. 157–74 in *Flora of North America*, vol. 4, New York: Oxford University Press.

———, and B. D. Parfitt. 2003b. *Ancistrocactus*. Pp. 209–11 in *Flora of North America*, vol. 4. New York: Oxford University Press.

——, C. Glass, R. Foster, and D. Pinkava. Forthcoming. Cactaceae. In *Chihuahuan Desert Flora*, by J. Henrickson and M. C. Johnston. Los Angeles: J. Henrickson.

——, and D. A. Zimmerman. 1977. A revision of the United States taxa of the *Mammillaria wrightii* complex with remarks upon the northern Mexican populations. *Cact. Succ. J. (U.S.)* 49:23–34, 51–62.

Zimmerman, D. A. 1972. The fruits of *Coryphantha hesteri*. *Cact. Succ. J. (U.S.)* 44:85.

Index

Boldface type indicates scientific names of Texas or closely adjacent species, preferred common names, and the page numbers where the main discussion of each taxon or name begins. Scientific names of synonyms and extraterritorial species are in italics.